In Vitro Cultivation of Parasitic Helminths

Editor
J. Desmond Smyth, Sc.D.
London School of Hygiene and Tropical Medicine
University of London
London, England

CRC Press
Taylor & Francis Group
Boca Raton London New York

CRC Press is an imprint of the
Taylor & Francis Group, an **informa** business

First published 1990 by CRC Press
Taylor & Francis Group
6000 Broken Sound Parkway NW, Suite 300
Boca Raton, FL 33487-2742

Reissued 2018 by CRC Press

© 1990 by Taylor & Francis
CRC Press is an imprint of Taylor & Francis Group, an Informa business

No claim to original U.S. Government works

A Library of Congress record exists under LC control number: 90001817

Publisher's Note
The publisher has gone to great lengths to ensure the quality of this reprint but points out that some imperfections in the original copies may be apparent.

Disclaimer
The publisher has made every effort to trace copyright holders and welcomes correspondence from those they have been unable to contact.

ISBN 13: 978-1-138-10596-6 (hbk)
ISBN 13: 978-1-138-56028-4 (pbk)
ISBN 13: 978-0-203-71179-8 (ebk)

Visit the Taylor & Francis Web site at http://www.taylorandfrancis.com and the
CRC Press Web site at http://www.crcpress.com

PREFACE

The successful *in vitro* cultivation of parasites has long been a major goal in parasitology as its realization allows investigations of the basic biology, biochemistry, and physiology of parasites to be carried out under controlled, reproducible conditions, free from interference by host-related factors. It also has considerable value as a tool for the collection of excretory/secretory (E/S) antigens for vaccination or diagnostic purposes. Success in this field also greatly furthers the replacement of laboratory animals by artificial systems — a very desirable end in its own right.

Although techniques for *in vitro* culture of parasitic helminths still lag a long way behind those for bacteria, fungi, and protozoa, major advances have been made within recent years and a number of species can now be cultured throughout part or the whole of their life cycles *in vitro*.

The aim of this volume is to provide details regarding the cultivation of representative species of parasitic helminths of vertebrates and to discuss the general problems involved and evaluate the success of available techniques. Enterophilic nematode parasites of insects are not dealt with as their life cycles are more akin to free-living forms. No attempt is made to review in detail all the species on which *in vitro* culture attempts have been made, but to deal largely with those species for which reasonably reliable and reproducible culture systems have been developed.

In order to be successful in this field, it is essential to be familiar with the morphology of the species in question and its various larval stage(s) so that the success of growth and/or differentiation during *in vitro* experiments can be assessed. This is especially important in dealing with trematodes and cestodes, many species of which have complex life cycles, but provide superb material for *in vitro* experiments. Such information is often difficult to obtain from current texts, being often hidden away in obscure publications. Attempts have been made to remedy this by providing, for these groups, very full, illustrative accounts of the life cycles of the species concerned, together with detailed descriptions of the morphology of the adult and larval stages.

With the availability of commercial culture media, sterile plastic ware, and, more recently, techniques for the cryopreservation of helminths, the horizons of helminth *in vitro* culture have been greatly expanded, and it is hoped that the contributions in this volume will encourage workers to take up the challenge of this difficult, but rewarding, field.

This volume makes no attempt to provide all the minor technical details for the preparation of widely used standard media, glassware, etc. Many such details can be obtained from laboratory manuals on tissue culture or from the excellent text, *In Vitro Methods for Parasite Cultivation*, by A. E. R. Taylor and J. R. Baker, 1987.

THE EDITOR

J. Desmond Smyth, Ph.D., Sc.D., is Emeritus Professor of Parasitology in the University of London and a Senior Research Fellow in the London School of Hygiene and Tropical Medicine, London, England. He holds the degrees of B.A., B.Sc. (1940), Ph.D. (1942), and Sc.D. (1958) from Trinity College, University of Dublin, Ireland. His previous appointments were Lecturer in Zoology, University College, Leicester, 1942 to 1945; University of Leeds, 1945 to 1947; and Trinity College, Dublin, 1947 to 1955; Professor of Experimental Biology, University of Dublin, 1955 to 1959; Professor of Zoology, Australian National University, Canberra, 1959 to 1970; Professor of Parasitology, Imperial College, University of London, 1971 to 1982; and Emeritus Professor of Parasitology, University of London, 1982 to present.

He is a past President of both the Australian and British Societies for Parasitology and is now an Honorary Fellow and is an Honorary Member of the American Society of Parasitology. He is the Foundation Editor of the *International Journal for Parasitology* and is on the Editorial Boards of several other international journals. In 1986 he was given a D.Sc. *Honoris Causa* by the University of Granada (Spain). Also in 1986, he was awarded the Elizabeth Frink Medal of the Zoological Society of London for his contributions to parasitology. He is the author of a number of books, including *An Introduction to Animal Parasitology*, 1962, 2nd ed. 1976; *The Physiology of Trematodes*, 1966, 2nd ed. 1983 (with D. W. Halton); *The Physiology of Cestodes*, 1969; *The Physiology and Biochemistry of Cestodes*, 1989 (with D. P. McManus); and *Frogs as Host-Parasite Systems*, 1978 (with M. M. Smyth).

He has worked for most of his career on the *in vitro* culture of cestodes and trematodes, being the first worker to cultivate a larval cestode (*Schistocephalus solidus*) to sexual maturity *in vitro* with the production of fertile eggs. This work led eventually to the successful culture of a number of other cestode and trematode species, including *Echinococcous granulosus* and *E. multilocularis*, the causative organisms of hydatid disease. The latter studies have resulted in a major expansion of research into the biochemistry, physiology, immunology, speciation, and ecological relationships of different strains of these pathogenic organisms throughout the world.

CONTRIBUTORS

Eric R. James, Ph.D.
Department of Ophthalmology
Medical University of South Carolina
Charleston, South Carolina

Jürgen Mössinger, M.Sc., Ph.D.
Institute of Tropical Medicine
University of Tübingen
Tübingen, Federal Republic of Germany

J. Desmond Smyth, Ph.D., Sc.D.
Department of Medical Parasitology
London School of Hygiene and Tropical
  Medicine
London, England

ACKNOWLEDGMENT

My best thanks are due to the London School of Hygiene and Tropical Medicine for providing the excellent working environment and library facilities which made the writing of this book possible.

TABLE OF CONTENTS

Chapter 1

CULTIVATING PARASITIC HELMINTHS *IN VITRO*: ADVANTAGES AND PROBLEMS

J. D. Smyth

TABLE OF CONTENTS

I. GENERAL INTRODUCTION

In vitro cultivation has a special contribution to make in the following areas:

1. Routine maintenance of helminth species
2. As a diagnostic and/or taxonomic tool
3. Biochemical and physiological studies
4. Immunological studies, antigen collection, immunodiagnosis, and vaccine development
5. Drug testing
6. Providing model systems for the study of differentiation, especially larval/adult transformation and asexual/sexual differentiation

The majority of helminth pathogens of man and domestic animals still defy attempts to culture them *in vitro* throughout their whole life cycle, but substantial advances have been made in all helminth phyla within recent years. Although there is an understandable tendency to concentrate research on helminths of man and domestic animals, it is important to stress that attempts to culture *any* species of parasitic helminth — whether of medical or economic importance — is worthwhile as, if successful, the techniques developed can often be applied to other species. An early example of this can be seen in the case of the cestode, *Schistocephalus solidus* — a bird tapeworm of no economic or medical significance — which was the first cestode to be successfully cultured to sexual maturity with the production of fertile eggs.[1,2] The experience gained with this species later led to the successful *in vitro* cultivation of a number of other cestode species (including the human pathogen, *Echinococcus granulosus*), as well as several species of trematodes.

II. TERMINOLOGY

The term *"in vitro"*, literally meaning "in glass", has long been used to describe a culture system involving a liquid or solid medium in a glass container. Although glass vessels have largely been replaced by plastic, the older term is still retained in the literature. More specific terms are also used: *axenic* (Greek a = free from; xenos = a stranger); hence an axenic culture of a species is one free from the presence of another species; *monoxenic* and *polyxenic* refer to conditions when one or more species of organisms are present.

III. SPECIFIC PROBLEMS OF HELMINTH CULTURE

The complex life cycles of many helminths, often involving three or more hosts, mean that *in vitro* culture of many species is exceptionally difficult. The chief problems involved are briefly summarized below.

A. HABITATS

Although the majority of parasitic helminths live in a restricted number of biological habitats, such as the intestine, lungs, liver, bile duct, or blood systems of vertebrate hosts, some species can occur in almost any part of a vertebrate or invertebrate host. In addition, some species have free-living stages, e.g., trematode cercaria and cestode coracidia. The physicochemical and nutritive characteristics of most of these habitats are very poorly known. The kinds of basic physiological data required, before the natural environment can be replicated *in vitro*, include pH, pO_2, pCO_2, Eh, viscosity, sugar and amino acid levels, osmotic pressure, and concentrations of the chief physiological ions. Except for some vertebrate hosts (e.g., man, dog), it is exceptionally difficult to obtain most of these data, as the process of measurement, involving instrumentation, when applied to a host habitat, is likely to interfere with the very parameter it is attempting to measure. The physicochemical and biological characteristics of some of the more common helminth habitats have been reviewed by Crompton,[3,4] Kennedy,[5] Smyth,[6] Smyth and Halton,[7] Smyth and McManus,[8] and Mettrick and Podesta.[9]

B. NUTRIENTS

The diet of parasitic helminths of vertebrates consists of a wide range of materials, e.g., blood, bile, mucus, tissue, secretions, cells, intestinal contents, etc., whose nutritional properties are difficult to replace by defined media. In invertebrates, the range of nutritive materials utilized by parasites is even more diverse. Like most organisms, some helminth species appear to require specific nutrients or growth factors before they will grow or reproduce normally; moreover, these may be required to be presented in a specific form (e.g., liquid, particulate, or solid, etc.). Trematodes, for example, appear to require some specific requirement (nutritional?; culture condition?) before eggs with normal eggshells are produced *in vitro* and this still remains the major unsolved problem in trematode development. Intraspecific nutritional differences, too, may be important. For example, in cestodes, although a large number of isolates of *Echinococcus granulosus* (from sheep, camel, goat, etc.) have been grown to sexual maturity *in vitro*, the horse isolate of this species has so far resisted all attempts to culture it, and it appears to have some very specific growth requirement at present not understood.[8]

C. STERILITY

Many stages of the life cycles of helminths involve living in nonsterile environments (such as the intestine) so that elimination of contaminating microorganisms is necessary before initial axenic culture can be achieved. A more satisfactory alternative, however, is to avoid this problem by commencing culture with stages (e.g., metacercariae, plerocercoids, and encysted nematodes) which occur in sterile habitats within the organs or tissues of vertebrates or invertebrates. Sterility used to be the major problem inhibiting workers from even attempting *in vitro* culture of parasitic helminths, but the advent of antibiotics has changed this situation and their application now makes it possible to obtain almost all stages of most species in a sterile condition.

D. ELIMINATION OF METABOLIC WASTE PRODUCTS

In the host habitat, the natural circulation of body fluids rapidly removes the metabolic waste products of a worm from the site of their production. For successful *in vitro* culture, the conditions provided must also allow for the removal of toxic waste products. Unless a continuous, unlimited flow system (from a reservoir) is available, this normally involves renewing media at intervals or using a complex recirculating system.

E. PROVISION OF "TRIGGER" STIMULI

The complex nature of many helminth life cycles means that each stage of development (e.g., the L_3 to L_4 molt in nematodes or the cercaria/schistosomulum transformation in trematodes) may require, as well as differing nutritional requirements, exposure to specific (usually physiological) triggers before stage transformation can take place. The provision of such triggers has represented a major challenge to workers in this field, and a wide range of such triggers has been identified. In cestodes, the following are examples of some which have proved to be necessary for triggering sexual differentiation: provision of a suitable substrate (*Echinococcus granulosus*);[10] presence of trypsin (*Mesocestoides lineatus*);[11] anaerobic conditions (*M. corti*);[12] and presence of hemin (*Hymenolepis microstoma*).[13] In trematodes, the cercaria/schistosomulum transformation can be brought about by a variety of methods, such as syringe passage, centrifuging, shaking vigorously, etc.[7] Similar triggers are likely to be required during many stages of nematode life cycles.

F. SPATIAL RELATIONSHIPS

In exceptional circumstances, the spatial relations of a worm with its *in vitro* environment may play an important role in development. For example, if cultured to sexual maturity free in a liquid medium, the cestode, *Schistocephalus solidus*, produces only infertile eggs, and (after some years of failure!) it was found that the organisms required compression during maturation[2] in order for the cirrus to be inserted into the vagina and to bring about insemination and fertilization (see Chapter 5, Figure 6). The insemination/fertilization problem, however, has never been solved in the case of some other cestode species, e.g., *E. granulosus* and *E. multilocularis* (which are also self-fertilizing),[14] and this also remains a major problem in many nematode and trematode *in vitro* culture systems.

IV. CRITERIA FOR ASSESSING GROWTH AND DEVELOPMENT

In estimating the success (or otherwise) of a particular *in vitro* technique, it is important to be able to distinguish between the mere "survival" of a worm and true "growth" and "development". These terms are interpreted here as follows.

Survival — Survival is the maintenance of an organism *in vitro* under conditions which allow the metabolism to operate at a level to keep the cells and tissues alive at their present level of differentiation.

Growth — Growth is the occurrence of cell division and tissue synthesis resulting in increase in cell numbers, but not necessarily in further differentiation.

Development — Development is the differentiation of cells, tissues, and organs and their resulting sequential transformation into the various stages of the life cycle from egg to adult.

Maturation — Maturation is the development of the reproductive system resulting in the production of viable spermatozoa and ova.

It cannot be too strongly urged that in order to obtain a reliable assessment of these processes *in vitro*, it is essential to work out in detail the pattern of growth, development, and maturation in the *normal* life cycle, so that appropriate criteria of progress can be established. It is surprising to find that even in some of the species widely used for experimental studies, that detailed knowledge of various basic aspects of the life cycle is still unknown. In this text, where appropriate, an attempt is made to provide details of the morphology of the adult parasite and its larval stages and to briefly describe its life cycle, with relevant data on the growth and development *in vivo*, if available.

V. LITERATURE REVIEWS ON HELMINTH *IN VITRO* CULTURE

The early literature on helminth *in vitro* culture has been reviewed by Taylor and Baker[15,16] and Smyth.[6] More recent work has been reviewed by Taylor and Baker,[17] Smyth and Halton,[7] and Smyth and McManus.[8]

REFERENCES

1. **Smyth, J. D.,** Studies on tapeworm physiology. I. Cultivation of *Schistocephalus solidus in vitro, J. Exp. Biol.,* 23, 47, 1946.
2. **Smyth, J. D.,** Studies on tapeworm physiology. VII. Fertilization of *Schistocephalus solidus in vitro, Exp. Parasitol.,* 3, 64, 1954.
3. **Crompton, D. W. T.,** *An Ecological Approach to Acanthocephalan Physiology,* Cambridge University Press, Cambridge, 1970.
4. **Crompton, D. W. T.,** The sites occupied by some parasitic helminths in the alimentary tract of vertebrates, *Biol. Rev.,* 48, 27, 1973.
5. **Kennedy, C. R., Ed.** *Ecological Aspects of Parasitology,* North-Holland, Amsterdam, 1979.
6. **Smyth, J. D.,** *An Introduction to Animal Parasitology,* 2nd ed., Edward Arnold, London, 1976.
7. **Smyth, J. D. and Halton, D. W.,** *The Physiology of Trematodes,* 2nd ed., Cambridge University Press, Cambridge, 1983.
8. **Smyth, J. D. and McManus, D. P.,** *The Physiology and Biochemistry of Cestodes,* 2nd ed., Cambridge University Press, Cambridge, 1989.
9. **Mettrick, D. F. and Podesta, R. B.,** Ecological and physiological aspects of helminth-host interactions in the mammalian gastrointestinal canal, *Adv. Parasitol.,* 12, 183, 1974.
10. **Smyth, J. D.,** *Echinococcus granulosus and E. multilocularis: in vitro* culture of the strobilar stages from protoscoleces, *Angew. Parasitol.,* 20, 137, 1979.
11. **Kawamoto, F., Fujioka, H., and Kumada, N.,** Studies on the post-larval development of cestodes of the genus *Mesocestoides:* trypsin-induced development of *M. lineatus in vitro, Int. J. Parasitol.,* 16, 333, 1986.
12. **Ong, S. J. and Smyth, J. D.,** The effects of some culture factors on sexual differentiation of *Mesocestoides corti* grown from tetrathyridia *in vitro, Int. J. Parasitol.,* 16, 361, 1986.
13. **Seidel, J. S.,** Hemin as a requirement in the development *in vitro* of *Hymenolepis microstoma, J. Parasitol.,* 57, 566, 1971.
14. **Smyth, J. D.,** The insemination-fertilization problem in cestodes cultured *in vitro,* in *Aspects of Parasitology,* Meerovitch, E., Ed., McGill University, Montreal, 1982, 393.
15. **Taylor, A. E. R. and Baker, J. R.,** *The Cultivation of Parasites In Vitro,* Blackwell Scientific, Oxford, 1968.
16. **Taylor, A. E. R. and Baker, J. R.,** *Methods of Cultivating Parasites In Vitro,* Academic Press, London, 1978.
17. **Taylor, A. E. R. and Baker, J. R.,** *In Vitro Methods for Parasite Cultivation,* Academic Press, London, 1987.

REFERENCES

Chapter 2

TREMATODA: BASIC PROBLEMS AND APPROACHES

J. D. Smyth

TABLE OF CONTENTS

I. GENERAL CONSIDERATIONS

A. SPECIES AND STAGES STUDIED

1. Metacercariae

The most successful *in vitro* culture of trematodes to date has been obtained using metacercariae (e.g., *Sphaeridiotrema globulus*) or, more rarely, cercariae (e.g., *Schistosoma mansoni*) as starting material. Culture attempts using adults have generally not been satisfactory and tend to be essentially records of how long a particular species "survived" under the conditions provided.

The metacercariae of some species exhibit *progenesis* (see Chapter 3). With the genitalia being partially differentiated, a simple stimulus, such as a rise in temperature, is sometimes sufficient to induce the completion of spermatogenesis, oogenesis, and vitellogenesis. Thus, the metacercariae of *Codonocephalus urniger* will mature in 48 h in saline alone, if cultured at 40°C.[1] However, these species are the exceptions and the metacercariae of many common species (e.g., *Fasciola* and *Diplostomum*) are not progenetic and are much more demanding in their culture requirements.

When seeking material for *in vitro* experiments, it should be remembered that many commonly available invertebrates and vertebrates (especially arthropods, fish, and amphibia) serve as intermediate hosts for trematodes and may contain metacercariae. It is thus worthwhile for workers in different countries to examine their local fauna for trematode larvae and, in this way, as yet untried species suitable for *in vitro* experimentation may be discovered.

2. Cercariae

Procedures for the artificial transformation of cercariae into metacercariae have made it possible to use cercariae as starting material, but most experimental work has been confined to metacercariae (see Chapters 3 and 4).

B. EXCYSTMENT OF METACERCARIAE

Depending on the species and its location in the host, metacercariae may be *free* (e.g., in North America, *Diplostomum baeri eucaliae* in the brain of the brook stickleback, *Eucalia inconstans*; in Europe, *Diplostomum phoxini* in the brain of the minnow) or *encysted* in its intermediate host (e.g., the cosmopolitan species, *Sphaeridiotrema globulus*) or on vegetation (e.g., *Fasciola hepatica*). It is common for the metacercarial cyst wall, when present, to consist of several layers, in which case consecutive treatments with both pepsin and trypsin are likely to be necessary to bring about excystment. The gas phase, especially the pCO_2, and the oxidation-reduction potential (Eh) may also need to be controlled and, as in the case of *Fasciola*, these conditions appear to play a part in stimulating the release of an endogenous secretion which assists the excystment process.[2] In some species (e.g., *Bucephaloides*, a common larva in the brain of marine fish), the thickness of the cyst wall differs in different sites; those from the nasal region and orbit have thicker walls (and hence require longer pepsin treatment) than those from the cranial fluid of the brain, which hatch more readily.[3]

In some encysted species (*Parvatrema timondavidi*, Chapter 3), the cyst wall is so thin that merely raising the temperature to 37 to 40°C is sufficient to bring about excystment.[4] Although the excystment techniques used by different authors are referred to under different species, it is likely that this technique can be varied, within reasonable limits, with equally satisfactory results.

C. OBTAINING STERILE METACERCARIAE FOR CULTURE

1. Without Using Antibiotics

It is possible to obtain metacercariae or tetratacotyles of some species in a sterile condition without using antibiotics, as, indeed, was the practice in the early pioneer days of *in vitro* culture before the advent of antibiotics.[5,6] This can be achieved by taking care to sterilize the surface of the host before dissecting out the larvae. This is best achieved by painting the host surface (e.g., the skin of a fish) with alcoholic iodine (1% in 90% ethanol), dissecting out the cysts or free larvae with sterile instruments, and transferring them to sterile saline.[6] As a further precaution, the larvae may be washed at least four times with sterile saline, either in tubes, by allowing them to settle, after shaking each time in tubes, or in a rotator, if available. It is convenient to hold small fish in a retort clamp during dissection; this leaves the hands free and enables larvae to be dissected out at eye level.

2. Using Antibiotics

Even if antibiotics are to be used, it is still advisable to minimize the risk of chance contamination by using the technique described above, or a similar one, to obtain the initial culture material as free from microorganisms as possible. Encysted metacercariae will clearly be much easier to surface sterilize if contaminated than, say, metacercariae of species (e.g., *Diplostomum*) which lie free in the host tissues. The use of acid pepsin for cyst excystment (see below) usually results in released larvae being in a sterile condition.

The antibiotics widely used are as follows: penicillin (100 IU/ml) + streptomycin (100 μg/ml) and/or gentamycin (100 μg/ml). Penicillin and streptomycin are most conveniently purchased as a concentrated solution of 5000 IU/ml and 5000 μg/ml, respectively, kept frozen at −20°C until required, and then stored at 4°C, and the addition of 2 ml/100 ml gives the required concentration shown above. Gentamycin is usually provided in a concentrated solution of 50 mg/ml. The addition of 2 ml/l of gentamycin gives the required concentration. The use of fungicides, such as myostatin or nyostatin, may be necessary, in some cases, if fungal infections are encountered.

NOTE: Workers in laboratories where antibiotics are not available should not be discouraged from attempting *in vitro* culture. Careful use of aseptic procedures and of the techniques outlined above in Section I.C.1., especially repeated washing, can readily enable sterile cultures to be obtained, although some experimentation may be necessary to overcome initial problems with different species.

D. CULTURE MEDIA
1. Commercially Prepared Media

A wide range of synthetic media is now available; those most used for helminth culture have been Parker 199, Parker 858, NCTC 135, NCTC 108, Eagle's Medium, RMPI 1640, and CMRL 1066. For details of the composition and source of these media, consult standard tissue culture texts.[7,8] Although various authors have used specific media for various species, it is often worthwhile to experiment with a substitute medium if the recommended one is not available.

2. Naturally Prepared Media

In laboratories which may not have the facilities (or funds) to obtain commercial media, it should be pointed out that it is quite often possible to prepare satisfactory alternative media by utilizing natural products such as blood, hen eggs, organ extracts (e.g., liver), or various "broths" as used by the early bacteriologists. The latter are prepared by digesting fresh meat in pepsin or trypsin, filtering and/or autoclaving the extract, and buffering it appropriately. One of the most widely used media was peptone broth, the preparation of which is given in standard microbiological texts.[9]

E. BUFFERS

All helminths produce quantities of acidic metabolic byproducts,[10] and trematodes are no exception, and efficient buffering is essential for successful cultures. Until recently, the standard buffers for culture media were $NaHCO_3$ and various phosphate buffers. These have now largely been replaced by organic buffers (of which the most widely used is HEPES which maintains a steady pH about 7.4) which are not subject to pH shift by CO_2 loss. However, as a source of CO_2 is often a metabolic requirement of helminth culture systems, a common practice is to use half the usual concentration of $NaHCO_3$ and make up the difference with HEPES. Recommended concentrations are 10 mM $NaHCO_3$ plus 20 mM HEPES.

F. CULTURE VESSELS

Almost any shape or size of vessel can be used for culturing helminths, the only criteria being that the size should be appropriate for the volume of medium used; screw-top vessels are convenient and are better for maintaining sterility, but rubber bungs or cotton wool plugs may also be used, with suitable precautions. Occasionally, certain types of glass may create problems by releasing ions (e.g., Na^+) which may upset the pH and/or the ionic balance. A wide range of presterilized plastic or glass culture vessels are now available commercially and are widely used in published techniques. However, even poorly equipped laboratories can often carry out rewarding experiments on *in vitro* culture by using a variety of vessels such as discarded medicine bottles or bottles used for domestic products. It should be pointed out that much of the early work in this field was carried out using simple equipment of this nature.

G. CRITERIA FOR ASSESSING GROWTH AND DEVELOPMENT
1. General Comment

As stressed in Chapter 1, an important prerequisite for *in vitro* culture attempts is to establish whether an organism is actually "developing" and not just "surviving". In order to determine this for a particular species, it is essential to be familiar with the normal pattern of development *in vivo*. In this text, details are provided for most of the species discussed. There are, however, a number of basic stages which are common to most species which can be used for assessing growth *in vitro*. These have been reviewed[2,10] and those appropriate to metacercariae development are listed in Table 1 and discussed further below. Some of these criteria may also be applied to pseudophyllidean cestodes (see Chapter 5) which, like trematodes, form sclerotin (quinone-tanned) eggshells.

2. Stages in Maturation[11] (Figure 1)

Stage 1: Cell multiplication — As in all embryonic development, a burst of mitotic activity characterizes the onset of growth and development. This stage cannot be recognized macroscopically, but, by inhibiting cell division in metaphase through the use of colchicine, mitoses can readily be detected in aceto-orcein squashes[11,12] (see Section II, Appendix, this chapter). To test for mitoses, colchicine is added to the culture medium, using a small sample, to produce a concentration of 10^{-4} and is incubated normally for 4 h and squashes are prepared. The inhibited mitoses can be counted under oil immersion.

NOTE: Colchicine is carcinogenic and should be handled with due precautions.

Stage 2: Body shaping — Cell division results in the lengthening of the body, particularly in the posterior region where the genitalia will develop.

TABLE 1

Criteria Recommended for the Recognition of Developmental Stages of Trematodes From Metacercaria to Adult (Based on *Diplostomum*) and Pseudophyllidean Cestodes From Plerocercoid to Adult (Based on *Diphyllobothrium*)

Phase	Time in host		Criterion recommended	Method of detection
	Diplostomum (h)	*Diphyllobothrium* (d)		
(1) Cell multiplication	0—24	0—1	Mitoses counts	Aceto-orcein squashes after colchicine treatment[a]
(2) Segmentation or body shaping	12—24	1—2	Division into proglottids (*Diphyllobothrium*) or regions (*Diplostomum*)	Direct observation on living material or aceto-orcein squashes
(3) Organogeny	24—48	2—3	Appearance of uterus and testes primordia	Squashes or whole mounts
(4) Early gametogeny	36—40	4—5	Appearance of "rosette" and "comma" stages in spermatogenesis	Squashes
(5) Late gametogeny	40—48	5—6	Appearance of mature spermatozoa	Squashes or unstained teases
(6) Eggshell formation and vitellogenesis	55—60	6—7	Presence of eggshell precursors in "vitelline" cells	Histochemical tests on whole specimens Diazo[+ve], catechol[+ve,b]
(7) Oviposition	60—72	7—8	Appearance of fully formed egg	Direct observations on living material or catechol-treated whole mounts

[a] For the preparation of aceto-orcein and its use in tissue "squashes", see Section II, Appendix, and Figure 2.

[b] In trematode species which form quinone-tanned eggs and in pseudophyllidean cestodes, the eggshell precursors give positive reactions with the diazo test (for phenols) and the catechol test (for phenol oxidase). For details see Section II, Appendix.

From Bell, E. J. and Smyth, J. D., *Parasitology*, 48, 131, 1958. With permission.

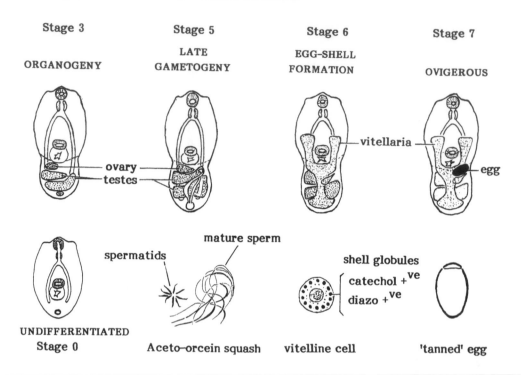

FIGURE 1.   Criteria for the recognition of the major stages in the development of an undifferentiated metacercaria to an adult trematode based on *Diplostomum phoxini* (also see Figure 1, Chapter 3). (Modified from Bell, E. J. and Smyth, J. D., *Parasitology*, 48, 131, 1958.)

Stage 3: Organogeny — Organogeny is characterized by the appearance of the anlagen of the genitalia (already present in progenetic forms) which can be seen in aceto-orcein squashes or whole mount preparations.

Stage 4: Early gametogeny — Early gametogeny, a characteristic and easily recognizable stage, is marked by the appearance of the beginning of spermatogenesis, easily recognizable in aceto-orcein squashes (Figure 1). The corresponding phases of oogenesis are less easy to recognize although developing ova have some affinity for neutral red (0.1% in BSS) in fresh preparations.

Stage 5: Late gametogeny — Late gametogeny is characterized by the appearance of mature spermatozoa in the testes.

Stage 6: Eggshell formation and vitellogenesis — Eggshell formation and vitellogenesis are characterized by the appearance of eggshell precursors in the vitelline cells. In species which form sclerotin eggshells (the majority of trematodes), the most readily identifiable precursor is a phenolic compound which gives strong color reactions with stable diazotates (e.g., Fast Red Salt B), a reaction easily recognizable in whole mounts or sections. Other precursors identifiable by cytochemical means are basic proteins and the enzyme, polyphenol oxidase. For technical details of applying these techniques,[11,13,14] see Section II, Appendix, this chapter.

Stage 7: Oviposition — Oviposition is characterized by the appearance of fully formed eggs.

Comparable stages for the dioecious schistosomes are given in Figure 13, chapter 4.

II. APPENDIX

A. ACETO-ORCEIN: PREPARATION

| Orcein (synthetic) | 1 g |
| 45% acetic acid | 100 ml |

a. Heat to boiling and simmer for 1 h in a fume cupboard, preferably under a reflux condenser.
b. Cool, filter, and store in a dark cupboard.
c. Before immediate use, filter just a small amount, as the solution is inclined to precipitate.
d. Tissues to be stained must be freed of saline or media by washing in water (see procedure below).

B. ACETO-ORCEIN: SQUASH TECHNIQUE (FIGURE 2)
1. Initial Procedure

a. Transfer metacercaria (or other tissue fragment) into a watch glass or deep cavity slide.
b. Draw off medium with a fine Pasteur pipette and replace it with basic saline (BSS).
c. Quickly repeat washing three times to remove all medium.
d. Finally, remove as much saline as possible and replace with freshly filtered aceto-orcein. If not washed sufficiently, precipitations may occur.
e. Cover cavity slide with a similar inverted slide or glass plate and incubate at 37 to 40°C. Very small metacercariae may be stained in 2 to 5 min; some take 5 to 15 min. Large, whole trematodes and cestode proglottids may require 15 + min, but appropriate staining times may vary substantially with species and can only be determined by trial and error.

2. Squashing

a. Transfer the stained tissue to a slide in a drop of aceto-orcein, add a cover glass (No. 0 or No. 1), and draw off surplus fluid with filter paper. Examine for basic morphology at low power.
b. To examine for cytological detail, place a piece of filter paper over the cover glass and squash firmly with the thumb. The tissues (now soft and macerated) will spread out and much cytological detail can be seen, especially at oil immersion level.
c. Such squashes only have a life of a few hours unless the edges are sealed. Nail varnish or stencil correcting fluid is excellent for this purpose, and preparations so sealed may be usable for 24 h or longer.

C. CYTOCHEMICAL TESTS FOR EGGSHELL PRECURSORS

The majority of trematodes form a quinone-tanned or sclerotin eggshell by the following mechanisms:[2]

$$\text{Protein} + \text{diphenol} \xrightarrow{\text{phenol oxidase}} \text{protein} + \text{quinone} = \text{tanned protein (sclerotin)}$$

The basic constituents of sclerotin are thus *protein*, *phenolic substances*, and the enzyme, *phenol oxidase*. From this it follows that the eggshell-forming material in the vitellaria (where it appears as globules), and later in the uterus, may be traced by means of cytological

1. Place a fragment of worm tissue in a deep cavity slide in a little culture medium.

6. Remove top covering slide, _before_ removing preparation; spillage can result in slides sticking together.

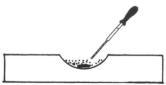

2. Withdraw medium with a fine pipette and replace with BSS. Leave 2–3 mins, withdraw liquid and replace with fresh BSS. Repeat for several rinsings.

7. Remove tissue in a little aceto-orcein and transfer to a flat slide. Cover with a No.0 or 1 cover glass, avoiding bubbles. Run in more aceto-orcein, if needed.

3. Withdraw BSS; replace by aceto-orcein (freshly filtered!). Leave for 2–3 mins and again replace with fresh aceto-orcein.

8. Observe preparation for gross morphology (testes etc.), before squashing for cytological details.

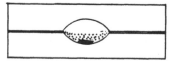

4. Incubate preparation by placing slide on _floor_ of 37–40°C incubator, and (not before) cover carefully with an inverted cavity slide to make a closed chamber.

5. Incubate preparation for:
2–5 mins – very small trematodes;
5–15 mins – small fragments of worms;
15–60 mins – large trematodes or whole proglottides of cestodes.

9. SQUASH: for cytology. Cover with filter paper and squash by pressing down firmly with a thumb. Further squashing pressure can be applied by rolling a round-bodied pencil on top.

FIGURE 2.   Technique for the preparation of aceto-orcein squashes of trematode or cestode tissues.

tests for these substances. Some species form keratin or other forms of eggshell which cannot be stained by these methods.[2]

1. Diazo Technique for Phenols [11,13,14]

Eggshell precursors first appear as globules in the vitelline cells and are readily detected by the Diazo technique. This method also works well in whole mounts after fixation in 70% ethanol.

Procedure

a.   Flatten and fix in 70% ethanol, 2 to 24 h.
b.   Wash in distilled water, 15 min.
c.   Place in 1% aqueous Fast Red Salt B (freshly prepared and filtered) in a watch glass or small petri dish and observe under a low-power binocular. In small trematodes (e.g., *Diplostomum*), the phenolic globules in the vitelline cells become colored in

0.25 to 3 min. Larger trematodes (e.g., *Fasciola* and *Schistosoma* may require 15 to 20 min.

d.   Wash in water, 15 min.

e.   (Optional) Counterstain lightly in Gower's Carmine, or similar stain; differentiate in acid alcohol if necessary.

f.   Dehydrate, clear, and mount.

Result

The color developed by the eggshell precursors depends on the nature of the phenolic compound(s) in that particular species. It ranges from yellow-orange to bright red, the vitellaria standing out against an almost colorless background. Because the phenol is converted to quinone in the fully formed sclerotin eggshell, the latter becomes diazo^{-ve} in the uterus.[11]

2. Catechol Technique for Phenol Oxidase[11,13,14]

When incubated in catechol (a di-phenol) as a substrate, phenol oxidase converts catechol into a quinone which "tans" the eggshell protein precursors and results in a yellow to brown color developing at the enzyme site. In some trematodes, e.g., *Diplostomum*, the reddish color which develops is almost as brilliant as that of the diazo technique; in others it is much paler. The catechol test[14] has the advantage over the diazo test in that it not only gives a color reaction with the eggshell precursors in the vitellaria, but also with the completed eggshell. This is useful for demonstrating the position and shape of eggs in the uterus and especially for revealing abnormalities in eggshell development.

Procedure

a.   Fix and flatten specimens in 70% ethanol, 2 to 12 h.

b.   Rinse in distilled water, 30 min.

c.   Incubate in 0.1% aqueous catechol (freshly prepared just before use) at 37°C for 5 to 30 min. Too long of an incubation is to be avoided as the catechol solution slowly self-oxidizes on exposure to air.

d.   Dehydrate, clear, and mount in balsam.

REFERENCES

1. **Dollfus, R. Ph., Timon-David, J., and Rebecq, J.,** Maturité gentitalé provoquée experimentalement chez *Codonocephalus urniger* (Rudolphi) (Trematoda, Strigeidae), *C. R. Acad. Sci.,* 242, 2997, 1956.
2. **Smyth, J. D. and Halton, D. W.,** *The Physiology of Trematodes,* 2nd ed., Cambridge University Press, Cambridge, 1983.
3. **Johnston, B. R. and Halton, D. W.,** Excystation *in vitro* of *Bucephaloides gracilescens* metacercaria (Trematoda: Bucephalidae), *Z. Parasitenkd.,* 65, 71, 1981.
4. **Yasuraoka, K., Kaiho, M., Hata, H., and Endo, T.,** Growth *in vitro* of *Parvatrema timondavidi* Bartoli 1963 (Trematoda: Gymnophallidae) from metacercarial stage to egg production, *Parasitology,* 68, 293, 1974.
5. **Ferguson, M. S.,** Excystment and sterilization of metacercariae of the avian strigeid trematode *Posthodiplostomum minimum,* and their development into adult worms in sterile conditions, *J. Parasitol.,* 26, 359, 1940.
6. **Smyth, J. D.,** Studies on tapeworm physiology. I. Cultivation of *Schistocephalus solidus in vitro, J. Exp. Biol.,* 23, 47, 1946.
7. **Paul, J.,** *Cell and Tissue Culture,* 5th ed., Churchill Livingstone, Edinburgh, 1975.

8. **Taylor, A. E. R. and Baker, J. R.**, *In Vitro Methods for Parasite Cultivation*, Academic Press, London, 1987.

9. **Collins, C. H. and Lyne, P. M.**, *Microbiological Methods*, 5th ed., Butterworths, London, 1984.

10. **Smyth, J. D.**, *An Introduction to Animal Parasitology*, Hodder & Stoughton, London, 1976.

11. **Bell, E. J. and Smyth, J. D.**, Cytological and histochemical criteria for evaluating development of trematodes and pseudophyllean cestodes *in vivo* and *in vitro*, *Parasitology*, 48, 131, 1958.

12. **Smyth, J. D.**, Studies on tapeworm physiology. VIII. Occurrence of somatic mitoses in *Diphyllobothrium* spp. and its use as a criterion for assessing growth, *in vitro*, *Exp. Parasitol.*, 5, 260, 1956.

13. **Johri, L. N. and Smyth, J. D.**, A histochemical approach to the study of helminth morphology, *Parasitology*, 47, 21, 1957.

14. **Smyth, J. D.**, A technique for the histochemical demonstration of polyphenol oxidase and its application to egg-shell formation in helminths and byssus formation in *Mytilus*, *Q. J. Microsc. Sci.*, 95, 139, 1954.

Chapter 3

TREMATODA: DIGENEA WITH PROGENETIC METACERCARIA

J. D. Smyth

TABLE OF CONTENTS

I. DEFINITIONS: PROGENESIS AND NEOTENY

Progenesis is defined as the advanced development of genitalia in a larval form, and a metacercaria which contains the anlagen (primordia) of the genitalia is said to be *progenetic*. The related term *neoteny* is used for the phenomenon in which the gonads reach maturity in a larval form. An advanced progenetic condition clearly borders on neoteny and the terms *neotentic* and *progenetic* are rather loosely used in the literature and are only definable and applicable within broad limits.

Based on the developmental stages in Table 1 (Chapter 2) and Figure 1, the degree of metacercarial "maturation" can be designated as Stage 0: *undifferentiated* (i.e., nonprogenetic); Stage 3: *organogeny* (i.e., progenetic), etc. The position of some of the species, whose *in vitro* culture is discussed in this text, is shown in Figure 1.

The occurrence of neotenic metacercariae (i.e., those which actually become ovigerous in the intermediate host) is more common than is generally appreciated. Buttner[1] has reviewed numerous species exhibiting this phenomenon. A good example is *Coitocaecum anaspidis* in the freshwater crustacean *Anaspides tasmaniae*.[2]

II. *POSTHODIPLOSTOMUM MINIMUM*

Cultured by: Ferguson[3]
Natural definitive hosts: numerous birds, especially herons
Experimental hosts: chicks[4]
Habitat: intestine
Prepatent period: 34 h[4]
Molluscan host: *Physa* spp.[4]
Second intermediate host: numerous freshwater fish
Distribution: cosmopolitan

A. LIFE CYCLE (FIGURE 2)
1. Adult (Figure 3)
P. minimum is a common strigeoid trematode parasitizing birds of the order Ciconiiformes, particularly herons, the great blue heron, *Ardea herodias*, being commonly infected.[5] It is, however, exceptionally nonhost specific and — as well as a wide range of birds — amphibia, reptiles, and mammals have been experimentally infected.[6] An unusual feature of the adult is its "morphological plasticity" and specimens from different hosts show variations in such features as the size and shape of the forebody, hindbody, suckers, and eggs.[5] This species may be part of a species complex and there is evidence that at least two physiological strains may occur.[7]

2. Molluscan Stages
The eggs are unembryonated when laid and the molluscan hosts are species of *Physa* (e.g., *P. sayii* and *P. gyrina*[7]) in which mother and daughter sporocysts develop.

3. Cercaria
The cercaria are of the longifurcate monostoma type and are phototropic.[4]

4. Metacercaria
Like the adult worm, the metacercaria is exceptionally nonhost specific and cysts have been recorded from 97 species of 18 families of freshwater fish.[7] Cysts show some predilection for the kidney, liver, pericardium, and spleen, but can be found in any organ. After cercarial penetration, genital anlagen appear first at 14 d and are well developed by 27 d.

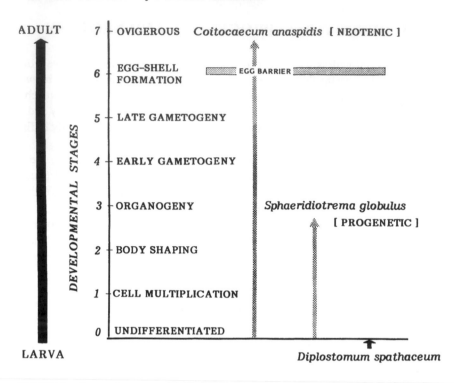

FIGURE 1.   Stages of metacercarial development reached by different species in the intermediate host. The metacercariae of *D. spathaceum* are entirely undifferentiated; those of *S. globulus* are *progenetic* and possess well-formed genital anlagen; those of *C. anaspidis* are *neotenic* and reach full sexual maturity in the "intermediate" crustacean host.

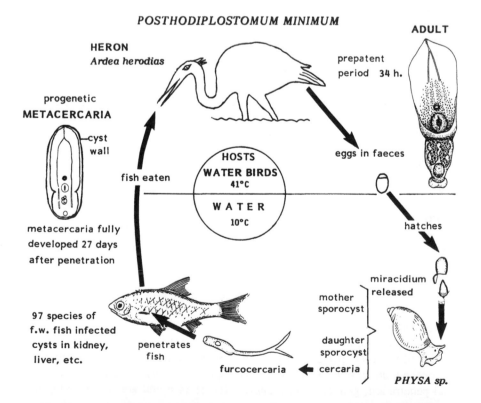

FIGURE 2.   *Posthodiplostomum minimum*: life cycle. (Based on data from Hoffman[7] and Miller.[4])

POSTHODIPLOSTOMUM MINIMUM

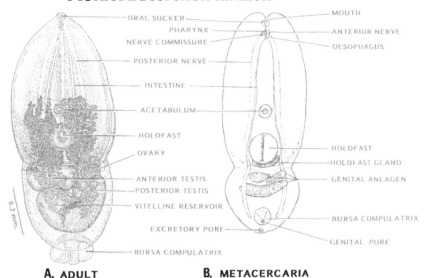

A. ADULT   B. METACERCARIA

FIGURE 3. *Posthodiplostomum minimum.* (A) Adult. (Modified from Ulmer, M. J., *J. Parasitol.*, 47, 608, 1961. With permission.) (B) Metacercaria. (Modified from Hughes, R. C., *Trans. Am. Microsc. Sci.*, 47, 320, 1928. With permission.)

A thin membranous cyst, of parasite origin, appears at about 19 to 21 d and is closely applied to the metacercaria; a host reaction cyst wall develops at about 29 d.[7]

B. *IN VITRO* CULTURE TECHNIQUE[3,7]

1. Source of Material

In the original experiment, metacercariae were obtained from the pumpkinseed sunfish (*Lepomis gibbosus*) and the liver and kidney were removed (nonaseptically) and digested in pepsin for 1 h. The concentration of this was not stated, but a 1% solution of pepsin in a basic saline (e.g., Hank's), adjusted to a pH of 2.0 (with HCl), is likely to be effective. After 1 h, cysts became free and some larvae excysted. After washing several times in sterile saline, larvae were finally excysted by a further treatment with sterile pepsin. A 5% solution of trypsin at pH 8.0 also resulted in effective excystment. The concentrations and times of exposure to both these enzymes are worth experimenting with.

2. Washing and Sterilization

The original technique[3] utilized gravity washing in repeated tubes of sterile saline, but saline + antibiotics and several rinses, with agitation on a roller tube system, would nowadays be more effective.

3. Culture Vessels

The original technique[3] used small tubes (13 × 100 mm) plugged with cotton wool and sealed with parafilm, but screw-top Leighton tubes (or similar) would clearly be more convenient.

4. Medium and Conditions

The original technique used 2 ml of Tyrode (diluted 5:3) plus three drops of yeast extract (concentration unstated). It is likely, however, that a well-balanced tissue culture medium (such as NCTC 109 or NCTC 1350) plus serum and yeast extract would be equally or more effective. The original gas phase was not stated (probably air?); culture temperature was 36°C.

5. Results

Although the results obtained with this species are relatively poor compared with species discussed later, they are included here for historical reasons as representing the first serious attempt to culture a larval trematode *in vitro*. In the best cultures, sexual maturation was achieved and eggs appeared in about 10 d. The eggs produced in culture, although normal in size, were abnormal in shape and appearance, with thin shells and abnormal yolk cell development.

C. EVALUATION

That the original culture system was unsatisfactory is reflected in the fact that only abnormal eggs were produced and the prepatent period of 10 d compares very unfavorably with that of 34 h *in vivo*. As with other species dealt with in this text, the failure to form normal eggshells is a common feature of trematode *in vitro* development.

III. *SPHAERIDIOTREMA GLOBULUS*

Cultured by: Berntzen and Macy[8]
Definitive hosts: swans, ducks, and other water fowl[9]
Experimental hosts: chickens,[9] ducklings,[8] and chick chorioallantois[10]
Prepatent period: 68 to 72 h
Location: lower small intestine and ceca
Molluscan hosts: species of *Bithynia, Fluminicola, Gabbia, Gonobiasis*,
  and *Oxytrema*
Second intermediate host: encysts in same molluscan host
Distribution: cosmopolitan

A. LIFE CYCLE (FIGURE 4)
1. Adult (Figure 5)

S. globulus (family Psilostomatidae) is a trematode with a worldwide distribution in a variety of water fowl such as the mute swan (*Cygnus ober*), the coot (*Fulica americana*), the lesser scamp (*Achya affinis*), and the muscovy duck (*Cairna moschata*). It can cause severe ulcerative, hemorrhagic enteritis, especially in swans.[9] Experimentally, it develops readily in ducks or chicks. The life cycle has been described by Szidat[11] and by Burns[12] (under the probable synonym of *S. spinoacetabulum*).

2. Molluscan Stages

Eggs are unembryonated when laid and measure 110 to 128 × 65 μm.[11] After passing through redial stages, cercariae are released.

3. Cercaria

This was described as *Cercaria helvetica XVII*.[11]

4. Metacercaria

Cercariae penetrate the same species of snails which serve as the first intermediate hosts and encyst on the inner side of the shell as well as between the shell and the mantle. The degree of progenesis reaches Stage 3 (see Figure 1), the anlagen of the cirrus sac, ovary, and testes being represented only by a cluster of cells.[8]

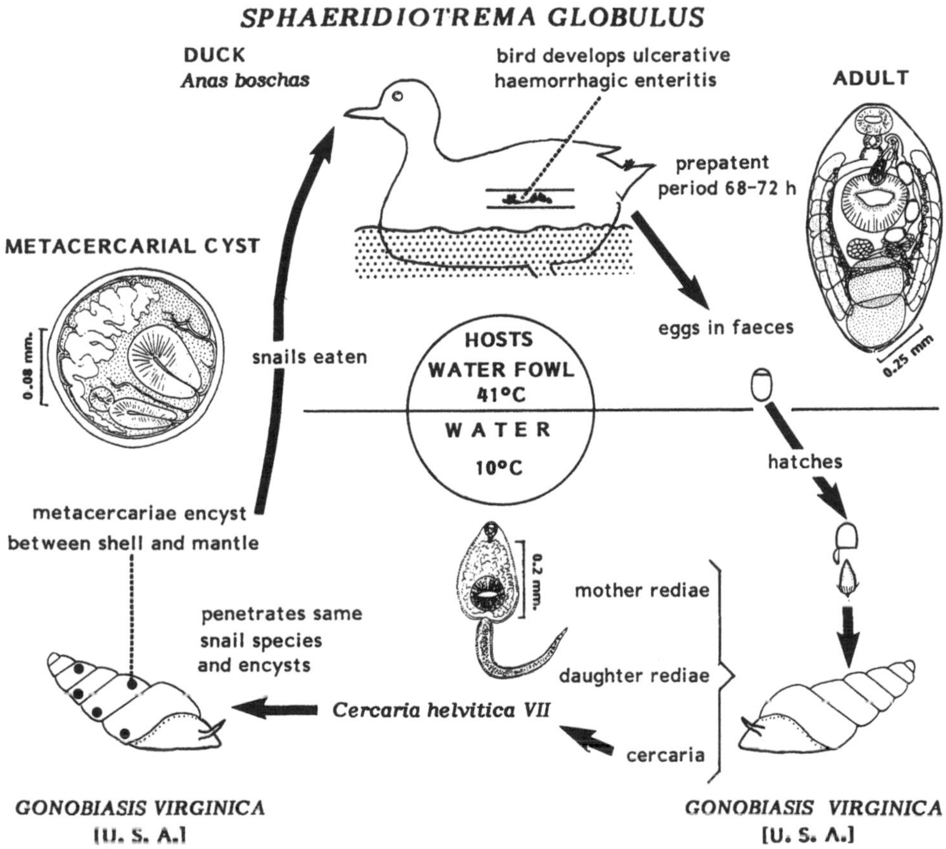

SPHAERIDIOTREMA GLOBULUS

FIGURE 4. *Sphaeridiotrema globulus*: life cycle. (Based on data from Fried and Huffman,[10] Szidat,[11] Burns,[12] and Macy and Ford.[13])

B. *IN VITRO* CULTURE TECHNIQUE[8]

1. Source of Material

Small numbers of metacercarial cysts can be obtained by crushing snails and dissecting the cysts out from the tissues.[14] For large numbers, digestion of whole snail tissues in 1% pepsin + 1% HCl in balanced saline is recommended.

2. Excystment

After isolation of cysts from snails, rinse in saline + antibiotics at 37°C for about 30 min. The original technique used 50 mg of streptomycin per milliliter and 25 mg of neomycin per milliliter, but basic antibiotic solutions are likely to be equally effective. After rinsing, aseptic procedures were followed. Cysts were excysted by treating with acid pepsin (as above) for 30 min, followed by 1% trypsin (1:250) in Earle's saline + $NaHCO_3$ at pH 8.3.

3. Culture Media and Conditions

The most satisfactory results were obtained in NCTC 109-y made up of 80% NCTC 109 + 20% egg yolk. Various methods of obtaining sterile egg yolk have been used by different workers; these authors (1) placed eggs in 70% ethanol for 30 min, (2) removed and opened the eggs under UV light, and (3) separated the yolk from the albumen, broke it up with a glass rod, and measured it into a measuring cylinder.

To each 20 ml of yolk, 80 ml of NCTC 109 was added and then centrifuged at 2000 × g for 1 h at room temperature. The resulting supernatant was labeled NCTC 109-

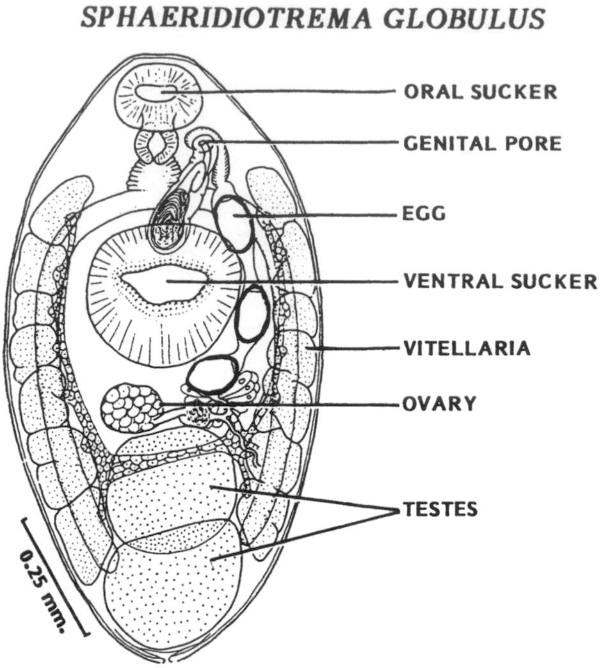

SPHAERIDIOTREMA GLOBULUS

FIGURE 5.   *Sphaeridiotrema globulus*: adult from ceca of domestic duck; experimental infection. (From Burns, J. P., *J. Parasitol.*, 47, 933, 1961. With permission.)

y. The pH was adjusted to 7.8 using 5% $NaHCO_3$. The best results were obtained with a gas phase of 10% O_2 + 10% CO_2 + 80% N_2. Metacercariae were cultured in 10 ml of medium in Kimax culture tubes (15 × 150 mm, screw top) at pH 7.8. The roller system was set to rotate for 5 min at 2-h intervals. Results at 42°C were better than at 37°C, which is not suprising considering that the adult is a bird parasite (see Figure 4).

4. Results

This is one of the few *in vitro* experiments in which detailed growth measurements *in vivo* were also made, so that the *in vitro* results could be critically compared with these directly. Although the size of the worms developed *in vitro* compared reasonably well with those *in vivo*, the gonads did not develop to the same extent.

Growth and development *in vitro* was slower than *in vivo* (Figure 6), worms becoming ovigerous *in vitro* at 126 h compared with 68 to 72 h *in vivo* in the duck. An average of 100 eggs per fluke per day were produced *in vitro* for periods up to 20 d, after which time degeneration set in. Of those produced by *in vitro* worms, some embryonated normally and hatched out (apparently) normal miracidia. Unfortunately, the ability of these miracidia to penetrate snails was not tested, nor was the percentage of viable eggs stated.

C. EVALUATION

This seems an excellent *in vitro* system with which to experiment further. The fact that apparently normal eggs were produced suggests that the eggshell formation system in this species may not be so nutritionally demanding as many other species.

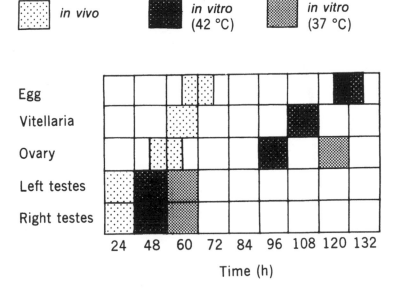

FIGURE 6. *Sphaeridiotrema globulus*: maturation of progenetic metacercariae *in vitro* at 37 and 42°C compared with that *in vivo*, in ducklings. (From Berntzen, A. K. and Macy, R. W., *J. Parasitol.*, 55, 136, 1969. With permission.)

IV. *PARVATREMA TIMONDAVIDI*

Cultured by: Yasuraoka et al.[15]
Definitive hosts: shore birds
Experimental hosts: mice[15] and herring gulls
Location: small intestine (ileum)
Pre patent period: 36 to 48 h
Molluscan host: *Tapes (Ruditapes) philippinarum,*
 Mytilus spp.
Second intermediate host: metacercariae "encyst" in same host
Distribution: Japan?

A. LIFE CYCLE (FIGURE 7)
1. Definitive and Intermediate Hosts

   P. timondavidi (family Gymnophallidae) is a parasite of shore birds throughout the world; these commonly become infected by eating mussels and clams containing metacercarial cysts. Yasuraoka et al.[15] failed to infect baby chicks or ducklings, but sexually mature worms (Figures 8 and 9) were recovered from mice 3 to 7 d postinfection (p.i.,), although eggs appeared as early as 36 h.

   Metacercariae "encyst" in the same species of mollusk and form contracted balls in gelatinous tissue in the inner part of the umbo. Although metacercariae are common in shore bivalves throughout the world, it is not known if there are one or two intermediate mollusk hosts. Bartoli[16] found numerous metacercariae in the mussel *Mytilus galloprovincialis*, but was unable to find sporocysts in this species or in numerous other lamellibranchs or prosobranchs in the same vicinity. It is likely then that there is only one intermediate host in which the cercariae "encyst" without being released from the mollusk,[17] but the earlier larval stages (sporocysts?; redia?) clearly must have rapid development phases.

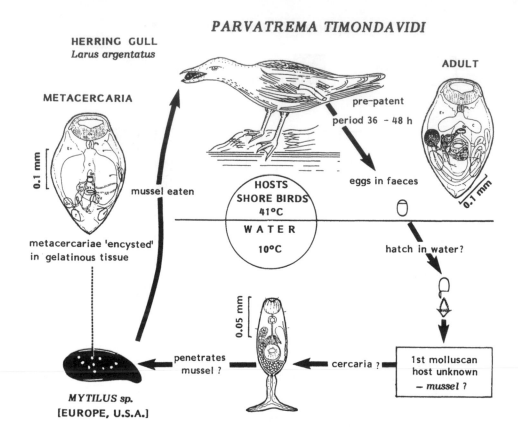

FIGURE 7.   *Parvatrema timondavidi*: life cycle. The molluscan stages are imperfectly known. Although metacercariae are common in several species of shore lamellibranchs throughout the world, it is not known if all the larval stages develop in the same species of mollusk intermediate host (and "encyst" there without escaping) or whether another species of mollusk is involved in the life cycle. The cercaria figured above is that of *Parvatrema borinquenae* which may be a synonym for *P. timondavidi*. (Based on data from Endo and Hoshina,[18] Yasuraoka et al.,[15] Bartoli,[16] and Cable.[17])

2. Metacercaria

The metacercaria (see Figure 8) is markedly progenetic,[18] reaching developmental stage 3 to 4 (see Figure 1) with well-developed (but immature) testes and ovary. The advanced state of progenesis in the testes is reflected in the fact that in the mouse, sperm appear as early as 24 h.[15]

B. *IN VITRO* CULTURE TECHNIQUE[15]

1. Source of Material

Metacercariae were extracted from the umbo region (adjacent to the hinge) of *T. philippinarum* where they are embedded in gelatinous tissue.

2. Excystment

Attempts to extract metacercariae from the viscous, gelatinous material surrounding them by using pepsin or trypsin failed. However, successful excystment was achieved by prolonged (8 h) incubation in a simple saline solution (0.85% sterile NaCl; pH 7.1) at 37°C, 70 to 80% excystment being achieved.

PARVATREMA TIMONDAVIDI

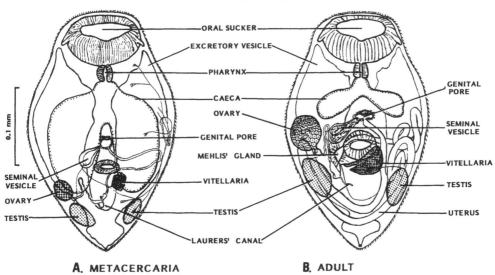

A. METACERCARIA   **B.** ADULT

FIGURE 8. *Parvatrema timondavidi*. (A) Metacercaria, from the clam, *Tapes (Ruditapes)* in Japan. (B) Adult, from experimental infection of mice. (Modified from Endo, T. and Hoshina, T., *Jpn. J. Parasitol.*, 23, 73, 1974. With permission.)

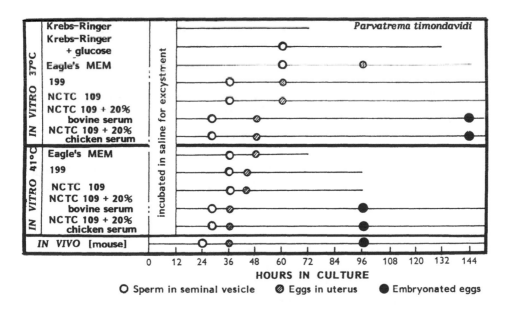

FIGURE 9. *Parvatrema timondavidi*: time of appearance of sperm and eggs *in vivo* and *in vitro*. (From Yasuraoka, K., Kaiho, M., Hata, H., and Endo, T., *Parasitology*, 68, 293, 1974. With permission.)

3. Culture Conditions and Media

Excysted metacercariae were washed in 5 to 10 ml of sterile Krebs-Ringer's-Trismalate (pH 7.1) in double-strength antibiotics (400 IU/ml of penicillin; 200 μg/ml of streptomycin) in 12 × 100 mm tubes. After settling, the washing process was repeated five times. Final cultures were set up in 2 ml of medium in 14 × 150 mm tubes, loosely capped, and incubated at 37 or 41°C in a gas phase of 8% CO_2. Tubes were placed on a rotator system set at 12 revolutions per hour. The media was replaced every second day. The following culture

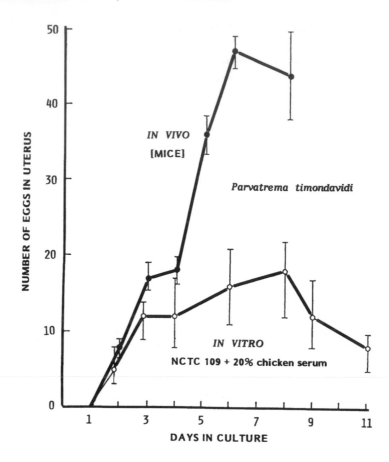

FIGURE 10. *Parvatrema timondavidi:* comparison of number of eggs in the uterus during development *in vivo* (in mice) and *in vitro*. Vertical bars represent 95% confidence limits. (Modified from Yasuraoka, K., Kaiho, M., Hata, H., and Endo, T., *Parasitology,* 68, 293, 1974. With permission.)

media were used: (1) Krebs-Ringer's-"Tris" solution with or without glucose, (2) Eagle's MEM, (3) Parker 199, (4) NCTC 109, and (5) NCTC supplemented with inactivated bovine or chick serum.

4. Results

Results are summarized in Figure 9. Best results were obtained in NCTC 109 with 20% bovine or chick serum, sperm appearing in 30 h in cultured worms compared with 24 h in *in vivo* worms. Growth at 41°C was much faster than at 37°C. The number of eggs in Medium 199 or NCTC 109 without serum was <20 eggs per uterus (Figure 10), whereas worms grown in mice contained 30 to 70 eggs per uterus. Some eggs (percent not stated) became embryonated and developed motile miracidia although (unfortunately) the ability of the latter to penetrate molluscan hosts was not tested.

C. Evaluation

The fact that some eggs embryonated and produced motile miracidia indicates that this is a promising result. Should the miracidia prove to be infective to snails, the species could be maintained without using an experimental animal definitive host, thus making it a valuable experimental system.

V. *MICROPHALLUS SIMILIS*

Cultured by: Davies and Smyth[19]
Chief definitive host: Common gull (*Larus canus*)
Experimental hosts: guinea pigs, rats, and mice[20]
Location: posterior small intestine
Pre patent period: 24 h — Stunkard;[21] 48 h — Davies and Smyth[19]
Molluscan host: *Littorina saxatalis, L.* spp.
Second intermediate host: metacercarial cysts in the "green" shore crab,
  Carcinus meanas (Europe), *Carcinides meanas* (U.S.)
Distribution: cosmopolitan

A. LIFE CYCLE (FIGURE 11)

1. Adult (Figure 12)

This species is a frequent intestinal parasite of the common gull in Europe and the U.S. and is probably cosmopolitan in distribution. It also occurs in terns and probably other shore birds. Its incidence, intensity, and distribution in the alimentary canal of gulls has been described in detail by Bakke.[22,23] The worm shows a tendency to establish itself in the posterior region of the intestine and in the rectum and cloaca.

2. Molluscan Stages

In Europe and North America, its first intermediate host is usually *L. saxatalis,* although it is also found in other *Littorina* spp. Eggs hatch only on ingestion by the snail and there are two generations of sporocysts. The infection causes parasitic castration of the snail.[24]

3. Metacercaria

The metacercariae (Figure 11), which occur chiefly in the digestive gland (hepatopancreas) of the crab, are markedly progenetic, reaching developmental stage 3 (see Figure 1) with well-formed, but immature, genital anlagen. Its ultrastucture has been investigated by Davies.[25] The advanced state of progenesis is reflected in the fact that, *in vivo,* eggs appear as early as 24 to 48 h p.i.

B. *IN VITRO* CULTURE TECHNIQUE[19]

1. Source of Material

Cysts were obtained from crabs (after pithing) by cutting away the carapace, removing the digestive gland, and washing in Hank's basic saline (BSS).

2. Excystment

Cysts readily excysted after treatment in 0.5% pepsin (pH 2.0) in BSS for 30 min followed by 0.5% trypsin (pH 7.0) in BSS for 10 to 30 min. Excysted metacercariae were washed in six changes of BSS + antibiotics (100 μg/ml of streptomycin; 100 IU of penicillin).

3. Culture Media and Conditions

Culture conditions included temperature, 38 to 41°C; gas phase, air, N_2, or $N_2 + CO_2$. Cultures were in Leighton tubes, 10 ml of medium and 50 to 100 organisms per tube. A very large number of media were tested (Table 1) and most culture conditions were capable of supporting spermatogenesis and vitellogenesis.

4. Results (Table 1)

Good results were obtained with a simple basic medium (BM) of NCTC 135 + 20%

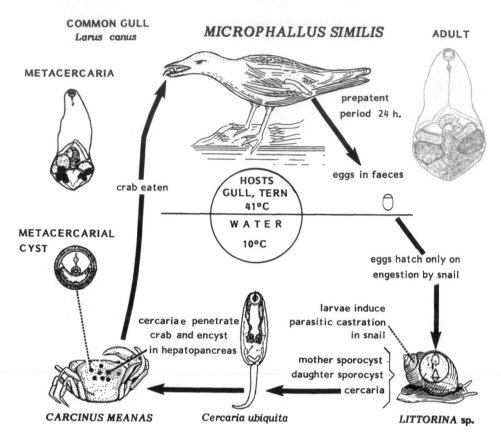

FIGURE 11.   *Microphallus similis*: life cycle. (Based on data from Stunkard[21] and Davies and Smyth.[19])

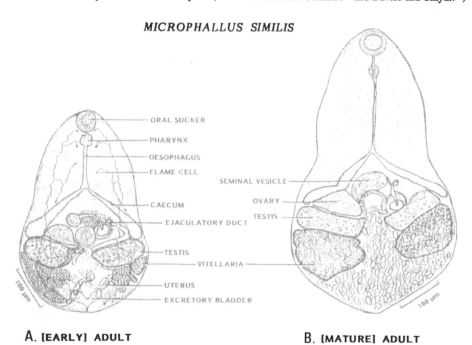

FIGURE 12.   *Microphallus similis*. (A) Early adult (24 h in mouse), eggs just appearing; (B) late adult, many days in mouse. (Modified from Stunkard, H. W., *Biol. Bull.*, 112, 254, 1957. With permission.)

TABLE 1

Microphallus similis: Egg Production *In Vitro* In Different Media

Culture medium[a]	% flukes with eggs	30 flukes examined on day 4			
		% flukes with some normal eggs	% flukes with some abnormal eggs	Max no. normal eggs/fluke	Max no. abnormal eggs/fluke
HBSS	All dead by day 3				
HBSS + 20% FCS (Flow)	86.7	30	86.7	61	117
NCTC 135	All dead by day 3				
NCTC 135 + 20% FCS (GIBCO)	83.3	50	83.3	17	28
NCTC + 20% FCS (Flow) = BM	100	96.7	100	44	59
BM + 10% MEE$_{50}$	All dead by day 3—no development				
BM + 10% CEE$_{50}$	All dead by day 3—no development				
BM + 1% BEE$_{50}$	56.7	0	56.7	0	52
BM + 5% BAF	83.3	30	83.4	24	86
BM + 5% yeast extract	80	36.7	80	17	102
BM + 0.02% DTT	Alive, but no development				
BM + 10% MIME·	All dead by day 3—no eggs				
BM + 50% egg macerate	86.7	23.3	86.7	21	30
Yolk medium + 20% FCS	All dead by day 3—no eggs				
BM + coag. NBCS with pits	96.7	66.7	96.7	46	168
BM + 0.2% rabbit r.b.c.	70	16.7	66.7	15	38
BM + 0.2% chick r.b.c.	73.3	10	73.3	10	49
BM + BHK monolayer	96.7	46.7	96.7	34	115
BM (anaerobic gas)	All dead by day 3—no development				

[a] FCS, fetal calf serum; BM, basic medium; MEE$_{50}$, mouse embryo extract; CEE$_{50}$, chick embryo extract; BEE$_{50}$, beef embryo extract; BAF, bovine amniotic fluid; DTT, dithiothreitol (Cleland's Reagent); MIME, mouse intestinal extract; NBCS, newborn calf serum; BHK, a cell line. For media preparation see original paper.

From Davies, C. and Smyth, J. D., *Int. J. Parasitol.*, 9, 261, 1979. With permission.

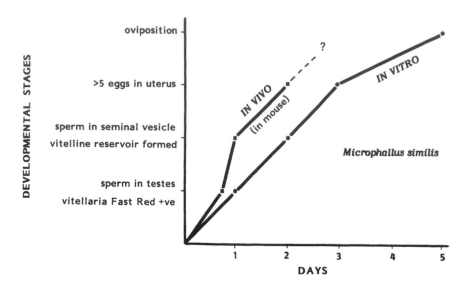

FIGURE 13.   *Microphallus similis*: rate of development *in vivo* (in mouse) compared with the best results *in vitro*. (From Davies, C. and Smyth, J. D., *Int. J. Parasitol.*, 9, 261, 1979. With permission.)

fetal calf serum, >80% of flukes producing eggs of which about 50% appeared to be normal. The addition of various supplements (such as chick embryo extract — CEE$_{50}$) did not improve results significantly and was, in fact, generally detrimental. Exceptions were the addition of a cell monolayer (BHK) or in a diphasic medium with substrate of coagulated newborn calf serum (NBCS) with pits in which the worms could burrow. In contrast to some species (e.g., *Diplostomum*), the addition of egg macerate did not appear to be advantageous.

The rate development *in vitro* was only slightly less than that *in vivo* (Figure 13). In the best results, most flukes contained some (apparently) morphologically normal eggs (N, Figure 14A and B), but abnormal eggs (A, Figure 14A) were also common. These abnormalities took various forms as follows:

1.   Tanned masses of vitelline cells with or without an associated ovum, but without a shell (Figure 14B, A$_1$)
2.   Small, shelled eggs containing vitelline cells, but no ovum (Figure 14B, A$_2$)
3.   Apparently normal eggs, but without an operculum or often with an irregularly thickened shell (Figure 14B, A$_3$)

Free ova were sometimes seen in the uterus. Attempts to embryonate eggs produced *in vitro* were unsuccessful and, after 2 weeks of incubation in sea water, the eggs still contained their original ova.

C. EVALUATION

The fact that eggs were produced in almost all media probably reflects the very advanced progenetic condition of the metacercaria. The failure of eggs to embryonate suggests that fertilization may not have taken place *in vitro*, a view supported by the fact that sperm were rarely seen in the uterus of cultured worms and then only in very small numbers. The production of abnormal eggshell is likely to be related to the occurrence of pretanning of the shell droplets in the vitellaria and vitelline ducts, a phenomenon reported in other species (e.g., *Fasciola hepatica* and *Microphalloides japonicus*).

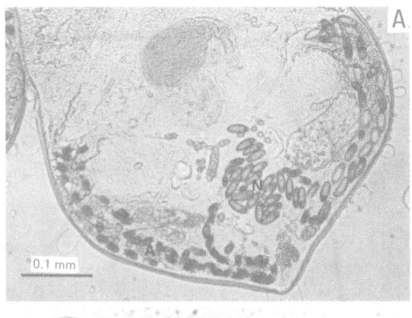

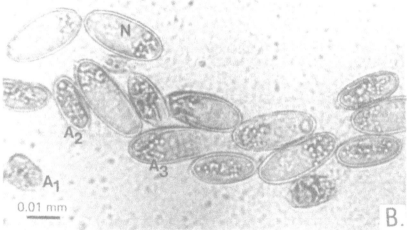

FIGURE 14. *Microphallus similis.* (A) 6 d *in vitro* culture in Hank's BSS + 20% fetal calf serum; apparently normal (N) and abnormal (A) eggs are present; (B) 4 d *in vitro* culture in NCTC 135 + 20% fetal calf serum; apparently normal (N) and various kinds of abnormal (A₁, A₂, A₃) eggs are present. (From Davies, C. and Smyth, J. D., *Int. J. Parasitol.*, 9, 261, 1979. With permission.)

VI. *MICROPHALLOIDES JAPONICUS*

Cultured by: Fujino et al.[26]
Definitive host: rat (*Rattus norvegicus*)
Experimental hosts: mouse, rat, guinea pig, dog, and sparrow[27]
Location: small intestine
Prepatent period: 12 to 24 h
Molluscan host: unknown
Second intermediate host: metacercarial cysts in the marsh crab, *Helice tridens tridens*
Distribution: Japan

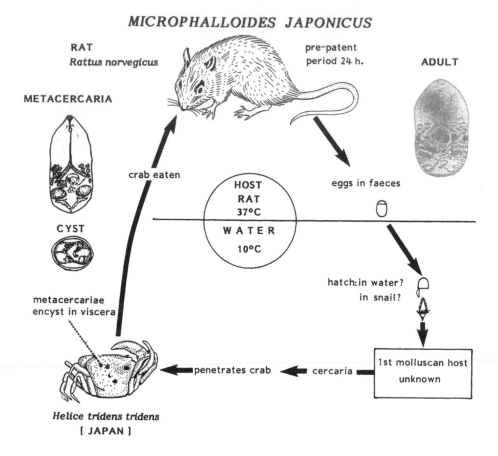

FIGURE 15. *Microphalloides japonicus*: life cycle. The diagram of the crab is notional. (Based on data from Fujino et al.[26] and Yoshida.[28])

A. LIFE CYCLE (FIGURE 15)

1. Adult

This species was found in the rats of Fukuoka City, Japan, but this may not be the original natural host as, experimentally, both birds and other mammals have been infected.

2. Molluscan Host

At present this is unknown.

3. Metacercaria

Metacercarial cysts, which are more oval than round, are very common in marsh crabs (see above) in Japan, cysts being found in the ovary, hypodermis lining of the carapace, on the stomach wall, and in other parts of the body cavity. Crabs of the genera *Chasmagnathus* and *Hemigraspus* can also serve as hosts.[27] The metacercaria (Figure 16) is highly progenetic and this condition is reflected in the fact that, in experimental mice, some incomplete eggs appear as early as 12 h p.i. and fully formed eggs appear in 24 h. The larva probably needs little (if any) nutrient in order to reach maturity and at 37°C, even in Hank's BSS, will produce a few eggs (see below).

B. *IN VITRO* CULTURE TECHNIQUE[26]

1. General Conditions

General conditions included temperature, 37°C; gas phase, air, pH 7.2 to 7.4.; and

MICROPHALLOIDES JAPONICUS

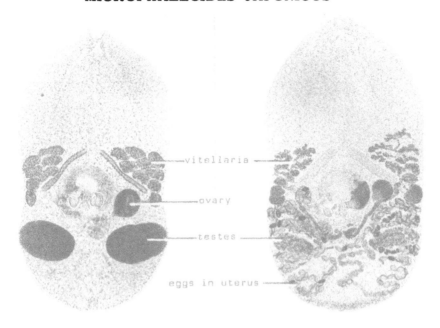

A. METACERCARIA   **B. ADULT**

FIGURE 16. *Microphalloides japonicus*. (A) Freshly excysted metacercaria; (B) adult developed after 48 h *in vitro* cultured in Krebs-Ringer. (From Fujino, T., Hamajima, F., Ishii, Y., and Mori, K., *J. Helminthol.*, 51, 125, 1977. With permission.)

culture tubes with rubber stoppers. Culture media included Hank's BSS, Eagle's MEM, MEM + (inactivated) 20% calf serum, NCTC 109, and NCTC 109 + 20% calf serum. Most media contained antibiotics (200 IU/ml of penicillin: 100 μg/ml of streptomycin). Media were unchanged throughout.

2. Excystment

Metacercariae fresh from crabs are enclosed in a thick transparent chitinous membrane. In physiological salines, within a range of temperatures of 31 to 41°C, the cyst wall becomes thin and larvae finally emerge so that enzyme treatment is not required. Most cysts excysted at 37°C within 1 h. Some cysts failed to excyst, but even these encysted worms, when cultured, produced some abnormal eggs.

3. Results

These are summarized in Table 2. Small numbers of eggs were produced even in Hank's BSS or Krebs-Ringer's. Best results were obtained in MEM + calf serum and NCTC 109 + calf serum, worms in the former producing most eggs. Sperm appeared in the seminal vesicle after about 12 h and after 24 h, eggs and/or vitelline masses were formed in the uterus. Both abnormal eggs and (apparently) normal eggs were produced. Unfortunately, tests to determine if the eggs were capable of embryonation were not carried out. No morphological differences between eggs produced *in vitro* and *in vivo* could be detected.

C. EVALUATION

This appears to be a promising system. The fact that some (apparently) normal eggs appeared indicates that some further experiments with this organism might result in larger numbers of normal eggs being produced.

TABLE 2

Microphalloides japonicus: **Egg Counts in Uterus Of Worms Cultured in Different Media**

Medium[a]	Days after cultivation											
	1			2			3			5		
	SF	ANE	(R)	SF	ANE	(R)	SF	ANE	(R)	SF	ANE	(R)
0.85% NaCl	0			0			0			0		
Krebs-Ringer's	15	0.3	(0—3)	15	16.4	(1—49)	15	25.0	(0—152)	15	22.5	(5—67)
Hanks' BSS	15	0.1	(0—1)	15	15.1	(0—45)	15	18.2	(0—46)	13	22.4	(3—41)
Eagle's MEM	14	1.4	(0—4)	15	25.6	(6—47)	15	55.3	(5—123)	13	44.9	(0—130)
NCTC 109	15	1.4	(0—3)	15	27.1	(4—100)	15	34.4	(3—90)	13	45.6	(0—162)
Eagle + serum	15	1.5	(0—7)	15	33.1	(0—103)	15	60.6	(14—122)	15	94.4	(17—192)
NCTC 109 + serum	15	2.9	(0—13)	15	10.8	(0—55)	15	25.7	(1—157)	15	45.2	(0—239)

[a] Abbreviations: SF, surviving flukes; ANE, average number of eggs per fluke; R, range (minimum number of eggs—maximum number of eggs).

From Fujino, T., Hamajina, F., Ishii, Y., and Mori, K. *J. Helminthol.*, 51, 125, 1977. With permission.

VII. *AMBLOSOMA SUWAENSE*

Cultured by: Schnier and Fried[29]
Definitive host: unknown, probably water birds
Experimental hosts: chick chorioallantoic membrane (CAM)[30] and chick eggs[31]
Location: alimentary canal?
Prepatent period: 4 to 5 d
Molluscan host: unknown
Second intermediate host: metacercariae (unencysted) in freshwater snails, *Sinotaia quadrata* (in Japan), *Campeloma decisum* (in the U.S.)
Distribution: Japan and the U.S.; may be cosmopolitan

A. LIFE CYCLE

1. Definitive Host

The life cycle is very imperfectly known, neither the definitive host nor the first molluscan host having yet been found. Fried and Schnier[32] failed to establish the worms in domestic chicks. However, the adult worm was first obtained[30] by growing it in developing chicken embryos using the elegant chorioallantoic membrane (CAM) technique developed by Fried;[33,34] the worms also can develop in the albumen of the egg.[31] This is the first time an adult worm has been identified by growing it to maturity utilizing this technique without the natural definitive host being known. Since the metacercariae are found in freshwater snails, it is likely that the natural hosts are water-dwelling birds.

2. Intermediate Hosts

As mentioned, the first intermediate host is unknown. Free metacercariae were first found in the freshwater snail, *Sinotaia quadrata,* in Japan[30] and later discovered in Wisconsin in the U.S., in another species of freshwater snail, *Campeloma decisum* Font.[35]

3. Metacercaria

The metacercariae (Figure 17), which are progenetic, occur between the epithelium and the shell. Two levels of progenesis appear to occur; in one, larvae have differentiated gonads and genital ducts, but no vitellaria; in the other, vitellaria are additionally present.[30]

AMBLOSOMA SUWAENSE

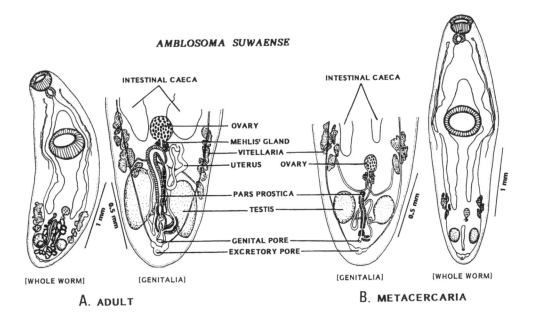

A. ADULT   B. METACERCARIA

FIGURE 17. *Amblosoma suwaense*. (A) Metacercaria; (B) adult worm grown in hen egg. (Modified from Schimazu, T., *Jpn. J. Parasitol.*, 23, 100, 1974. With permission.)

B. *IN VITRO* CULTURE TECHNIQUE[29]

1. Source of Material

The free metacercariae were dissected out from the epithelium of the snail, *C. decisum*, and rinsed in Locke's BSS + antibiotics (200 IU/ml of penicillin + 200 μg/ml of streptomycin).

2. Culture Conditions and Media

General conditions included temperature, 37.5°C; 12 × 75 mm plastic culture tubes; stationary upright cultures; 2 ml of medium per tube; and gas phase, air. Half of the medium was replaced daily. A wide range of media were tested: 0.85% NaCl, Hank's BSS, NCTC 135, NCTC 135 + 50% chicken serum (normal), NCTC 135 + 50% chicken serum (inactivated), NCTC 135 + 20% egg yolk, and NCTC 135 + 20% albumen.

3. Results

Best results were obtained in NCTC 135 + 20% egg yolk, the same medium successfully used for *Sphaeridiotrema globulus*.[8] Worms were ovigerous in 4 d (Figure 17). These authors also reported that results with normal (i.e., not inactivated) serum were better than with inactivated serum. Although some (apparently) normal eggs developed, no attempts were made to determine their viability. Three types of abnormal eggs were obtained, small, unshelled, and incompletely shelled.

C. EVALUATION

This is clearly a useful experimental organism, especially as it can be grown *"in vivo"* by the CAM technique, thus providing a valuable method for comparison with *in vitro* methods. However, as with many other species discussed here (i.e., *Microphallus similis*), the eggs were largely abnormal.

VIII. *METAGONIMUS YOKOGAWAI*

Cultured by: Yasuroaka and Kojima[36]
Definitive hosts: man, cats, dogs, and fish-eating mammals
Experimental hosts: mice, rats, and hamsters[37,38]
Location: duodenum
Prepatent period: 5 d (in mice)
Molluscan host: *Semisulcospira* spp.
Second intermediate host: numerous fish species, e.g., the sweet fish,
  Plecoglossus altivelis, in Japan
Distribution: Far East, U.S.S.R., and Balkans

A. LIFE CYCLE (FIGURE 18)
1. Adult
   M. yokogawai is a well-known human pathogen, man becoming infected by eating fish containing metacercariae. Its general biology has been reviewed in detail by Ito.[39] Cats and dogs serve as active reservoir hosts, but it has also been reported from many other fish-eating mammals and even birds (pelicans).

2. Molluscan Stages
   Eggs hatch only on ingestion by a suitable snail host (e.g., *Semisulcospira* sp.) in which sporocysts and rediae develop. The cercaria is of the parapleurolophocercous type.[40]

3. Metacercaria
   The cercariae readily penetrate numerous fish species where the metacercariae normally encyst under the scales, but can occasionally be found in the muscles. The metacercaria is not highly progenetic, but contains the anlagen of the ovary and testes. It becomes infective to kittens at 15 d.[41]

B. *IN VITRO* CULTURE TECHNIQUE[36]
1. Source of Material
   Cysts were separated from scales of infected fish with a scalpel and rinsed thoroughly in 0.4% NaCl.

2. Excystment
   Cysts were treated with 0.03% pepsin (1:3000) in 0.7% HCl for 3 h at 39 to 40°C. After rinsing twice in sterile 0.8% NaCl, they were treated in trypsin (20,000 Hb units per milliliter) at pH 7.1 for 30 min at 37°C.
   The action of pepsin readily freed the cysts from the scales, but excystment only occurred after subsequent treatment with trypsin, 80 to 90% of metacercariae excysting within 20 min, escape taking place through a small hole which appeared in the cyst. Excysted metacercariae were rinsed in Krebs-Ringer's-"Tris" solution containing 200 IU/ml of penicillin and 100 µg/ml of streptomycin.

3. Culture Conditions and Media
   General conditions included temperature, 37.5°C; 14 × 150 mm culture tubes or 25-mm Carrel flasks (both with Morton closures); 2 ml of medium per culture; 50 to 100 metacercariae per culture; gas phase, 8% CO_2 in air. All media contained antibiotics (concentrations as above).

METAGONIMUS YOKOGAWAI

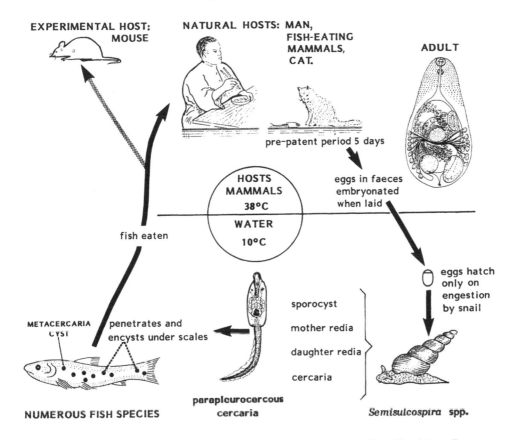

FIGURE 18.   *Metagonimus yokogawai*: life cycle. (Based on Smyth,[38] Ito,[39] and Saito.[40])

4. Results

A variety of media were tested, best results being obtained in Medium C (Table 3) containing NCTC 109, chick embryo extract, and human serum in a ratio of 3:4:3. Worms became ovigerous in 12 d (Figure 19) with only abnormal eggs being formed. The vitellaria stained poorly with Fast Red Salt B indicating that eggshell formation was not normal. That *in vitro* development was inhibited was further indicated by the fact the worms became ovigerous *in vivo* (in mice) in 5 d and *in vivo* worms also grew in size at a much faster rate (Figure 20).

The addition of various supplements, liver concentrates, yeast extract, vitamins, or hen egg yolk, failed to be beneficial to development *in vitro*.

C. EVALUATION

The fact that this species can mature in the mouse within 5 d makes it a useful organism for *in vivo* and *in vitro* laboratory studies. The failure to form normal vitellaria, and therefore normal eggshell, reflects a common pattern seen in many of the species discussed in this text and suggests a nutrient deficiency in the medium.

TABLE 3

Metagonimus yokogawai: **Components of Culture Media Utilized by Yasuraoka
and Kojima**

Medium	CEE$_{100}$	Human serum	Liver concentrate (1%)	Yeast extract (0.5%)	Vitamin mixture	Y-109	NCTC 109
A	—	—	—	—	—	—	10
B	—	2	—	—	—	—	8
C	4	3	—	—	—	—	3
D	4	3	0.5	—	—	—	2.5
E	4	3	—	1	—	—	2
F	4	3	0.5	—	0.5	—	2
G	4	3	0.5	1	—	—	1.5
H	4	3	—	—	—	3	—
I	—	—	—	—	—	10	—

Note: Concentrations are given in parts per 10 ml.

From Yasuraoka, K. and Kojima, K., *Jpn. J. Med. Sci. Biol.,* 23, 199, 1970. With permission.

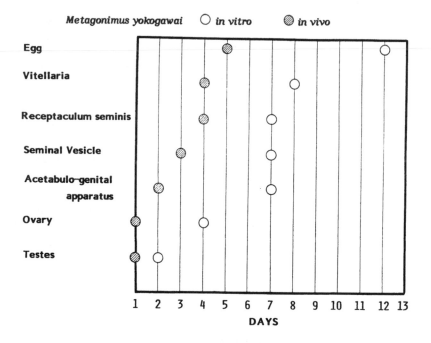

FIGURE 19.   *Metagonimus yokogawai:* time of appearance of the various organs in
worms grown *in vivo* (in mice) and *in vitro* in Medium C. (From Yasuraoka, K. and
Kojima, K., *Jpn. J. Med. Sci. Biol.,* 23, 199, 1970. With permission.)

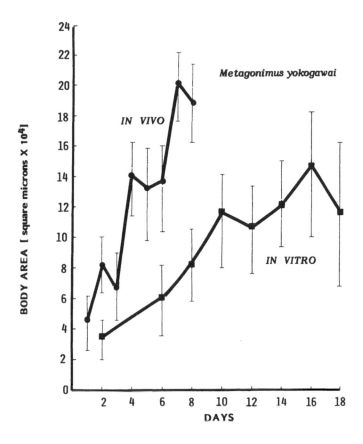

FIGURE 20. *Metagonimus yokogawai*: comparison of relative body worm area in worms grown *in vivo* (in mice) and *in vitro*. Each point represents the mean of the best ten flukes in a sample. (Modified from Yasuraoka, K. and Kojima, K., *Jpn. J. Med. Sci. Biol.*, 23, 199, 1970. With permission.)

REFERENCES

1. **Buttner, A.**, La Progénèse chez les Trématodes digénétiques, *Ann. Parasitol. Hum. Comp.*, 25, 376, 1950.
2. **Hickman, V. V.**, On *Coitocaecum anaspidis* sp.nov., a trematode exhibiting progenesis in the fresh-water crustacean *Anaspides tasmaniae* Thompson, *Parasitology*, 26, 121, 1934.
3. **Ferguson, M. S.**, Excystment and sterilization of metacercariae of the avian strigeid trematode, *Posthodiplostomum minimum* and their development into adult worms in sterile cultures, *J. Parasitol.*, 26, 359, 1940.
4. **Miller, J. H.**, Studies on the life history of *Posthodiplostomum minimum* (MacCallum 1921), *J. Parasitol.*, 40, 255, 1954.
5. **Ulmer, M. J.**, Passerine birds as experimental hosts for *Posthodiplostomum minimum* (Trematoda: Diplostomidae), *J. Parasitol.*, 47, 608, 1961.
6. **Palmieri, J. R.**, Additional natural and experimental hosts and interspecific variation in *Posthodiplostomum minimum* (Trematoda: Diplostomatidae), *J. Parasitol.*, 59, 744, 1973.
7. **Hoffman, G. L.**, Experimental studies on the cercaria and metacercaria of a strigeoid trematode, *Posthodiplostomum minimum, Exp. Parasitol.*, 7, 23, 1958.
8. **Berntzen, A. K. and Macy, R. W.**, *In vitro* cultivation of the digenetic trematode *Sphaeridiotrema globulus* (Rudolphi) from the metacecarial stage to egg production, *J. Parasitol.*, 55, 136, 1969.
9. **Huffman, J. E., Fried, B., Roscoe, D. E., and Cali, A.**, Comparative pathologic features and development of *Sphaeridiotrema globulus* (Trematoda) infections in the mute swan and domestic chicken and chicken allantois, *Am. J. Vet. Res.*, 45, 387, 1984.

10. **Fried, B. and Huffman, J. E.**, Excystation and development in the chick and on the chick chorioallantois of the metacercaria of *Sphaeridiotrema globulus* (Trematoda), *Int. J. Parasitol.*, 12, 427, 1982.

11. **Szidat, L.**, Uber die Entwicklungsgeschichte von *Spaeridiotrema globulus* Rud. 1814 und die Stellung der Psilostomatidae Odhner im natürlichen System. I, *Z. Parasitenkd.*, 9, 529, 1937.

12. **Burns, J. P.**, The life cycle of *Sphaeridiotrema spinoacetabulum* sp.n. Trematoda: Psilostomidae from the ceca of ducks, *J. Parasitol.*, 47, 933, 1961.

13. **Macy, R. W. and Ford, J. R.**, The psilostome trematode *Sphaeridiotrema globulus* (Rud.) in Oregon, *J. Parasitol.*, 50, 93, 1964.

14. **Macy, R. W., Berntzen, A. K., and Benz, M.**, *In vitro* excystation of *Sphaeridiotrema globulus* metacercariae, structure of cyst, and the relationship to host specificity, *J. Parasitol.*, 54, 28, 1968.

15. **Yasuraoka, K., Kaiho, M., Hata, H., and Endo, T.**, Growth *in vitro* of *Parvatrema timondavidi* Bartoli 1963 (Trematoda: Gymnophallidae) from the metacercarial stage to egg production, *Parasitology*, 68, 293, 1974.

16. **Bartoli, P.**, Données nouvelles sur la morphologie et la biologie de *Parvatrema timondavidi* Bartoli 1963 (Trematoda: Digenea), *Ann. Parasitol. Hum. Comp.*, 40, 155, 1965.

17. **Cable, R. M.**, The life cycle of *Parvatrema boriquenae* gen. et sp. nov. (Trematoda: Digenea) and the systematic position of the Subfamily Gymnophallinae, *J. Parasitol.*, 39, 408, 1953.

18. **Endo, T. and Hoshina, T.**, Redescription and identification of a gymnophallid trematode in a brackish water clam, *Tapes (Ruditapes) phillippinarum*, *Jpn. J. Parasitol.*, 23, 73, 1974.

19. **Davies, C. and Smyth, J. D.**, The development of the metacercariae of *Microphallus similis in vitro* and in the mouse, *Int. J. Parasitol.*, 9, 261, 1979.

20. **James, B. L.**, Host selection and ecology of marine digenean larvae, in *Marine Larvae*, Crisp, D. J., Ed., Cambridge University Press, Cambridge, 1971, 179.

21. **Stunkard, H. W.**, The morphology and life history of the digenetic trematode *Microphallus similis* (Jaegerskiold, 1900), Baer, 1943, *Biol. Bull.*, 112, 254, 1957.

22. **Bakke, T. A.**, Studies on the helminth fauna of Norway. XXII. The common gull, *Larus canus* L., as final host for *Digenea* (Platyhelminthes). I. The ecology of the common gull and the infection in relation to season and the gull's habitat, together with the distribution of the parasites in the intestine, *Norw. J. Zool.*, 20, 165, 1972a.

23. **Bakke, T. A.**, Studies on the helminth fauna of Norway. XXIII. The common gull, *Larus canus* L., as a final host for *Digenea* (Platyhelminthes). II. The relationship between infection and weight, sex and age of the common gull, *Norw. J. Zool.*, 20, 189, 1972b.

24. **Combescot-Lang, C.**, Étude des Trématodes parasites *Littorina saxatilis* (Olivi) et de leurs effects sur cet hôte, *Ann. Parasitol. Hum. Comp.*, 51, 27, 1976.

25. **Davies, C.**, The forebody glands and surface features of the metacercariae of adults of *Microphallus similis*, *Int. J. Parasitol.*, 9, 553, 1979.

26. **Fujino, T., Hamajima, F., Ishii, Y., and Mori, K.**, Development of *Microphalloides japonicus* (Osborn, 1919) metacercariae *in vitro* (Trematoda: Microphallidae), *J. Helminthol.*, 51, 125, 1977.

27. **Yamaguti, S.**, *Systema Helminthum*, Vol. 1, *The Digenetic Trematodes of Vertebrates*, Interscience, New York, 1958.

28. **Yoshida, S.**, On a trematode larva encysted in a crab, *Helice tridens* (de Haan), *J. Parasitol.*, 3, 76, 1917.

29. **Schnier, M. S. and Fried, B.**, *In vitro* cultivation of *Amblosoma suwaense* (Trematoda: Bracylaimidae) from the metacercaria to the ovigerous adult, *Int. J. Parasitol.*, 10, 391, 1980.

30. **Schimazu, T.**, A new digenetic trematode, *Amblosoma suwaense* sp. nov., the morphology of its adult and metacercaria (Trematoda: Brachylaimidae), *Jpn. J. Parasitol.*, 23, 100, 1974.

31. **Fried, B., Heyer, B. L., and Pinski, A. K.**, Cultivation of *Amblosoma suwaense* (Trematoda: Brachylaimidae) in chick embryos, *J. Parasitol.*, 67, 50, 1981.

32. **Fried, B. and Schnier, M. S.**, Infectivity of *Amblosoma suwaense* (Trematoda: Brachylaimidae) in the domestic chick, *Proc. Helminthol. Soc. Wash.*, 48, 83, 1981.

33. **Fried, B.**, Growth of *Philophthalmus* sp. (Trematoda) on the chorioallantois of the chick, *J. Parasitol.*, 48, 545, 1962.

34. **Fried, B.**, Transplantation of trematodes to the chick cholioallantois, *Proc. Pa. Acad. Sci.*, 43, 232, 1969.

35. **Font, W. F.**, *Amblosoma suwaense* (Trematoda: Brachylaimidae; Leucochloridiomorphinae) from *Campeloma decisum* in Wisconsin, *J. Parasitol.*, 66, 861, 1980.

36. **Yasuraoka, K. and Kojima, K.**, *In vitro* cultivation of the heterophid trematode *Metagonimus yokogawai*, from the metacercaria to adult, *Jpn. J. Med. Sci. Biol.*, 23, 199, 1970.

37. **Yokogawa, M. and Sano, M.**, Studies on intestinal flukes. IV. The development of the worm in experimentally infected animals with metacercariae of *Metagonimus yokogawai*, *Jpn. J. Parasitol.*, 17, 540, 1968 (in Japanese; English summary).

38. **Smyth, J. D.**, *An Introduction to Animal Parasitology*, 2nd ed., Edward Arnold, London, 1976.

39. **Ito, I.**, *Metagonimus* and other human heterophyid trematodes, *Prog. Med. Parasitol. Jpn.*, 1, 315, 1964.

40. **Saito, S.,** On the differences between *Metagonimus yokogawai and Metagonimus takahashii.* I. Morphological comparisons, *Jpn. J. Parasitol.,* 21, 449, 1972 (in Japanese; English summary).
41. **Shatrov, A. S.,** The biology of *Metagonimus yokogawai* Katsurada, 1912 in the Upper Priamurje, *Parazitologia (Leng.),* 8, 196, 1974.

Chapter 4

TREMATODA: DIGENEA WITH NON-PROGENETIC METACERCARIA

J. D. Smyth

TABLE OF CONTENTS

I. *COTYLURUS ERRATICUS*

Cultured by: Mitchell, Halton, and Smyth[1]
Definitive hosts: various species of gulls
Experimental host: gull
Location: small intestine
Prepatent period: 96 h[2]
Molluscan hosts: *Valvata* spp., *Lymnaea* spp., and *Physa fontinalis*
Second intermediate host: metacercaria (tetracotyle) in numerous salmonid
  fish
Distribution: cosmopolitan

A. LIFE CYCLE (FIGURE 1)
1. Adult
This species is a common intestinal parasite of fish-eating birds, especially gulls, throughout the world. It has been reported in *Larus californicus* in the U.S. and in *L. ridibundus*,[3] *L. canus*, and *L. fuscus*[4] in Europe. Details for maintaining the life cycle in the laboratory are given by Olson;[2] its morphology has been described (Figure 2A and B) by Guberlet.[4]

2. Molluscan Stages
The chief molluscan host appears to be species of *Valvata, V. lewisi*, in the U.S.[2] and (probably) *V. piscinalis* in Europe, although it has also been reported from species of *Lymnaea* and *Physa*.[6] There are two sporocyst generations and the average time for cercariae to emerge after snail infection is 35 d.[2]

3. Metacercaria (Figure 2C)
Cercaria penetrate and encyst in a wide range of salmonoid fish (Salmonidae) throughout the world, such as trout, graying, whitefish, cisco, and various varieties of salmon.[2,3,5,7] The metacercariae were originally described[7] under the larval generic name, *Tetracotyle intermedia*, a term which is still widely used.

Metacercariae occur almost exclusively in the pericardial cavity, becoming encysted in 2 to 3 weeks at 21°C. They occasionally occur in the mesenteries of the pyloric caeca in close proximity to the pericardium. The wall of the cyst is composed of an outer, fibrous, spongy layer and an uneven, thin, transparent layer. Their size range is 0.48 to 1.48 × 0.41 to 1.11 mm.[3]

B. *IN VITRO* CULTURE TECHNIQUE[1]
1. Source of Material
Cysts were removed with aseptic precautions from the pericardial cavity and transferred to Young's teleost saline[8] + antibiotics (streptomycin, 1 mg/ml; benzylpenicillin, 100 IU/ml).

2. Excystment
Cysts were washed in Hank's BSS and then readily excysted by incubating them at 41°C in 0.5% trypsin + 0.2% sodium taurocholate at pH 7.5. Excystment began after 30 to 45 min and was usually completed within 1 to 2 h.

3. Culture Media and Conditions
General conditions included temperature, 41°C; gas phase, air; and pH 7.4 (?). Culture vessels were Leighton tubes (Bellco) with 5 ml of media + antibiotics (as above). The culture media included the basic medium, Morgan 199, or NCTC 135 + inactivated chicken serum (Flow Laboratory, Irvine, Scotland).

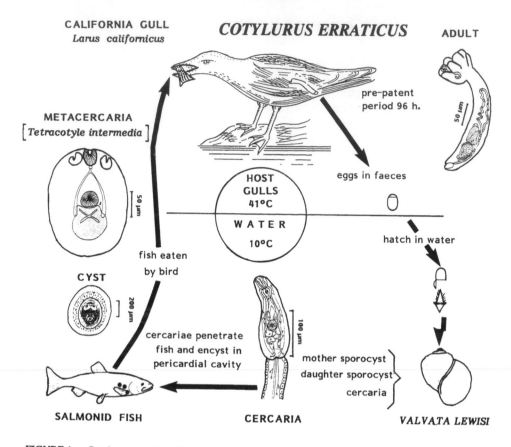

FIGURE 1.   *Cotylurus erraticus*: life cycle. (Based on data from Olson[2] and Niewiadomska and Kozicka.[3])

COTYLURUS ERRATICUS

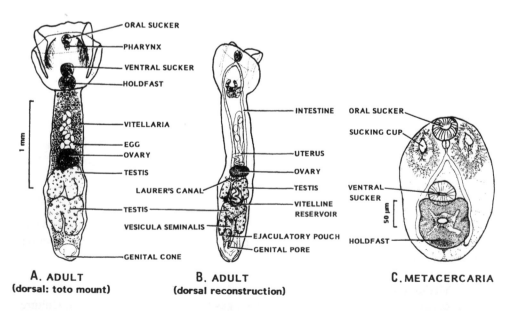

FIGURE 2.   *Cotylurus erraticus*. (A) and (B) Adult (After Guberlet, J. E., *J. Parasitol.*, 9, 6, 1922. With permission.) (C) Metacercaria. (After Niewiadomska, K. and Kozicka, J., *Acta Parasitol. Pol.*, 18, 487, 1970. With permission.)

TABLE 1
Cotylurus erraticus: In Vitro Culture; Development in Different Media Compared with that *In Vivo*

Treatment	Period in culture/*in vivo* (d)	First egg production (d)
In vitro		
M199	6 (died)	—
NCTC135	6 (died)	—
M199 + 20% chicken serum	9	9
M199 + 40%	9	7
NCTC135 + 40%	9	8
M199 + 80%	9	6
NCTC135 + 80%	8	6
In vivo		
Grown in *Larus ridibundus*	5	4—5

From Mitchell, J. S., Halton, D. W., and Smyth, J. D., *Int. J. Parasitol.*, 8, 390, 1978. With permission.

4. Results

Little development took place in M 199 or NCTC alone. The addition of chicken serum had a marked effect (Table 1). The best results were obtained in NCTC 135 + 80% serum in which the hind body developed rapidly and paired testes could be identified after 3 d. Eggs appeared in the uterus at 6 to 9 d, compared with 4 to 5 d *in vivo* (Figure 3).[2,9] Ultrastructure studies showed that the development of the testes and spermatozoa *in vitro* compared favorably with that *in vivo* (see Figure 2A and B). In the ovary and vitellaria, cellular differentiation *in vitro* compared well with that *in vivo*, although fewer fully differentiated cells were present in *in vitro* worms. However, the eggs produced *in vitro* were abnormal both in size and shape (compare *Microphallus similis*). This result is in contrast to that obtained with *Cotylurus lutzi* which, when cultured[10] in NCTC 135 + 40% chicken serum, but supplemented with chicken gut extract, produced normal eggs. The degree of abnormality in *C. erraticus in vitro* eggs was further emphasized by the observation that no ova were found in TEM sections of eggs.

C. EVALUATION

The fact that only abnormal eggs were produced *in vitro* is likely to be related to (1) pretanning of eggshell material, (2) failure of insemination and fertilization to occur, and (3) failure of ova to be released into the ootype, possibly due to the lack of the appropriate trigger. Nevertheless, the organism is useful as an experimental model to study the developmental biology of trematodes.

II. *COTYLURUS LUTZI*

Cultured by: Basch et al.;[10] Voge and Jeong[11]
Definitive hosts: unknown; probably birds
Experimental hosts: finches, canaries, and chicks (poor hosts)[1]
Location: small intestine
Prepatent period: 7 d
Molluscan hosts: *Biomphalaria glabrata*
Second intermediate host: *B. glabrata* in which metacercariae (tetracotyles) encyst in ovotestis
Distribution: Brazil

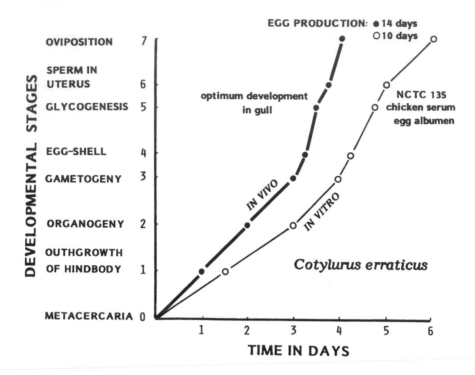

FIGURE 3. *Cotylurus erraticus:* comparative development *in vivo* (in gulls) and *in vitro*. (After Halton, D. W. and Mitchell, J. S., *Z. Parasitenkd.*, 70, 515, 1984. With permission.)

A. LIFE CYCLE (FIGURE 4)
1. Definitive Host

This species was first described by Basch[12] from the metacercarial (tetracotyle) stage in *Biomphalaria glabrata* in Brazil. Its natural definitive host is still unknown. The fact that finches, canaries, and chickens can act as experimental hosts may suggest that terrestrial rather than aquatic birds may be the definitive hosts.

2. Molluscan Stages

B. glabrata appears to be the only molluscan host so far identified. It can be readily infected experimentally at 25 to 27°C, furcocercariae emerging 25 d postinfection (p.i.).

3. Metacercaria

Cercaria penetrate the same snail species (*B. glabrata*) and develop into tetracotyles (metacercariae; Figure 5A) in the ovotestis. "Encysted" tetracotyles contract, but do not form cyst walls. The tetracotyle is only slightly progenetic, the anlagen of the genitalia being present as a small group of undifferentiated cells (see Figure 5A).

B. *IN VITRO* CULTURE TECHNIQUE[10]
1. Source of Material

The material was obtained from young adult *Biomphalaria glabrata* exposed to large numbers of cercariae. After 1 to 2 months, snails were washed and their shells gently crushed in 0.34% NaCl. The ovotestis containing "encysted" metacercariae (tetracotyles) were teased in sterile 0.34% NaCl + antibiotics (10 μg of gentamycin, 500 U of penicillin, and 500 μg of streptomycin per milliliter) and the freed organisms were rinsed four times before culturing. Since a cyst wall is not formed, enzymes or bile salts are not necessary. However, it has been shown that a temperature of 41°C was essential for activation as this did not occur at 37°C.[11]

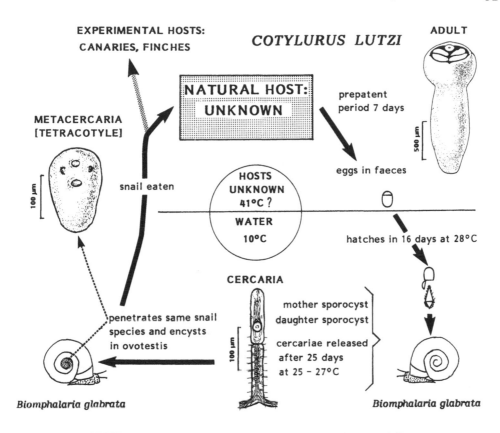

FIGURE 4. *Cotylurus lutzi:* life cycle. (Based on data from Basch.[12])

COTYLURUS LUTZI

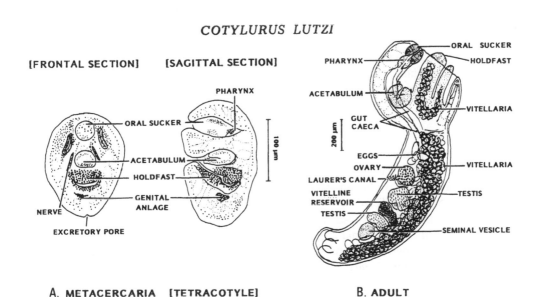

A. METACERCARIA [TETRACOTYLE]

B. ADULT

FIGURE 5. *Cotylurus lutzi.* (A) Sections of "encysted" metacercaria (tetracotyle) from the snail *Biomphalaria glabrata;* (B) adult, from experimental infection of a zebra finch, *Taeniopygia castanotus.* (After Basch, P. F., *J. Parasitol.,* 55, 527, 1969. With permission.)

2. Culture Media and Conditions

C. lutzi was first cultured to sexual maturity by Voge and Jeong,[11] who experimented with a number of different culture conditions, including the gas phase. In the best of these conditions, adult worms developed, but only eggs with abnormal shells, which failed to embryonate, were produced. Best results were obtained in NCTC 135 + 50% chick serum at 39 to 41°C in a gas phase of air.

Basch et al.[10] later improved this technique by the addition of chick intestine mucosal extract and obtained eggs which gave rise to normal miracidia (see below). These workers used NCTC 135 + 40% chick serum + antibiotics (100 U of penicillin, 100 μg of streptomycin, and 10 μg of gentamycin per milliliter) supplemented with a number of extracts of chicken tissues. The most successful of these was an extract of chicken mucosa prepared as follows.

The small intestine was cut at the gizzard and at the beginning of the large intestine. This portion (approximately 70 cm long) was cut into two equal lengths and labeled "upper" and "lower" small intestine. Each half was washed in several changes of cold (5°C) normal saline and then the mucosal layer was scraped off with a microscope slide and homogenized in a glass tissue grinder with a Teflon plunger. These extracts were squeezed through silk bolting cloth, brought up to 20 ml with normal saline, and centrifuged for 60 min at 3000 rpm at 5°C. The supernatant was filtered twice and then sterilized by passage through 0.65 and 0.45 μm Millipore filters. Dialyzed intestinal extracts were also prepared.

Cultivation was carried out in 25 × 150 mm glass tubes, containing 15 larvae in 5 ml of medium; tubes were closed with silicone rubber stoppers and sealed with parafilm. The gas phase was air.

3. Results

Intestinal extract was the only extract which was beneficial in that all cultures containing it produced some eggs with normal eggshells. However, miracidia only developed in eggs from cultures containing the undialyzed "upper" small intestine mucosal extract. In the best results, the first eggs appeared at 5 d, a day earlier than the prepatent period reported *in vivo*.[12] Normal (embryonated) egg production continued for about 24 d and then fell away.

In vitro-produced eggs were allowed to hatch and, of the 72 snails exposed to the miracidia, 5 became infected. Well-encysted tetracotyles were found in the ovotestis several weeks later. These larvae proved to be infective to zebra finches (*Taeniopygia castanotus*) which later produced *C. lutzi* eggs, thus completing the life cycle.

C. EVALUATION

This is clearly a most successful *in vitro* system in that it represents the only culture system (to date) in which normal eggs, which gave rise to active miracida, were produced. Moreover, the infectivity of the miracida to snails was proved experimentally. Compare this result with that of *Sphaeridiotrema globulus* in which embryonated eggs were obtained, but their hatchability and infectivity was not further tested.

The variation in the nutritional requirements of even closely related species is reflected in the fact that tetracotyles of *Cotylurus strigeoides*, when cultured in the same medium which was successful for *C. lutzi*, produced only abnormal eggs and when grown on the chick chorioallantois, only showed minimal postmetacercarial development.[13]

III. *DIPLOSTOMUM SPATHACEUM*

Cultured by: Kannangara and Smyth[14]
Definitive hosts: fish-eating birds, especially gulls
Experimental host: gull

Location: small intestine

Prepatent period: 93 to 104 h[15]

Molluscan hosts: *Lymnaea* spp., especially *L. stagnalis*, *L. palustris*, and
L. peregra[15,16]

Second intermediate host: metacercaria (diplostomulum) in numerous salmonid fish

Distribution: cosmopolitan, especially Europe and the U.S.

A. LIFE CYCLE (FIGURE 6)

1. Adult (Figure 7A)

There are a number of synonyms of *D. spathaceum*, the most common being *D. volvens*, *D. huronense*, and *D. flexicaudum*.[17] Because of the occurrence of the metacercaria in the eye, it is commonly referred to as the "eye fluke". Its biology, ecology, speciation, and distribution has been comprehensively reviewed by Chubb[17] and Dubois.[18] Some authors recognize the existence of subspecies.

The species is a common parasite of fish-eating birds, especially gulls, *Larus* spp., and 38 species of birds from 7 families have been reported as definitive hosts.[16] Its morphology is shown in Figure 7A.

2. Molluscan Stages

The chief molluscan hosts are species of *Lymnaea*, *L. stagnalis* (cosmopolitan) and *L. peregra* (Europe) being commonly infected. Mother and daughter sporocysts develop (see Figure 6) and cercaria are released. Cercaria hang in the water in a very characteristic position, their furca spread at an angle of about 120° with the tail stem.[15]

3. Metacercaria (Figure 7B)

Cercaria penetrate a wide range of fish hosts, and the metacercaria has been reported in 105 species of freshwater fish in Europe and the U.S.; at least 23 fish species have been reported to be infected in the U.K.[19] The South African clawed toad, *Xenopus laevis*, has also been experimentally infected, and natural infections have been found in wild amphibia, reptiles, and mammals.[16] Several cases of human eye infections have been reported.[16] The cercariae are remarkable for showing a powerful taxis toward the eye after penetration. Thus, although they may penetrate all over the body, within 12 h, 80% are found in the eye. Most fish lens have from 1 to 20 metacercariae, but up to 100 may be present. Infections (*diplostomiasis*) result in some degree of blindness and stunted growth and this increases with accumulation of larvae with age. The metacercaria — often referred to as a *diplosto-mulum* — is nonprogenetic and also remains unencysted. It should be noted, however, that there are occasional records of small numbers of progenetic metacercaria found in snails, and these may mature directly in gulls without passing through the fish intermediate stage.[20]

B. *IN VITRO* CULTURE TECHNIQUE[14]

1. Source of Material

Fish (*Rutilus rutilus*) were killed by severing the spinal cord just behind the head and holding the head in a retort clamp for easy dissection under a strong light. The eyes were painted with alcoholic iodine (1% in 90% ethanol) and left to dry. The lens was extracted by incising the cornea and the lens teased out in sterile NCTC 135 to free the contained metacercariae.

It should be noted that in the U.S. another species, *D. adamsi*, is found in the retina.[21] In Europe, a closely allied species, *D. gasterostei*,[22] occurs in the retina in the three-spined stickleback, *Gasterosteus aculeatus*, and another species, *Tylodelphys clavata*, occurs within the humor of the eye of trout.[23] These species should not be confused with the present species.

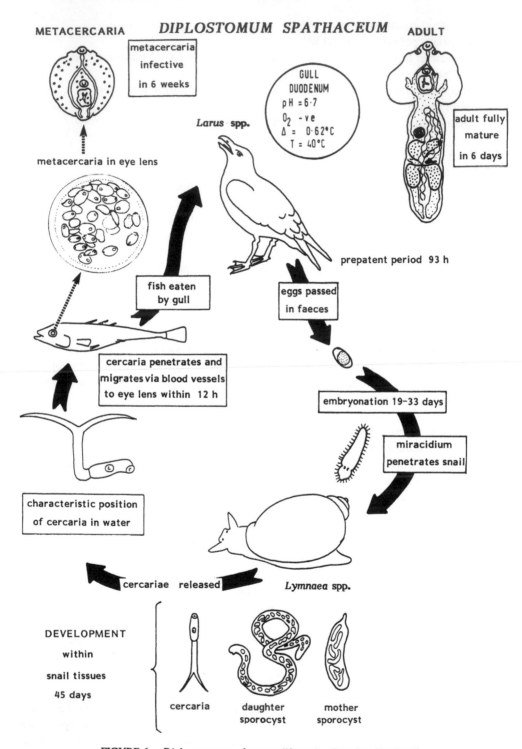

METACERCARIA   *DIPLOSTOMUM SPATHACEUM*   ADULT

metacercaria infective in 6 weeks

GULL DUODENUM
pH = 6·7
O_2 − v e
Δ = 0·62°C
T = 40°C

Larus spp.

adult fully mature in 6 days

metacercaria in eye lens

prepatent period 93 h

fish eaten by gull

eggs passed in faeces

cercaria penetrates and migrates via blood vessels to eye lens within 12 h

embryonation 19–33 days

miracidium penetrates snail

characteristic position of cercaria in water

cercariae released   *Lymnaea* spp.

DEVELOPMENT within snail tissues 45 days

cercaria   daughter sporocyst   mother sporocyst

FIGURE 6.   *Diplostomum spathaceum*: life cycle. (Based on Heatley.[15])

2. Culture Media and Conditions

Media — On the basis that metacercariae of the closely related species, *D. phoxini*, from the brain of the minnow can be grown to sexual maturity in a simple yolk/albumen mixture,[24,25] it was expected that the same sort of semisolid medium would suffice for *D. spathaceum*. In preliminary experiments, this was found not to be the case, and the organism

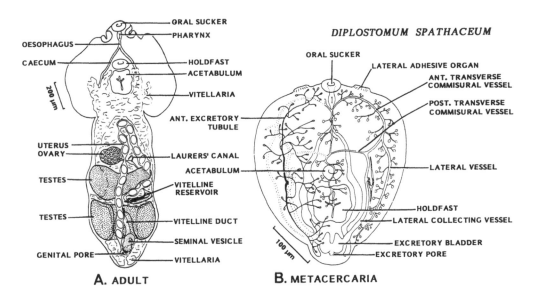

FIGURE 7. *Diplostomum spathaceum*: (A) Adult, from gull; (B) metacercaria, from the three-spined stickleback, *Gasterosteus aculeatus*. (After Heatley, W., M.Sc. thesis, University of Dublin, Dublin, Ireland, 1958.)

proved to be very refractory to *in vitro* culture. It was eventually found that the composition, consistency, and (probably) the particulate nature of the egg-saline medium was crucial to development.[14]

Of the various types of egg macerates prepared, two were more successful than others, although sexual maturity was achieved in all media. These two media were (1) NCTC 135 to yolk-albumen-macerate, 5:1; and (2) NCTC 135 to cooked egg whole macerate, 5:1. Details of preparation are given below.

Preparation of egg medium — Components from an egg (opened aseptically) were coagulated by heating to 80°C for 1 1/4 h in a covered sterile beaker. The coagulated material was transferred to a sterile glass homogenizer with an equal volume of NCTC 135 containing 2% yeast extract (GIBCO), 1% glucose, and 20% fetal calf serum (Flow) with penicillin and streptomycin at 100 U/ml. The mixture was homogenized for 15 s at slow speed in a sterile Waring blender and stored in sterile bottles at 40°C. A number of other liquid or diphasic media were tried without success (Table 2)

Culture conditions — Culture vessels used were Falcon flasks, Leighton tubes, universal containers, and 11-ml (disposable) plastic tubes; the last proved most suitable. The gas phase was 5% CO_2 + 15% O_2 + 50% N_2 or air. No difference was found between these two, and air was routinely used. Cultures were agitated gently in an agitating incubator at 40 rpm at 41°C.

3. Results

These are summarized in Table 2. Full sexual maturity up to oviposition was only achieved in egg media (Figure 8) and, in media containing 15 to 20% or less egg yolk in NCTC 135, gonadal development did not proceed beyond the stage of genital development.

In the best results, i.e., in cooked whole egg macerate (see Table 2, 5d) and 5:1 yolk/albumen macerate (see Table 2, 5e), the earliest time that eggs appeared was 10 to 12 d, worms in the cooked macerate being slightly bigger. However, all eggs produced were abnormal, and the vitellaria were not as well developed as in *in vivo* matured worms. Moreover, egg appearance at 10 to 12 d compared very unfavorably with an *in vivo* oviposition time of 6 d when eggs can be seen in the uterus; Heatley[15] reports that a single egg was found in the uterus as early as 93 h. However, as shown below, the *in vitro* prepatent period is similar to that for worms grown on the chick chorioallantois.

TABLE 2

Diplostomum spathaceum and *D. phoxini*: *In Vitro* Culture; Degree of Development in Different Media

Classification of Media According to the Degree of Development (*Diplostomum* spp.)

Medium		Mitoses	Somatic growth	Genital rudiment	Developed testes	Follicular ovary	Vitellaria	Eggs
1. Salines	Hanks' saline	±	—	—	—	—	—	—
2. Tissue culture media	NCTC 135	++	±	—	—	—	—	—
3. A. Diphasic media								
B. "Double culture tube" technique		++	+	±	—	—	—	—
C. Circulatory system								
4. A. Animal sera								
a. 35% Horse serum								
b. 40% bovine serum		++	+	±	—	—	—	—
c. 25% fetal calf serum								
B. Tissue culture media + yeast extract								
5. Egg media								
a. Egg yolk		++	++	++	++	++	+	++
b. Egg yolk and albumen		++	++	++	++	++	++	++
c. Fresh whole egg macerate		++	++	+	++	++	++	++
d. Cooked whole egg macerate		++	+++	+	+	+	++	++
e. 5:1 yolk/albumen macerate		++	+++	+			++	++

From Kannangara, D. W. W. and Smyth, J. D., *Int. J. Parasitol.*, 4, 669, 1974. With permission.

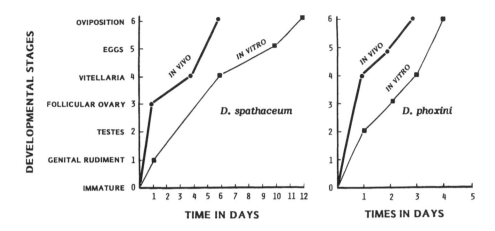

FIGURE 8. Optimum growth of *Diplostomum spathaceum* and *D. phoxini in vitro*, compared with growth *in vivo*. (From Kannangara, D. W. W. and Smyth, J. D., *Int. J. Parasitol.*, 4, 667, 1974. With permission.)

C. CULTURE ON THE CHICK CHORIOALLANTOIS

It is worth recording that, in an elegant experiment, Leno and Holloway[26] have also successfully grown this species to sexual maturity (Figure 9) using the valuable CAM (chorioallantois) technique developed by Fried.[27] Although only a small proportion of worms produced eggs, some of these proved to be normal, viable, and fertilized and developed viable miracidia, eye spots appearing 15 d postembryonation. Thus, this valuable technique can be used as an adjunct to *in vitro* experiments to obtain viable stages of developing worms without the need to use experimental birds as hosts.

D. EVALUATION

The fact that this species has a very wide cosmopolitan distribution and its mollusk stages occur in species of *Lymnaea* make it an especially useful laboratory model for *in vitro* culture experiments. Although only abnormal eggs were produced, the fact that worms grew on the chick chorioallantois suggests that nutritional factor(s) are missing from the *in vitro* medium used. It should also to be noted that results in liquid media were very poor, which suggests that (like *D. phoxini*) this species has a feeding mechanism that readily takes in food in semisolid, but not liquid form.

IV. *DIPLOSTOMUM PHOXINI*

Cultured by: Bell and Smyth[24]
Definitive hosts: fish-eating birds
Experimental hosts: ducks[28]
Location: duodenum
Prepatent period: 84 h[28]
Molluscan hosts: *Lymnaea* spp., especially *L. peregra* and *L. auricularia*
Second intermediate host: metacercaria (diplostomulum) in brain of minnow
Distribution: Europe and the U.S.S.R.[17]

NOTE: Because the life cycle, morphology, and *in vitro* technique for this species are so similar to that of *D. spathaceum*, only brief details are provided.

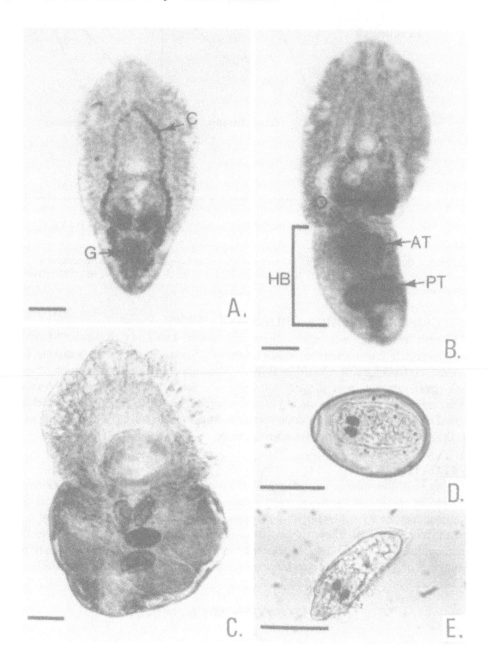

FIGURE 9. *Diplostomum spathaceum* (in U.S., syn. *D. flexicaudum?*[17]): development on the chick allantois. (A) Early stage, showing genital anlagen (G) and ingested blood (C), bar = 100 μm; (B) the hindbody (HB) becomes apparent and the anterior (AT) and posterior (PT) testes develop and the ovary anlage, bar = 100 μm; (C) sexual maturation with appearance of eggs, bar = 100 μm; (D) egg, after embryonation, containing miracidium with eyespots, bar = 50 μm; (E) free miracidium; bar = 100 μm. (From Leno, G. H. and Holloway, H. L., *J. Parasitol.*, 72, 555, 1986. With permission.)

A. LIFE CYCLE AND MORPHOLOGY

Details of the life cycle are given in Figure 10. The common definitive hosts are various species of ducks, but the metacercarial stage is very host specific and appears to be confined to the European minnow (*Phoxinus phoxinus*) in whose brain the (unencysted) metacercariae are found. In the U.K., up to 60% of minnows are commonly infected. Details of the morphology of the various stages and of the life cycle are given by Bell and Hopkins,[28] Blair,[29] Rees,[30,31] and Smyth.[32]

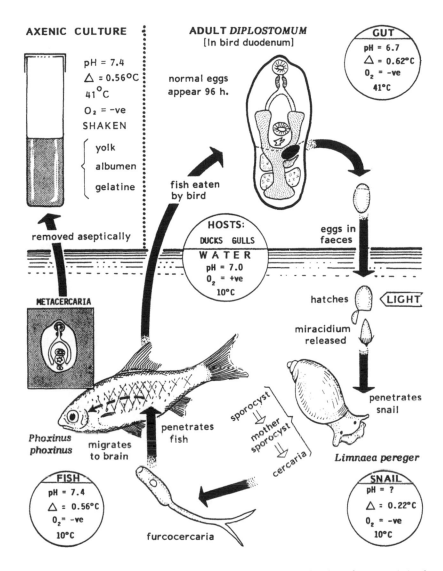

FIGURE 10. *Diplostomum phoxini:* life cycle. (From Smyth, J. D., *Introduction to Animal Parasitology*, 2nd ed., Edward Arnold, London, 1976. With permission.)

B. *IN VITRO* CULTURE TECHNIQUE

The general culture technique and conditions were the same as those used for *D. spathaceum.*

1. Results

The most satisfactory results were obtained with semisolid egg media, liquid media being generally unsatisfactory. This was first used by Bell and Smyth[24] and Smyth,[25] who found that egg yolk, slightly diluted with saline, stimulated mitosis, and development to spermatogenesis was obtained. The addition of albumen to the medium and the use of continuous agitation finally led to complete maturation leading to oviposition. The eggs produced, however, were abnormal, being imperfectly formed with thin shells. Later, Kannangera and Smyth[14] found that in a medium consisting of 2 ml of egg yolk, 0.5% yeast extract, and 0.5 ml of albumen, worms produced eggs on the fourth day (see Figure 8). Cooked egg macerate or yolk/albumen (5:1) as used for *D. spathaceum,* proved to be somewhat better. A number of workers have attempted to replace the egg medium with defined constituents, but without success.[33,34]

C. EVALUATION

As with *D. spathaceum*, the fact that only abnormal eggs were produced points to the lack of a (nutritional?) factor or condition required for eggshell production. It should be noted that this organism is especially useful for teaching and research purposes, as the metacercaria will still produce eggs after storage in a refrigerator for at least a month.[14]

V. FASCIOLA HEPATICA

Cultured by: Smith and Clegg[35]
Definitive hosts: commonly sheep and cattle, but also other mammals
Experimental hosts: mice[36]
Location: bile duct
Prepatent period: 90 weeks (sheep)[32]
Molluscan hosts: commonly *Lymnaea* spp., more rarely in species of *Pseudosuccinea, Fossaria, Physa,* and *Similiminaea*[37]
Metacercaria: encysted on pasture
Distribution: cosmopolitan

A. LIFE CYCLE AND MORPHOLOGY

The life cycle of this species is so well known that its life cycle and morphology will not be discussed here.

B. EXPERIMENTAL HOSTS

Development readily takes place in the mouse, egg production commencing at about 5 to 6 weeks.[36]

C. *IN VITRO* CULTURE: EARLY WORK

In spite of the fact that *F. hepatica* causes significant disease in livestock throughout the world, very few attempts have been made to culture either the adult or larval stages. Early work, which was largely unproductive, has been reviewed by Fried.[40] More recently, some limited success was achieved by Davies and Smyth[41] who found that in a medium of NCTC 135 + 50% chick serum + red blood cells, growth in size was comparable to that *in vivo* (in mice), but only a trace of genital rudiment developed, whereas *in vivo*, cirrus, ovary, and testes anlagen had developed by that time. An improved technique was later developed by Smith and Clegg[35] and this method is described below.

D. *IN VITRO* CULTURE: TECHNIQUE[35]
1. Source of Materials

Cercarial emergence from infected snails can be enhanced by placing the snails in a glass dish and lowering the temperature to (say) 14°C by adding ice. As the temperature rises, cercariae emerge *en masse* and encyst on the glass walls from which they can be scraped off, after allowing time for the metacercarial cyst wall to harden (about 24 h).[35] This can equally well be carried out in small tubes with the walls lined with cellophane on which the encysting metacercariae can adhere. This latter technique has the advantage that specific numbers of cysts can be cut out at a time and used for culturing or feeding experiments.

2. Excystment

Numerous techniques have been described for the excystment of the metacercaria. One of the simplest is as follows:[38]

a. Place cysts in N/20 HCl, at 39°C.
b. Add an equal volume of a solution of 1% $NaHCO_3$, 0.8% NaCl, and 20% bile (sheep, dog, or pig).
c. Metacercaria (70 to 80%) will excyst within 4 to 5 h (1.2% crude tauroglycocholate or pure synthetic sodium taurocholate can be used instead of the bile).

A similar method for large-scale excystment of metacercariae is also available.[39] This is useful for metabolic experiments and involves incubation for 1 h under 60% CO_2/40% N_2, followed by the addition of 10% sterilized sheep bile (or taurocholate) which stimulates excystment.

3. Culture Medium and Conditions
Medium — All media contained 50 U/ml of penicillin + 50 μg/ml of streptomycin. Best results were obtained in RMPI 1640 + 50% human serum and 2% human red blood cells. The serum was prepared from type A blood allowed to clot at room temperature, centrifuged at 10,000 × g for 30 min, and filtered through a 0.2-μm cellulose membrane. Lower concentrations of serum gave inferior results.

Conditions — Culture vessels were plastic, screw-top, Leighton culture tubes, sealed with parafilm; 2 ml of medium were gassed for 30 s with 8% O_2 in air. Temperature was 37°C, and the medium was renewed every 3 to 4 d.

4. Results
Results are summarized in Figure 11. Most worms grew at a relatively linear rate for up to 14 weeks before finally degenerating. However, a few worms (seven in four experiments) grew more rapidly during the 7th week in culture and by 14 weeks had grown to 6 to 7 mm by 3.0 to 3.5 diameter — a size comparable to that of sexually mature *Fasciola* grown in mice.[36] These larger worms showed extensive development of the uterus and a morphologically normal cirrus and cirrus sac. Testes were present in all specimens and mature spermatozoa in some. The vitellaria showed extensive development, but only a rudimentary ovary was present and no eggs were formed.

E. EVALUATION
Perhaps the most interesting feature of these results was the sudden development and differentiation of a small number of worms. Since the nutrient provided to all worms was the same, this small number were either initially physiologically different from the rest or (more likely) this small number was somehow subjected to specific (chance!) culture conditions in the system which triggered their differentiation. This system clearly has considerable potential for further experimental work.

VI. SCHISTOSOMES OF MAN

A. GENERAL REVIEW
1. Early Work: Pre-approximately 1975
The successful culture of adult schistosomes from cercariae has long been a desirable aim of parasitologists, although considering the importance of schistosomes and the number of persons working on them, surprisingly little effort (or resources) has been directed towards this end. Most early experiments were carried out on *Schistosoma mansoni*, the most successful of this early work being that of Clegg.[42] Using a relatively simple medium based on lactalbumen hydrolosate, glucose, and rabbit red blood cells (RBC), he transformed

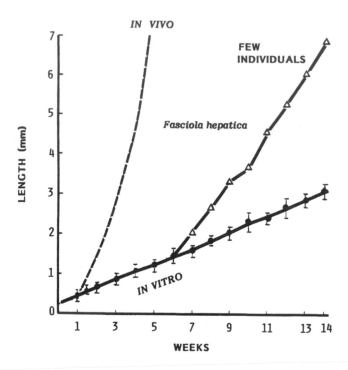

FIGURE 11.   *Fasciola hepatica:* growth of metacercariae in RPMI 1640, 50% human serum, and 2% RBC. The growth rate is based on 96 worms (16 cultures in 8 experiments). A few individuals (seven in four experiments) showed an accelerated growth rate. The approximate growth rate is based on the growth in mice. (After Smith, M. A. and Clegg, C. A., *Z. Parasitenkd.,* 66, 9, 1981. With permission.)

cercariae to schistosomula by allowing them to penetrate mouse skin. Most *in vitro* grown males produced spermatozoa, but females failed to produce eggs except in one, apparently abberant, worm in which sexual maturity was reached.

During this period, too, a number of workers developed what can be described as "maintenance" techniques which enabled adult schistosomes, removed from laboratory hosts, to survive *in vitro* for periods long enough for metabolic experiments to be carried out — up to 12 d in some instances. Details of these are given in Fried,[40] Hansen and Hansen,[43] and Smyth and Halton.[44] Most of these early experiments were carried out on *S. mansoni,* but a few investigators used *S. haematobiumn, S. japonicum,* and the bird schistosome, *Trichobilharzia ocellata,* again with very limited success.

2. Recent Work: Post-approximately 1975

More recent work in the field has been reviewed by Clegg and Smith[45] and Smyth and Halton.[44]

The cercarial-schistosomule transformation — One of the major difficulties facing earlier workers was the supply of culture material. Schistosomula for *in vitro* culture were normally obtained from the lungs of a laboratory host as 7-d-old worms which were difficult to obtain in quantity, although they could also be obtained (in small numbers) by allowing them to penetrate a membrane.[42] This situation was greatly improved by the development of a number of techniques for artificially transforming cercariae to schistosomula. These techniques have been reviewed by a number of workers;[44-46] they fall into two categories, incubation and mechanical methods.

Incubation methods

1. Incubation in tissue culture (Rose) chambers + tissue explants[44]
2. Incubation in diffusion chambers in the peritoneal cavity[44]
3. Incubation in saline containing specific substances such as lecithin, serum, ELAC (lactalbumen hydrolysate in Earle's), etc. or various combinations of these[44,47-51]

Mechanical methods

1. High-speed centrifugation[52]
2. Whorling in a vortex mixer[53]
3. Repetitive syringing through hypodermic needles[54]

Not all of these methods (Table 3) produce identical results and to what extent the cercarial-schistosomule transformation is completed by them has been much examined.[50,51,55] Both the mechanical and incubation methods have produced good results in the hands of different workers. The incubation method, using ELAC, has been described as the method which is " . . . least traumatic for the organisms. Furthermore, it provides almost limitless numbers of schistosomules with the least work and equipment."[50]

The mechanical method has also been shown to be very efficient, and the successful culture of *S. mansoni* by Basch,[56,57] described below, made use of an ingenious double-headed syringe which enables cercariae to be passed from one syringe to another (Figure 12).

The development of these methods proved to be a great stimulus for further work. All three species of human schistosomes, *S. mansoni*,[56-58] *S. japonicum*,[48,49,59,60] and *S. haematobium*[61] plus one species of rodent schistosome, *Schistosomatium douthitti*,[62] have been grown to sexual maturity *in vitro*, with pairing taking place, although eggs were not produced in all species. The most successful of these techniques are reviewed below.

B. *IN VITRO* CULTURE OF *SCHISTOSOMA MANSONI*

> Cultured by: Basch[56,57]
> Definitive hosts: man
> Experimental hosts: mice and other rodents
> Location: hepatic portal system
> Prepatent period: 34 to 48 d[32]
> Molluscan host: commonly *Biomphalaria glabrata*, but, occasionally, other
>   species
> Distribution: tropics

1. Life Cycle and Morphology

The life cycle and morphology of this species are so well known that they will not be discussed here.

2. Growth Rate

In order to establish suitable criteria for comparing growth *in vitro* with that *in vivo*, it is essential to have available reliable data on the normal growth rate in an experimental host. The following were used by Clegg[42] as representing significant stages in development which could be readily identified:

●   Stage 1 (7 d) — Lung forms; no cell division detectable (tested by colchicine-aceto-orcein)

TABLE 3
Schistosoma mansoni: **Assessment of Different Techniques for the Transformation of Cercariae into Schistosomules Evaluated Against the Characteristics of 1-h *In Vivo* Schistosomules**

	Digestive system	Nuclear condition	Glycocalyx	CHR[a]
Cercariae	Granules in esophageal glands	Heterochromatic	Present	Positive
In vivo schistosomules	Granules in mucosal epithelium	Euchromatic	Absent	Negative
In vitro schistosomules				
Dried rat skin	1 h*	6—24 h	1—3 h	3 h
Centrifuged/vortexed	1—3 h	24—48 h	3—6 h	48 h
Syringe-passaged	1—3 h	24—48 h	6—24 h	72 h
Omnimixed	1—3 h	24—48 h	6—24 h	48 h
Centrifuged/incubated	1—3 h	24—48 h	6—24 h	72 h
Serum incubated	1 h	6—24 h	3—6 h	24 h
ELAC incubated	1—3 h	24—48 h	6—24 h	48 h

Note: Asterisk indicates time at which *in vitro* schistosomules are comparable to 1-h *in vivo* schistosomules.

[a] CHR, Cercarienhüllen reaktion.

From Cousin, C. E., Stirewalt, M. A., Dorsey, C. H., and Watson, I. P., *J. Parasitol.*, 72, 606, 1986. With permission.

Syringe A   Syringe B

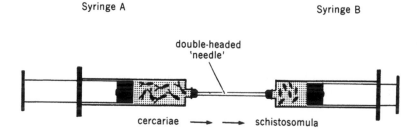

FIGURE 12.   Double syringe system used by Basch[56] for the passage of cercariae of *Schistosoma mansoni*, backward and forward, to transform them into schistosomula. (From Smyth, J. D. and Halton, D. W., *The Physiology of Trematodes*, 2nd ed., Cambridge University Press, Cambridge, 1983. With permission.)

- Stage 2 (15 d) — Gut ceca joined behind ventral sucker
- Stage 3 (28 d) — Males, two testes; females, narrow uterus
- Stage 4 (28 d) — Male, eight testes, anterior two with spermatozoa; female, small ovary; pairing begins
- Stage 5 (30 d) — Vitellaria give a positive reaction for eggshell phenol materials with diazo salt (e.g., fast red salt B)
- Stage 6 (34 to 35 d) — First eggs produced

These criteria, or variations of them, have been used by a number of workers. Basch[56,57] (see below) based his assessment of growth on the stages drawn up by Faust et al.[63]

3. *In Vitro* Culture Technique[56,57]
Production of Schistosomula — The source of material was a Puerto Rican strain of *S. mansoni* maintained in PR-albino *Biomphalaria glabrata* and golden hamsters.

1.   Cercariae were released from snails by exposure to a 75-W bulb.

2. Water containing cercariae was collected through a Sartorius SM 13420 nylon mesh, size 0.1 mm, into 50-ml screw-top glass centrifuge tubes and cooled to 10°C.

3. Cercariae were concentrated by centrifuging for 30 to 60 s at 1000 rpm.

4. The upper 90% water was discarded and 20 ml of wash medium was added. This consisted of BME medium with added 16 mM HEPES and 300 U/ml of penicillin, 300 µg/ml of streptomycin, and 160 µg/ml of gentamycin.

5. The tubes were centrifuged again to form a pellet of cercariae. All subsequent steps were carried out in a bacteriological hood.

6. Cercariae from the two tubes were pooled and resuspended into two new tubes of 25 ml of wash media and again centrifuged.

7. The two pellets were removed separately by Pasteur pipettes to fresh tubes with wash media. These pairs of tubes were clipped into a SI-150 rotator (Scientific Industries, Springfield, MA), tilted to keep the fluid below the tube lip, and rotated at sufficient speed to keep the cercariae remaining in suspension in the antibiotic wash. After 10 min, the tubes were centrifuged and the process repeated with two new tubes.

8. A pellet of washed cercariae was then drawn into a 5-ml glass syringe with a double passage system and a six gauge needle (see Figure 12), and the cercariae passed backwards and forwards causing shearing of the tails.

9. The cercaria were poured into a petri dish and, with a rotatory movement, the tails separated from the bodies.

10. The tails were withdrawn with a Pasteur pipette, fresh wash medium supplemented with 5% serum being added when necessary.

11. The cercarial bodies were transferred to a 15-ml centrifuge tube containing 9 ml of medium + serum and the process of 10-min rotation, centrifugation, and transfer repeated three times.

12. The cercarial bodies at this point were considered to be schistosomula.

Establishment of cultures — Sedimented schistosomula were distributed, with a Pasteur pipette, equally into 8 to 12, 35 × 10 mm, petri dishes each containing 2.0 to 2.5 ml of prewarmed and pregassed culture medium. Eight small petri dishes were placed in larger (150 × 15 mm) plastic petri dishes and incubated at 36°C in a humidified, flowing CO_2 atmosphere.

Culture medium — The composition of the culture medium (Medium 169) is given in Table 4. It was stored at 4°C, in 10-ml volumes, in 16 × 125 mm tubes with slightly loosened caps, under 5% CO_2. Dilution of the medium to 275 mOsm/kg, to compensate for evaporation, was said to improve growth somewhat. A drop of sterile, washed RBC was added to each culture after 24 to 48 h. The RBC (type O), obtained from the local blood transfusion service, were washed twice by centrifugation. The culture media and the RBC were renewed twice weekly. When worms reached (Faust's) Stage 18 (Figure 13), five worms of each sex were transferred to a glass Leighton tube in 1.5 mm medium. A drop of RBC was also added.

4. Results

This technique achieved a considerable degree of success. Schistosomula developed pigment in the gut by the 5th day, gut fusion began on day 11, and pairing was first observed during the 7th week of culture. Large numbers of adult worms were eventually obtained, about 10% of which produced eggs after more than two months culture. Most females produced 15 to 20 eggs, but up to 40 were obtained in some worms.

Unfortunately, the eggs produced by these experiments were abnormal, being only about half the normal size (Figure 14) with the spine typically thin and poorly formed; they were assumed to be infertile. Although the vitelline glands, ovaries, and testes were relatively

TABLE 4
Schistosoma mansoni: **Culture Medium 169 of Basch[56]**

	Component[a]	Stock conc	Add stock	Working conc
1 (i)	BME (E) liquid[b]		ll	
	or			
1 (ii)	BME (E) powder[c]		1 packet	
	plus			
1 (iii)	Water 3 × distilled[c]		ll	
2	Lactalbumin hydrolysate		1 g	1 g/l
3	Glucose		1 g	11.1 mmol/l
4	Hypoxanthine	10^{-3} mol/l[d]	0.5 ml	5×10^{-7} mol/l
5	Serotonin	10^{-3} mol/l[d]	1 ml	10^{-6} mol/l
6 (i)	Insulin U-100[e]	100 U/ml	2 ml	0.2 U/ml
	or			
6 (ii)	Insulin, crystalline[f]	8 mg/ml[d]	1 ml	8 μg/ml
7	Hydrocortisone	10^{-3} M[d]	1 ml	10^{-6} mol/l
8	Triiodothyronine	2×10^{-4} M[d]	1 ml	2×10^{-7} mol/l
9	MEM vitamins[g]	100 ×[d]	5 ml	0.5 ×
10	Schneider's medium[h]	1 ×	50 ml	5%
11	HEPES		2.4 g	10 mmol/l
12	Serum	1 ×[d]	100 ml	10%
13	NaOH	5 mol/l	q.s.	pH 7.4
14	NaHCO₃ (for 1, ii, only)		2.2 g	26 mmol/l
15	Water, 3 × distilled		q.s.	to 275 mosmol/k

Note: q.s., *quantum sufficit.*

[a] Biochemicals from Sigma, St. Louis, MO.
[b] Grand Island Biological (GIBCO), New York, 320—1015.
[c] GIBCO 420—1100.
[d] Stocks frozen at −20°C. Working concentration is based upon a theoretical total volume of 1 l, dilution to 275 mosmol/k and addition of other components increases the volume and reduces the nominal working concentrations.
[e] Lilly Iletin NDC 0002—1135—01.
[f] Bovine origin; Sigma 1—5500.
[g] GIBCO 320—1120.
[h] GIBCO 350—1720. (This medium was supplemented by human blood cells, Type O.)

From Basch, P. F., *J. Parasitol.*, 67, 179, 1981. With permission.

poorly developed, electron microscope studies[64] showed that the structure of the vitelline cells and reproductive tracts was similar to that of normal *in vivo* worms.

An extension of this work was a series of experiments in which paired *in vitro* worms were injected or transplanted into the veins of mice.[65] These worms failed to develop further. However, some normal eggs *were* produced by the use of any of the following:

1. *Ex vivo* worms, i.e., those grown *in vivo* to pairing and then cultured *in vitro* for 34 to 53 d
2. Schistosomula grown *in vitro* for up to 13 d before injection into mice
3. Worms grown up to prepairing stage (see Figure 13) before injection into mice

C. EVALUATION

Although normal, fertile eggs were not obtained by this technique, the results of Basch and colleagues must be considered to represent a major step forward in schistosome culture. The fact that paired worms grown entirely *in vitro* failed to produce normal eggs, whereas

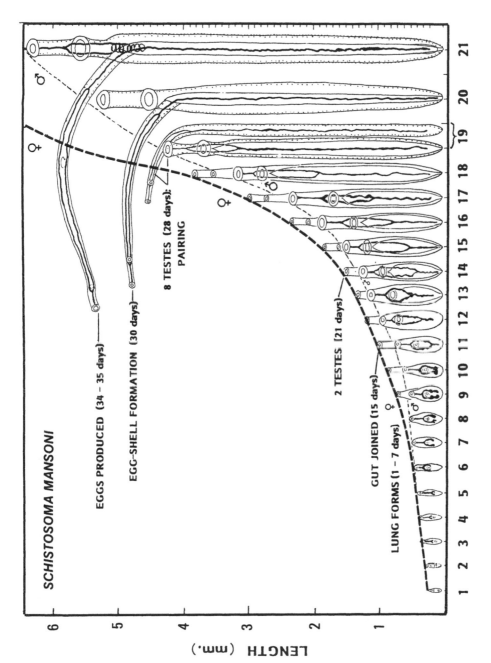

FIGURE 13. *Schistosoma mansoni*: semidiagrammatic representation of the successive growth stages in the mammalian host as used by Faust et al.[63] and Basch.[56] Time scale is based on that of Clegg.[42]

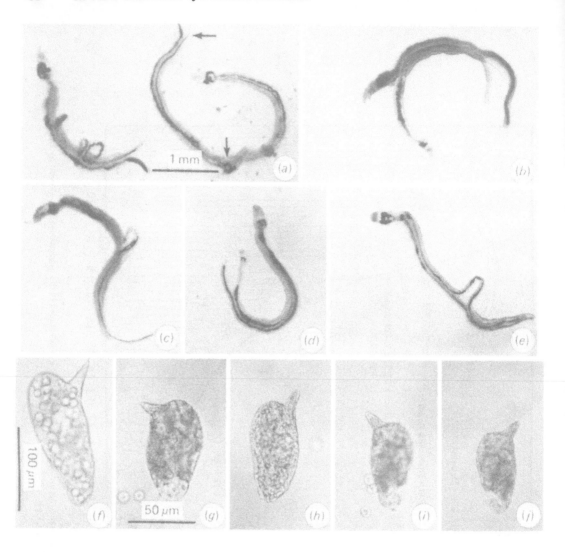

FIGURE 14.   *Schistosoma mansoni:* (a) to (e) paired adults grown from transformed cercariae *in vitro.* (a) One female (arrows) clasped by two males; (f) normal egg from worm grown *in vivo* in hamster; (g) to (j) abnormal *in vitro* eggs (about half normal size). (From Basch, P. F., *J. Parasitol.,* 67, 186, 1981. With permission.)

in vitro worms injected into mice, before pairing, produced normal eggs, suggests that *in vitro* development only becomes abnormal at, or near, the pairing stage — perhaps due to the lack of an appropriate stimulus from the male and/or suboptiminal *in vitro* environment.

This view is supported by later work[66] in which various combinations of *ex vivo*, cultured, and unisexual male and female worms were implanted into hamsters. It was found that females of *any* type (*ex vivo* or cultured) coimplanted with *ex vivo* males, were able to develop to sexual maturity, and produced viable eggs. Many of the cultured males did not induce growth and maturation of females, but in a few cases viable miracidia were produced when cultured males plus *ex-vivo* or unisexual females were implanted. It was concluded[66] '' . . . that both cultured males and females have the potential for full growth and reproductive maturation, but are retarded by inadequate culture conditions.''

D. OTHER *SCHISTOSOMA* SPECIES

Only limited success has been achieved with the other two species of *Schistosoma* infecting man, *S. haematobium* and *S. japonicum*. Although *in vitro* techniques for both these species have been established, only limited development has been achieved, and eggs have not been obtained. For these reasons, these species are only briefly mentioned here.

1. *Schistosoma haematobium*

Using cercariae (from *Bulinus truncatus*), Smith et al.[61] transformed them to schistosomula by passing them through mouse skin and culturing them in Leighton tubes in 2 ml of medium. The medium consisted of 50% human serum and 50% Earle's BSS + antibiotics in a gas phase of 8% CO_2 in air; 1% RBC (type O) were added later. In the best results, worms grew at a rate approaching that *in vivo* and males produced sperm, but females failed to develop vitellaria or eggs. For technical details, the original paper[61] or the summary by Fried[40] should be consulted.

2. *Schistosoma japonicum*

Experiments with this species have been somewhat more successful. Both recently transformed schistosomula[48,49,60] and 17-d liver forms[59] have been grown *in vitro* to near maturity with pairing, and abnormal eggs were produced in one instance.[60] Growth *in vitro* was also considerably less than that *in vivo*.

VII. SCHISTOSOMES OF ANIMALS

Several species of schistosomes infecting animals develop much more rapidly than those in man, and it is surprising that, until recently, they have been little used as experimental models for *in vitro* studies. The best known species are *Trichobilharzia ocellata*[67] and *Schistosomatium douthitti*.[62] Most success has been achieved with the latter species, whose culture is described briefly below.

A. *SCHISTOSOMATIUM DOUTHITTI*

Cultured by: Basch and O'Toole[62]
Definitive hosts: numerous wild rodent species (see below)
Experimental hosts: mice and hamsters[32]
Location: hepatic portal system
Prepatent period: 11 d[68]
Molluscan hosts: species of *Lymnaea*, *Physa*, *Stagnicola*, and *Pseudosuccinea*
Distribution: Canada and the U.S.

1. Life Cycle

Details for maintaining the life cycle in the laboratory are given in Smyth[32] and Kagan et al.[68] In North America, the common hosts are the field vole (*Microtus pennsylvanicus*) and the musk rat (*Ondatra zibethica*), but field mice, nutria, and other rodent species may also serve as natural hosts. The worms normally occur in the blood vessels of the mesenteries of the small intestine. Occasionally, they occur in the lungs.

The males and females are paired until the latter are ready to oviposit. The females then separate from the males and migrate into the blood vessels and deposit their eggs. Unisexual female infections[69] can also produce parthenogenetic eggs so that (unlike *S. mansoni*) a stimulus from the male is not required for sexual maturation to be achieved.

The eggs have no spine, but penetrate the blood vessels and gut wall and are embryonated when they reach the intestinal lumen. Eggs are passed out in the feces and hatch on contact with water.

2. Molluscan Development

The most common molluscan host is probably *L. stagnalis*, but several other species and genera are infected.[70] There are 2 sporocyst generations and as many as 40,000 to 60,000 cercariae may arise from a single miracidium. The incubation period for snails is about 37 to 52 d.

3. Adult (Figure 15)

The morphology and general biology of the adult worms and cercariae have been described by Price[71] and Malek.[70]

Male — Length is 1.9 to 6.3 mm. The body is divided into a forebody, which is flattened, and a hind-body, which forms the gynecophoric canal. There are 14 to 16 testes.

Female — Length is 1.1 to 5.4 mm. The essential female genitalia are present except that a Laurer's canal has not been described. When *in copula*, the female is held in the gynecophoric canal of the male with only its anterior end protruding.

4. *In Vitro* Culture Technique[62]

General procedure — The source of material was *S. douthitti* maintained in laboratory mice or Syrian golden hamsters with the larval stages in *Stagnicola emarginata angulata*. The general culture protocol was based on that used successfully by Basch[56] for *S. mansoni*, with minor modifications, summarized below.

1. Cercariae were processed and converted to schistosomula by the double syringe method (see Figure 12).
2. Schistosomula were placed, approximately 500 per dish, into 10 × 35 mm petri dishes with 2.5 ml of medium 169 (see Table 4).
3. The medium was changed after 2 to 3 d and washed human RBC (Type O) were added.
4. The medium was further renewed every 2 to 3 d for 2 weeks, when about 10 to 12 schistosomula were transferred to glass Leighton tubes containing 1 to 5 ml of medium + cells.
5. The medium was renewed every 2 to 3 d thereafter.

Results — Growth of *S. douthitti* was much more rapid than that of *Schistosoma mansoni*, but when the cultured worms reached a similar stage of maximal development, again the eggs, which were produced in 3 to 4 weeks *in vitro*, were not viable. Eggs were much more numerous and were even produced in single female cultures, as in *in vivo* unisexual infections.

B. EVALUATION

The fact this species develops more rapidly *in vivo* and *in vitro* than *S. mansoni* makes it a particularly useful laboratory model. The additional observation that unpaired females (unlike *S. mansoni*) will produce viable eggs is also an added advantage, and comparison of the physiology of the two species could possibly throw some light on the nature of the "trigger" for sexual maturation of the female *S. mansoni* provided by the male during pairing.

VIII. MISCELLANEOUS TREMATODE SPECIES

A. GENERAL COMMENT

This text has concentrated on giving practical details of the *in vitro* culture of those species which (1) have been most successfully cultured and/or (2) those which are more likely to be easily available from local sources in different countries. However, for com-

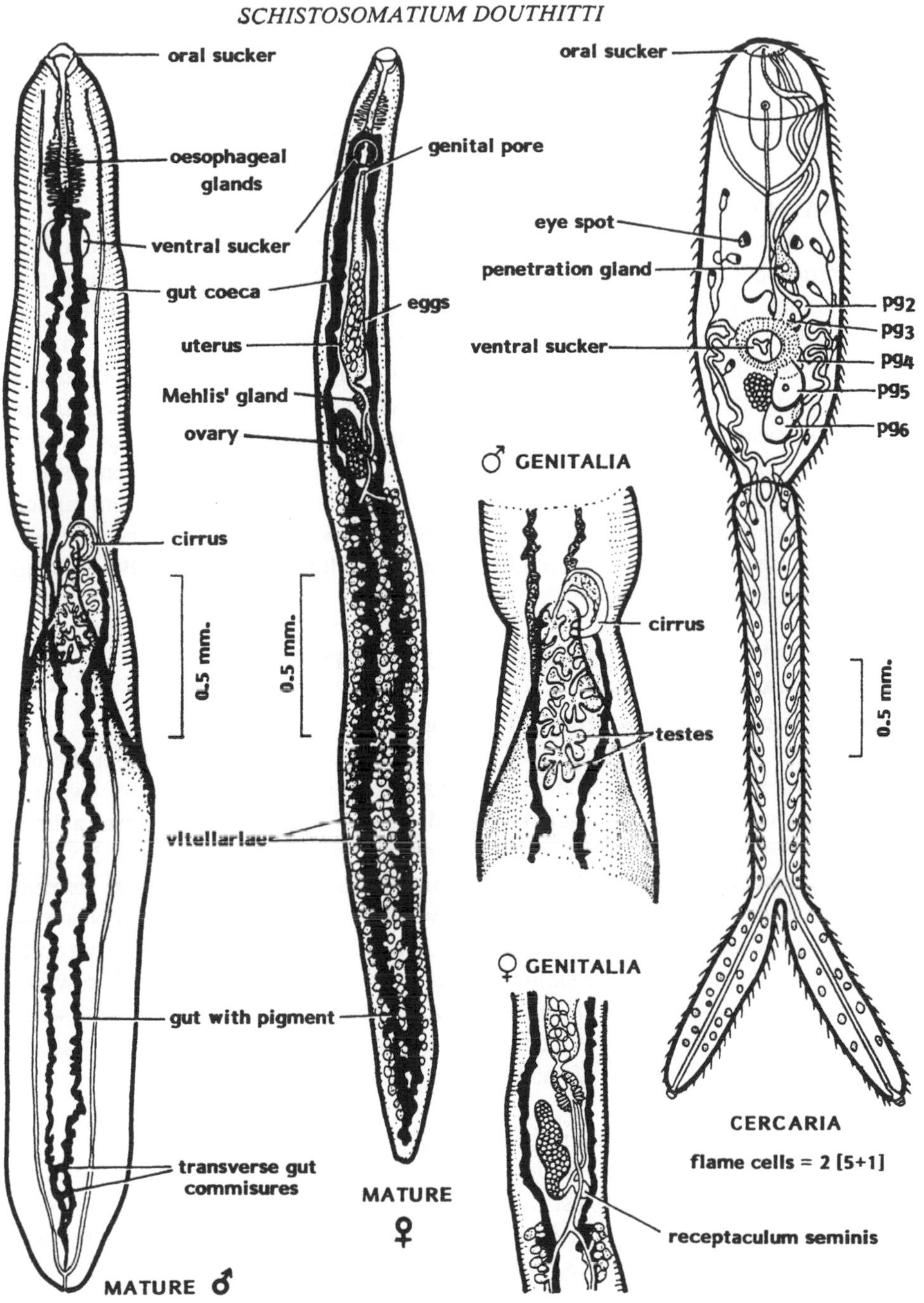

SCHISTOSOMATIUM DOUTHITTI

FIGURE 15. *Schistosomatium douthitti:* morphology of adult and cercaria. (From Smyth, J. D., *An Introduction to Animal Parasitology*, 2nd ed., Edward Arnold, London, 1976. With permission.)

pleteness, it is worthwhile listing some other species which have been cultured with limited degrees of success, which are species not generally available, or where results were published too recently for inclusion.

B. SPECIES CULTURED

These species are as follows: Fasciolidae, *Fasciolopsis buski*;[72] Echinostomatidae, *Echinoparyphium serratum*[73] and *Echinostoma malayanum*;[74] Brachylaemidae, *Leucochloridi-*

omorpha constantiae;[75] Troglotrematidae, *Paragonimus westermani,*[76] *P. miyazakii,* and *P. ohira;*[77] Opisthorchiidae, *Opisthorchis (= Clonorchis) sinensis;*[78] Isoparorchiidae, *Isoparochis hypselobagri;*[79] and Amphistomatidae, *Orthocoelium scoliocoelium.*[80]

IX. TREMATODA: INTRAMOLLUSCAN STAGES

A. GENERAL COMMENT

Relatively little success has been achieved in the *in vitro* culture of the intramolluscan stages of trematodes and yet it is self-evident that the successful culture of such stages would, by eliminating the need to maintain snails in the laboratory, greatly facilitate research in trematode diseases, especially schistosomiasis. Most experiments have been carried out on schistosomes of man with some on the bird schistosome, *Trichobilharzia ocellata.* Although the transformation of the miracidia to sporocysts can readily be carried out *in vitro,* further development has been very limited, and the production of viable sporocysts or cercariae from miracidia has not yet been achieved. So little progress has been made that only a brief account is given below, and the *in vitro* cultivation of cercariae from miracidia remains a major challenge to parasitologists. Various aspects of the field have been reviewed by Hansen and Hansen,[43] Clegg and Smith,[45] Fried,[40] Mellink and van den Bovenkamp,[81] and Smyth and Halton.[44]

B. MIRACIDIUM/MOTHER SPOROCYST/DAUGHTER SPOROCYST TRANSFORMATIONS

1. The Miracidium/Mother Sporocyst Transformation

A number of workers have used relatively simple techniques to bring about the miracidium-sporocyst transformation. The stage is characterized by the cessation of swimming and the shedding of ciliated plates from the miracidium. Di Conza and Basch[82] used a defined medium + 20% human serum for *S. mansoni* and reported mother sporocists developing and increasing three to four times their length, but no daughter sporocysts were formed.

If cultures were started from daughter sporocysts (see below), some growth occurred in that some embryos were formed internally and microvilli developed externally, but further development did not occur. Rather similar results were obtained by other workers with *S. mansoni,*[83,84] *S. japonicum,*[84] and *Trichobilharzia ocellata.*[81] Earlier work was reviewed by Hansen and Hansen.[43] Most of the media used in the above experiments were based on Schneider's *Drosophila* medium plus additives or modifications.

2. The Mother Sporocyst/Daughter Sporocyst Transformation

Daughter sporocysts of *S. mansoni* develop in snails in 10 to 18-d-old mother sporocysts, breaking through the mother sporocyst wall and migrating to the hepatopancreas where they give rise to a further generation of daughter sporocysts.[85] Cultures started from this stage have been more successful than those from the earlier stages, and a second generation of sporocysts has been produced *in vitro.*[86,87]

A controlling factor in these later, successful cultures appears to be the maintenance of a low Eh which was achieved by the addition of Cleland's Reagent (dithiothreitol) or cysteine plus glutathione, together with low levels of oxygen.[86,87] In Schneider's medium, at the appropriate Eh, and supplemented with galactose, reducing agents, serum, and a low pO_2 and a high pCO_2, a second generation of sporocysts developed which emerged over 7 to 11 d culture.[44] A summary of the limited achievements of all stages of larval *in vitro* culture is given in Table 5. In general, the techniques are not yet sufficiently established to be used routinely and are therefore not given here. For details, the original papers should be consulted.

TABLE 5
Schistosoma mansoni: Cultivation of Larval Stages *In Vitro*

Stage	Period of cultivation (25—27°C)
Miracidium to mother sporocyst	10 d
Mother sporocyst to daughter	Not accomplished
Daughter sporocyst	3 weeks
Daughter to secondary daughter	7 d
Secondary daughter sporocyst	11 d
Sporocyst to cercaria	Not accomplished

REFERENCES

1. **Mitchell, J. S., Halton, D. W., and Smyth, J. D.,** Observations on the *in vitro* culture of *Cotylurus erraticus* (Trematoda: Strigeidae), *Int. J. Parasitol.*, 8, 389, 1978.

2. **Olson, R. E.,** The life cycle of *Cotylurus erraticus* (Rudolphi, 1809) Szidat, 1928 (Trematoda: Strigeidae), *J. Parasitol.*, 56, 55, 1970.

3. **Niewiadomska, K. and Kozicka, J.,** Remarks on the occurrence and biology of *Cotylurus erraticus* (Rudolphi, 1809) (Strigeidae) from the Mazuriaw Lakes, *Acta Parasitol. Pol.*, 18, 487, 1970.

4. **Guberlet, J. E.,** Three new species of Holostomidae, *J. Parasitol.*, 9, 6, 1922.

5. **Wooten, R.,** Occurrence of the metacercariae of *Cotylurus erraticus* (Rudolphi, 1809) Szidat, 1928 (Digenea: Strigeidae) in brown trout, *Salmo trutta* and the rainbow trout, *S. gairdneri* Richardson 1836, from Hanningfield Reservoir, Essex, *J. Helminthol.*, 47, 389, 1973.

6. **Zajicek, D.,** Contribution to the knowledge of the life-history of trematodes of the genus *Cotylurus* Szidat 1928 in Czechoslovakia, in *Proc. Symp. Parasitic Worms and Aquatic Conditions*, Czechoslovak Academy of Sciences, Prague, 1964, 131.

7. **Hughes, R. C.,** Studies on the trematode family Strigeidae (Holostomidae). XIII. Three new species of Tetracotyle, *Trans. Am. Microsc. Soc.*, 47, 414, 1928.

8. **Hale, L. J.,** *Biological Laboratory Data*, Methuen, London, 1958.

9. **Halton, D. W. and Mitchell, J. S.,** *Cotylurus erraticus:* section autoradiography of thymidine incorporation during growth and development *in vitro*, *Z. Parasitenkd.*, 70, 515, 1984.

10. **Basch, P. F., Di Conza, J. J., and Johnson, B. E.,** Strigeid trematodes (*Cotylurus lutzi*) cultured *in vitro:* production of normal eggs with continuance of life cycle, *J. Parasitol.*, 59, 319, 1973.

11. **Voge, M. and Jeong, K.,** Growth *in vitro* of *Cotylurus lutzi* Basch 1969 (Trematoda: Strigeidae), from tetracotyle to patent adult, *Int. J. Parasitol.*, 1, 139, 1971.

12. **Basch, P. F.,** *Cotylurus lutzi* sp.n. (Trematoda: Strigeidae) and its life cycle, *J. Parasitol.*, 55, 527, 1969.

13. **Fried, B., Barber, L. W., and Butler, M. S.,** Growth and development of the tetracotyle of *Cotylurus strigeoides* (Trematoda) in the chick, on the chorioallantois and *in vitro*, *Proc. Helminthol. Soc. Wash.*, 45, 162, 1978.

14. **Kannangara, D. W. W. and Smyth, J. D.,** *In vitro* culture of *Diplostomum spathaceum* and *Diplostomum phoxini* metacercariae, *Int. J. Parasitol.*, 4, 667, 1974.

15. **Heatley, W.,** Studies on the Development of Helminths. Observations of the Development of *Diplostomum spathaceum* (Rud.) *In Vivo* and *In Vitro*, M.Sc. thesis, University of Dublin, Dublin, Ireland, 1958.

16. **Palmieri, J. R., Heckmann, R. A., and Evans, R. S.,** Life history and habitat analysis of the eye fluke *Diplostomum spathaceum* (Trematoda: Diplostomatidae) in Utah, *J. Parasitol.*, 63, 427, 1977.

17. **Chubb, J. C.,** Seasonal occurrence of helminths in freshwater fish. II. Trematodes, *Adv. Parasitol.*, 17, 141, 1979.

18. **Dubois, G.,** Synopsis des Strigeidae et des Diplostomatidae (Trematoda). Deuxieme partie, *Mem. Soc. Neuchatel. Sci. Nat.*, 10, 259, 1970.

19. **Sweeting, R. A.,** Investigations into natural and experimental infections of freshwater fish by the common eye-fluke *Diplostomum spathaceum* Rud., *Parasitology*, 69, 291, 1974.

20. **Lester, R. J. G. and Lee, T. D. G.,** Infectivity of progenetic metacercariae of *Diplostomum spathaceum*, *J. Parasitol.*, 62, 832, 1976.

21. **Lester, R. J. G. and Freeman, R. S.,** Survival of two trematode parasites (*Diplostomum* spp.) in mammalian eyes and associated pathology, *Can. J. Ophthalmol.,* 11, 229, 1976.
22. **Williams, M.,** Studies on the morphology and life-history of *Diplostomulum (Diplostomum) gasterostei* (Strigeida: Trematoda), *Parasitology,* 56, 693, 1966.
23. **Kennedy, C. R.,** The use of frequency distributions in an attempt to detect host mortality induced by infections of diplostomatid metacercariae, *Parasitology,* 89, 209,1984.
24. **Bell, E. J. and Smyth, J. D.,** Cytological and histochemical criteria for evaluating development of trematodes and pseudophyllidean cestodes *in vivo* and *in vitro, Parasitology,* 48, 131, 1958.
25. **Smyth, J. D.,** Maturation of larval pseudophyllidean cestodes and strigeid trematodes under axenic conditions: the significance of nutritional levels in platyhelminth development, *Ann. N.Y. Acad. Sci.,* 77, 25, 1959.
26. **Leno, G. H. and Holloway, H. L.,** The culture of *Diplostomum spathaceum* metacercariae on the chick chorioallantois, *J. Parasitol.,* 72, 555, 1986.
27. **Fried, B.,** Growth of *Philophthalmus* sp. (Trematoda) on the chorioallantois of the chick, *J. Parasitol.,* 48, 545, 1962.
28. **Bell, E. J. and Hopkins, C. A.,** The development of *Diplostomum phoxini* (Strigeida, Trematoda), *Ann. Trop. Med. Parasitol.,* 50, 275, 1956.
29. **Blair, D.,** A key to cercariae of British strigeoids (Digenea) for which the life cycles are known, and notes on the characters used, *J. Helminthol.,* 51, 155, 1977.
30. **Rees, F. G.,** The adult and diplostomulum stage (*Diplostomulum phoxini* (Faust)) of *Diplostomum pelmatoides* Dubois, and an experimental demonstration of part of the life cycle, *Parasitology,* 45, 295, 1955.
31. **Rees, F. G.,** Cercaria *Diplostomum phoxini* (Faust), a furcocercaria which develops into *Diplostomum phoxini* in the brain of the minnow, *Parasitology,* 47, 126, 1957.
32. **Smyth, J. D.,** *Introduction to Animal Parasitology,* 2nd ed., Edward Arnold, London, 1976.
33. **Williams, M. O., Hopkins, C. A., and Wyllie, M. R.,** The *in vitro* culture of strigeid trematodes. III. Yeast as a medium constituent, *Exp. Parasitol.,* 11, 121, 1961.
34. **Wyllie, M. R., Williams, M. O., and Hopkins, C. A.,** The *in vitro* culture of strigeid trematodes. II. Replacement of a yolk medium, *Exp. Parasitol.,* 10, 51, 1960.
35. **Smith, M. A. and Clegg, C. A.,** Improved culture of *Fasciola hepatica in vitro, Z. Parasitenkd.,* 66, 9, 1981.
36. **Dawes, B.,** On the growth and maturation of *Fasciola hepatica* L. in the mouse, *J. Helminthol.,* 36, 11, 1962.
37. **Yamaguti, S.,** *Systema Helminthum,* Vol. I., *The Digenetic Trematodes of Vertebrates — Part 1,* Interscience, New York, 1958.
38. **Sewell, M. M. H. and Purvis, G. M.,** *Fasciola hepatica:* the stimulation of excystation, *Parasitology,* 59, 4P, 1969.
39. **Tielens, A. G. M., Van der Meer, P., and Van den Meer, S. G.,** *Fasciola hepatica:* simple, large scale, *in vitro* excystment of metacercariae and subsequent isolation of juvenile liver flukes, *Exp. Parasitol.,* 52, 8, 1981.
40. **Fried, B.,** Trematoda, in *Methods of Cultivating Parasites in Vitro,* Taylor, A. E. R. and Baker, J. R., Eds., Academic Press, London, 1978, 151.
41. **Davies, C. and Smyth, J. D.,** *In vitro* cultivation of *Fasciola hepatica* metacercariae and of partly developed flukes from mice, *Int. J. Parasitol.,* 8, 125, 1978.
42. **Clegg, J. A.,** *In vitro* cultivation of *Schistosoma mansoni, Exp. Parasitol.,* 16, 133, 1965.
43. **Hansen, E. L. and Hansen, J. W.,** Present state of culture of schistosomes; limitations, improvements, applications to control, in *The In Vitro Cultivation of the Pathogens of Tropical Diseases, Tropical Disease Research Series,* 3, Schwabe, Basel, 1980, 325.
44. **Smyth, J. D. and Halton, D. W.,** *The Physiology of Trematodes,* Cambridge University Press, Cambridge, 1983.
45. **Clegg, J. A. and Smith, M.,** Trematoda, in *In Vitro Methods for Parasite Cultivation,* Taylor, A. E. R. and Baker, J. R., Eds., Academic Press, London, 1987, 254.
46. **Stirewalt, M. A., Cousin, C. E., and Dorsey, C. H.,** *Schistosoma mansoni:* stimulus and transformation of cercariae into schistosomules, *Exp. Parasitol.,* 56, 358, 1983.
47. **Eveland, L. K. and Morse, S. I.,** *Schistosoma mansoni: in vitro* conversion of cercariae to schistosomula, *Parasitology,* 71, 327, 1975.
48. **Yasuraoka, K., Irie, Y., Hata, H., and Shimomura, H.,** Culture of *Schistosoma japonica* (Philippine strain) from the cercarial stage and the effects of immune rabbit and human sera *in vitro:* a preliminary report, *Southeast Asian J. Trop. Med. Public Health,* 7, 197, 1976.
49. **Yasuraoka, K., Irie, Y., and Hata, H.,** Conversion of schistosome cercariae to schistosomula in serum-supplemented media and subsequent culture *in vitro, Jpn. J. Exp. Med.,* 48, 53, 1978.
50. **Cousin, C. E., Stirewalt, M. A., and Dorsey, C. H.,** *Schistosoma mansoni:* transformation of cercariae to schistosomules in ELAC, saline and phosphate-buffered saline, *J. Parasitol.,* 72, 609, 1986.

51. **Cousin, C. E., Stirewalt, M. A., Dorsey, C. H., and Watson, L. P.,** *Schistosoma mansoni:* comparative development of schistosomules produced by artificial techniques, *J. Parasitol.,* 72, 606, 1986.

52. **Gazzinelli, G., De Oliveira, C., Figueiredo, E. A., Pereira, L. H., Coelho, P. M. Z., and Pellegrino, J.,** *Schistosoma mansoni:* biochemical evidence for morphogenetic change from cercaria to schistosomule, *Exp. Parasitol.,* 34, 181, 1973.

53. **Ramalho-Pinto, F. J., Gazzinelli, G., Howells, R. E., Mota-Santos, T. A., Figueiredo, E. A., and Pellegrino, J.,** *Schistosoma mansoni:* defined system for stepwise transformation of cercaria to schistosomule *in vitro, Exp. Parasitol.,* 36, 360, 1974.

54. **Colley, D. C. and Wikel, S. K.,** *Schistosoma mansoni:* simplified method for the production of schistosomules, *Exp. Parasitol.,* 35, 44, 1974.

55. **James, E. R. and Taylor, M. G.,** Transformation of cercariae to schistosomula: a quantitative comparison of transformation techniques and of infectivity by different injection routes of the organisms produced, *J. Helminthol.,* 50, 223, 1976.

56. **Basch, P. F.,** Cultivation of *Schistosoma mansoni in vitro.* I. Establishment of cultures from cercariae and development until pairing, *J. Parasitol.,* 67, 179, 1981.

57. **Basch, P. F.,** Cultivation of *Schistosoma mansoni in vitro.* II. Production of infertile eggs by worm pairs cultured from cercariae, *J. Parasitol.,* 67, 186, 1981.

58. **Michalick, M. S. M., Gazzinelli, G., and Pellegrino, J.,** Cultivation of *Schistosoma mansoni* cercarial bodies to adult worms, *Rev. Inst. Med. Trop. Sao Paulo,* 21, 115, 1979.

59. **Fu, H.-M., Chow, K., and Chiu, J.-K.,** *In vitro* cultivation of *Schistosoma japonicum, Int. J. Zoonoses,* 3, 105, 1976.

60. **Wang, W. and Zhou, S. L.,** Studies on the cultivation *in vitro* of *Schistosoma japonicum* schistosomula transformed from cercariae by artificial methods, *Acta Zool. Sin.,* 33, 144, 1987 (in Chinese); *Helminthol. Abstr.,* 57, 508, 1988.

61. **Smith, M., Clegg, J. A., and Webbe, G.,** Culture of *Schistosoma haematobium in vivo* and *in vitro, Ann. Trop. Med. Parasitol.,* 70, 101, 1976.

62. **Basch, P. F. and O'Toole, M. L.,** Cultivation *in vitro* of *Schistosomatium douthitti* (Trematoda: Schistosomatidae), *Int. J. Parasitol.,* 12, 541, 1982.

63. **Faust, E. C., Jones, C. A., and Hoffman, W. A.,** Studies on schistosomiasis mansoni in Puerto Rico. III. Biological studies. 2. The mammalian phase of the life cycle, *P. R. J. Public Health Trop. Med.,* 10, 133, 1934.

64. **Irie, Y., Basch, P. F., and Beach, N.,** Reproductive ultrastructure of adult *Schistosoma mansoni* grown, *in vitro, J. Parasitol.,* 69, 559, 1983.

65. **Basch, P. F. and Humbert, R.,** Cultivation of *Schistosoma mansoni in vitro.* III. Implantation of cultured worms into mouse mesenteric veins, *J. Parasitol.,* 67, 191, 1981.

66. **Basch, P. F. and Rhine, W. D.,** *Schistosoma mansoni:* reproductive potential of male and female worms cultured *in vitro, J. Parasitol.,* 69, 567, 1983.

67. **Howell, M. J. and Bourns, T. K. R.,** *In vitro* culture of *Trichobilharzia ocellata, Int. J. Parasitol.,* 4, 471, 1974.

68. **Kagan, I. G., Short, R. B., and Nez, M. M.,** Maintenance of *Schistosomatium douthitti* (Cort, 1914) in the laboratory, *J. Parasitol.,* 40, 1, 1954.

69. **Short, R. B.,** Sex studies on *Schistosomatium douthitti* (Cort, 1914) Price, 1931 (Trematoda: Schistosomatidae), *Am. Mid. Nat.,* 48, 1, 1942.

70. **Malek, E. A.,** Geographical distribution, hosts and biology of *Schistosomatium douthitti* (Cort, 1914) Price, 1931, *Can. J. Zool.,* 55, 661, 1977.

71. **Price, H. E.,** Life history of *Schistosomatium douthitti* (Cort), *Am. J. Hyg.,* 13, 685, 1931.

72. **Lo, C. T. and Cross, J. H.,** *In vitro* cultivation of *Fasciolopsis buski, Southeast Asian J. Trop. Med. Public Health,* 5, 252, 1974.

73. **Howell, M. J.,** Excystment and *in vitro* cultivation of *Echinoparyphium serratum, Parasitology,* 58, 583, 1968.

74. **Jaw, C. Y. and Low, C. T.,** *In vitro* cultivation of *Echinostoma malayanum* Leiper, 1911, *Chin. J. Microbiol.,* 7, 157, 1974.

75. **Fried, B. and Contos, N.,** *In vitro* cultivation of *Leucochloridiomorpha constantiae* (Trematoda) from the metacercaria to the ovigerous adult, *J. Parasitol.,* 59, 936, 1973.

76. **Kannangara, D. W. W.,** *In vitro* cultivation of the metacercariae of the human lung fluke *Paragonimus westermani, Int. J. Parasitol.,* 4, 675, 1974.

77. **Hata, H., Yokogawa, M., Kobayashi, M., Niimura, M., and Kozima, S.,** *In vitro* cultivation of *Paragonimus miyazakii* and *P. ohirai, J. Parasitol.,* 73, 792, 1987.

78. **Sun, T.,** Maintenance of adult *Clonorchis sinensis in vitro, Ann. Trop. Med. Parasitol.,* 63, 399, 1969.

79. **Nizami, W. A. and Siddiqi, A. H.,** Studies on the *in vitro* survival of *Isoparorchis hypselobagri* (Digenea: Trematoda), *Z. Parasitenkd.,* 45, 263, 1975.

80. **Sharma, A. N. and Sharma, P. N.,** Preliminary experiments on the *in vitro* cultivation of an amphistome *Orthocoelium scolicoelium* (Trematoda; Digenea), *J. Helminthol.*, 56, 169, 1982.
81. **Mellnik, J. J. and van den Bovenkamp, W.,** *In vitro* culture of intramolluscan stages of the avian schistosome *Trichobilharzia ocellata*, *Z. Parasitenkd.*, 71, 337, 1985.
82. **Di Conza, J. J. and Basch, P. F.,** Axenic cultivation of *Schistosoma mansoni* daughter sporocysts, *J. Parasitol.*, 60, 757, 1974.
83. **Benex, J. and Jacobelli, G.,** Évolution *in vitro* des miracidiums de *Schistosoma mansoni*, *Ann. Parasitol. Hum. Comp.*, 56, 57, 1981.
84. **Voge, M. and Seidel, J. S.,** Transformation *in vitro* of miracidia of *Schistosoma mansoni* and *S. japonicum* into young sporocysts, *J. Parasitol.*, 58, 699, 1972.
85. **Hansen, E. L.,** Secondary daughter sporocysts of *Schistosoma mansoni:* their occurrence and cultivation, *Ann. N.Y. Acad. Sci.*, 266, 426, 1975.
86. **Buecher, E. J., Perez-Mendez, G., Hansen, E. L., and Yarwood, E.,** Sulfhydryl compounds under controlled gas in culture of *Schistosoma mansoni* sporocysts, *Proc. Soc. Exp. Biol. Med.*, 146, 1101, 1974.
87. **Hansen, E. L., Perez-Mendez, G., Yarwood, E., and Buecher, E. J.,** Second generation daughter sporocysts of *Schistosoma mansoni* in axenic culture, *J. Parasitol.*, 60, 371, 1974.

Chapter 5

CESTODA

J. D. Smyth

TABLE OF CONTENTS

I. BASIC PROBLEMS OF CESTODE *IN VITRO* CULTURE

A. GENERAL COMMENTS

Substantial progress has been made with the *in vitro* cultivation of this group and a number of species can now be cultured to maturity or near maturity and some (e.g., *Hymenolepis diminuta*) have been cultured through their entire life cycle *in vitro*. Early work in this field has been reviewed by Smyth,[1,2] Taylor and Baker,[3] and Voge.[4] More recent work has been reviewed by Arme,[5] Evans,[6] Howell,[7] and Smyth and McManus.[8]

B. SPECIFIC PROBLEMS

Many of the problems of cestode cultivation are shared with other parasitic helminths, such as trematodes or nematodes. There are, however, some problems which are unique to cestodes. A major one is the fact that cestodes have no gut so that all nutrient must be taken in through the tegument and all waste material must be excreted through it. This clearly places a limit on the kind of molecules which can pass through the tegument and greatly limits the composition of the culture media which can be used. A further complication, which inhibited *in vitro* culture of cestodes for many years, is the fact that adult cestodes normally inhabit the intestines of vertebrates which contain a rich microflora, so that obtaining adult worms in a sterile condition represented a major problem. The availability of antibiotics such as gentamycin, penicillin, and streptomycin has largely eliminated this difficulty, and there is no reason why any adult tapeworm cannot be obtained in a sterile condition by the use of appropriate washing and sterilizing techniques. Although adult worms can now be obtained in a sterile condition, it has been found that the more satisfactory approach has been to use the larval stages as starting material for *in vitro* culture attempts, as — at least in vertebrates — they occur in sterile tissue sites from which larvae can readily be dissected in a sterile condition. This approach has been very successful with many species of the better-known genera such as *Schistocephalus*, *Ligula*, and *Diphyllobothrium* among the Pseudophyllidea and *Hymenolepis*, *Taenia*, and *Echinococcus* among the Cyclophyllidea.

C. CONDITIONS OF CULTIVATION

Once sterility has been established, further problems arise, the chief of which are considered below.

1. Physicochemical Conditions

A major difficulty is that the precise conditions pertaining in the alimentary canal of a particular host at a particular time are very poorly known. Moreover, the nutrient content and general composition of the intestinal contents will fluctuate with the feeding patterns of the hosts. The situation is further complicated by the fact that some species, e.g., *Hymenolepis diminuta*, undergo complex migratory movements relative to the feeding patterns of the host[9] — all situations difficult to reproduce *in vitro*. Again, although the broad characteristics of the vertebrate gut are well known, little information is available on the conditions within specific sites, such as the crypts of Lieberkühn, for example, in which some cestode species embed their scoleces. Much more information is available on the biological environment within tissue sites inhabited by larvae, such as liver, body cavity, muscles, blood stream, bile ducts, brain, etc.,[2,10,11] although data for hosts other than man and laboratory animals are relatively scarce.

Ideally, to provide appropriate conditions for *in vitro* culture, data should be available on the pH, pO_2, pCO_2, Eh, amino acid and carbohydrate levels, temperature, osmotic pressure, and concentrations of the common physiological ions of the natural habitat. Unfortunately, for most species, very little of this data is available and much more information is needed in this area of research.

2. Nutritional Requirements

The nutritional materials available to cestodes in intestinal or tissue sites are complex in composition and difficult to replace by defined media. Moreover, individual species have been shown to have very specific nutritional requirements and what is satisfactory for one species may not be so for another, closely related, species *even though it utilizes (apparently) the same site in the same host*! A classical example of this is the case of *Echinococcus granulosus* in which protoscoleces from sheep hydatid cysts (sheep "isolates") grow readily to sexual maturity *in vitro*, whereas those from horse hydatid cysts, although surviving for long periods *in vitro* fail to undergo sexual differentiation.[12] Since both isolates grow to sexual maturity in the gut of a dog, it is clear that the worms are selecting different nutrients and/or conditions in the intestine for their growth processes. The requirements of the sheep isolate are clearly being provided for *in vitro* and those for the horse isolate are not. This simple experiment led to the discovery of different nutritional (metabolic?) strains of *E. granulosus*.[13] Little precise information is known regarding the special nutritional requirements of cestodes, apart from the basic metabolic requirements such as carbohydrates, proteins, and lipids,[8] and most culture media used have been derived empirically, utilizing the many synthetic media now commercially available.

3. Removal of Metabolic Waste

In the natural habitat, the metabolic waste products of adult and larval cestodes are readily removed from the vicinity of the organisms as a result of the natural circulation of body fluids or associated body movements. As with all helminth culture, a successful *in vitro* technique must similarly provide conditions which allow for the rapid removal of toxic waste products. Some early authors have used complex continuous flow systems,[14] although later workers have found that roller tube systems, with reasonably frequent media renewals, are equally effective.

4. Provision of Differentiation Stimuli

Cestodes have complex life cycles often involving up to three hosts (two intermediate and one definitive) and it is well established that specific stimuli may be required to "trigger" the differentiation of one stage to another. The nature of these triggers has been resolved for very few species; examples of well-recognized triggers are change in temperature (e.g., *Shistocephalus*, see Section II) and presence of bile (e.g., *Spirometra*, see Section IV).

5. Host-Parasite Spatial Relationships

The spatial relationship between an adult tapeworm and the intestinal mucosa of its host is difficult to reproduce artificially *in vitro*, and in some cases these relationships play a vital role in the developmental biology of the worm. Thus, it is well established that compression of the strobila is necessary for insemination in species such as *Ligula* and *Schistocephalus* and probably other species.[8] Other specific spatial requirements (attachment?, surface contact?) may be necessary in other instances, but little is understood in this area of cestode biology.

II. *SCHISTOCEPHALUS SOLIDUS*

Cultured by: Smyth[15,20,21]

Definitive hosts: numerous fish-eating birds; more rarely, fish-eating mammals (e.g., otters)

Experimental hosts: ducks, chicks, pigeons, rats, hamsters

Prepatent period: 36 to 48 h

First intermediate host: procercoid in fresh-water copepods

Second intermediate host: plerocercoid in three-spined stickleback, *Gasterosteus aculeatus*, and related species

Distribution: Europe, Russia, North America

This species is of some historic importance as being the first cestode to be cultured to sexual maturity *in vitro* with the production of fertile eggs.[15,20]

A. GENERAL BIOLOGY AND MORPHOLOGY
1. Adult

The adults have been reported from a wide range of fish-eating birds, but because the plerocercoid is highly progenetic (see below) and matures within 36 to 48 h in the bird, it may rarely be found on autopsy, even in areas where fish infected with plerocercoids abound. It has occasionally being reported from fish-eating mammals, such as otters. The morphology of the adult has been described in detail by Hopkins and Smyth[22] (Figures 1 to 3). It is a lanceolate-shaped worm measuring 50 to 80 by 10 mm. The bothria on the scolex are little more than shallow grooves which result in the worm having little adhesive powers to attach to the intestinal wall. The lack of powerful bothria is compensated for by the fact that the plerocercoid is highly progenetic, being already segmented with well-formed genital anlagen (see Figure 1) and can mature and commence producing eggs as early as 36 h postinfection. The species is also unusual, however, in possessing an additional band of circular muscles

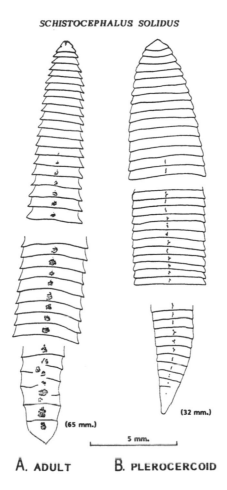

A. ADULT   B. PLEROCERCOID

FIGURE 1. *Schistocephalus solidus*. (A) Adult from experimental infection in ducks; details of the genitalia are given in Figure 2; (B) plerocercoid from body cavity of stickleback, *Gasterosteus aculeatus*. Note the advanced state of progenesis, as evidenced by the segmented condition and the presence of genital anlagen (see Figure 3) in all except the most anterior proglottides. (From Hopkins, C. A. and Smyth, J. D., *Parasitology*, 41, 283, 1951. With permission.)

which may help it to brace itself against the peristalsis of the gut and so retain its position. This could explain how in some experimental hosts (e.g., hamsters) it has been found to remain (exceptionally) for as long as 18 d.[23]

2. Plerocercoid

The plerocercoid (Figures 1 and 4) has the main features of the adult, with 62 to 92 proglottides and progenetic genitalia, containing spermatocytes and oocytes, but no spermatozoa or ova. It occurs in the body cavity of the fresh-water and marine forms of the three-spined stickleback, *Gasterosteus aculeatus*. The ten-spined stickleback, *Pungitius (Pygosteus) pungitius,* has been infected experimentally[24] and may be a natural host in Russia. The siting of the plerocercoid in the (sterile) body cavity makes it readily removed in a sterile condition for *in vitro* culture. When the body cavity is slit open with a fine scalpel, the larva(e) emerges immediately (Figure 4); often only a single large larva is present.

3. Life Cycle (Figure 5)

The life cycle follows the typical pseudophyllidean pattern.[11] The eggs need a warm

SCHISTOCEPHALUS SOLIDUS

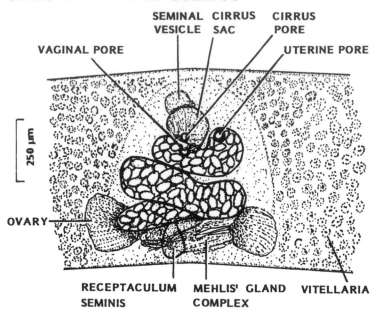

FIGURE 2. *Schistocephalus solidus*, dorsal view of adult genitalia as seen in whole mount preparation. (From Hopkins, C. A. and Smyth, J. D., *Parasitology*, 41, 283, 1951. With permission.)

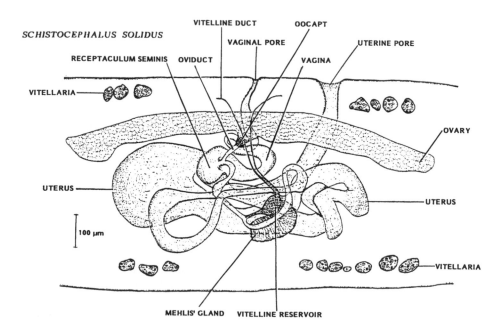

FIGURE 3. *Schistocephalus solidus*, anatomy of female genitalia; reconstruction from serial sections. (From Hopkins, C. A. and Smyth, J. D., *Parasitology*, 41, 283, 1951. With permission.)

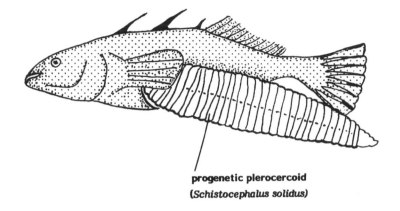

progenetic plerocercoid
(Schistocephalus solidus)

FIGURE 4. *Gasterosteus aculeatus* (three-spined stickleback) with body cavity cut open to release a plerocercoid of *Schistocephalus solidus* which may itself weigh >90% of the (unopened) fish. (Drawn from an original photograph.)

temperature and oxygen to embryonate, requiring about 8 d at 26°C. In the laboratory, this can readily be achieved by placing washed eggs on cellophane disks on watch glasses enclosed in petri dishes, with occasional changes of water. Eggs hatch on exposure to light and develop into procercoids if eaten by a suitable copepod species, such as *Eucyclops* (= *Cyclops?*) *agilis*,[25] becoming infective in about 10 d. Fish become infective by ingesting infected copepods, and birds become infective by eating infected fish.

4. Laboratory Maintenance: Experimental Hosts

Practical details regarding the maintenance of this species in the laboratory have been comprehensively reviewed by Orr and Hopkins.[25] Birds and mammals (rats, hamsters) have proved to be effective experimental hosts, and it is interesting to note that the position taken up in the host gut, as well as the longevity, varies substantially from host to host. The most satisfactory bird laboratory hosts appear to be 1 to 4-week-old ducks or 2 to 5-week-old chicks in which 40 to 50% become established.[11] Pigeons can also be infected, but they require to be fed fish or plerocercoids enclosed in gelatine capsules; feeding free plerocercoids to pigeons is generally unsatisfactory. Intracoelomic infection can also be readily carried out in mice, which is a particularly useful technique for providing a ready supply of fertile eggs.[26,27]

B. *IN VITRO* CULTURE TECHNIQUE: PLEROCERCOID TO ADULT[15,20,21]

1. Source of Material

With practice it is relatively easy to remove the plerocercoids of this species from the body cavity of the stickleback which should be killed by pithing. It is convenient to leave the needle in the fish so that it can be handled without touching the skin (as pressure on the swollen abdomen of an infected fish may result in premature release of the larva, (see Figure 4, with subsequent risk of infection).

In the original technique,[15] the surface of the fish was painted with alcoholic iodine (1% iodine, w/v, in 90% ethanol), before dissection with sterile instruments. The fish was held in a retort stand at eye level for easy dissection and on removal, by sterile forceps or platinum loop, larvae were placed in sterile saline in petri dishes and rinsed several times in sterile saline before culturing. This early work was carried out without the benefit of antibiotics, but the basic technique was sufficiently successful to allow a substantial number of sterile cultures to be established, over a number of years, and most of the basic problems of cultivation were satisfactorily solved.[20] Nowadays, workers would be well advised to rinse

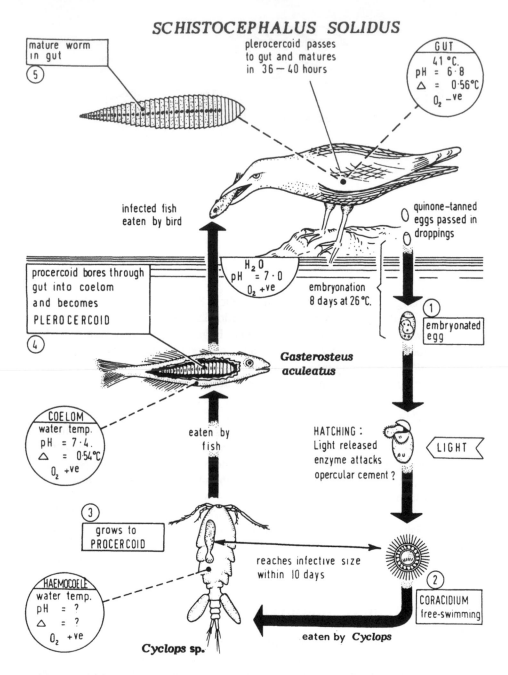

FIGURE 5.   *Schistocephalus solidus*: life cycle and some physiological factors relating to it. (From Smyth, J. D., *An Introduction to Animal Parasitology*, 2nd ed., Edward Arnold, London, 1976, 255. With permission.)

plerocercoids, after removal from the fish, in BSS containing the usual antibiotic mixtures to assure sterility before culturing.

2. Culture Media and Conditions

(commercial tissue culture media not then being available). The plerocercoid has exceptionally high food reserves, such as glycogen (50% of the dry weight[8]), so that probably little, if any, nutrient uptake appears to be necessary to achieve sexual maturation; prolonged maintenance would, however, clearly need some additional nutrients.

Although eggs were produced in these early experiments (see below), these proved to be infertile. In addition, cytological examination of cultured worms revealed that the testes showed some cytological abnormalities and insemination — as shown by the absence of spermatozoa from the receptaculum seminis — had not taken place. Subsequent work[20] showed that for plerocercoids to develop to normal sexually mature adults and for insemination and fertilization to occur *in vitro,* the following culture conditions were essential:

1. Use of well-buffered media to counteract the toxic effects of acidic metabolic waste products — Nutrient broth was replaced by 100% horse serum (which has strong buffering powers) with considerable success, but most of the modern tissue culture media, such as Morgan 199, NCTC 135, NCTC 109, and RPMI 1640, now available, buffered with added serum and, perhaps, HEPES, are likely to be equally successful.
2. Cultivation under anaerobic conditions or at a pO_2 sufficiently low to prevent premature oxidation of the phenolic egg-shell precursors in the vitellaria — this can be achieved by using tall culture tubes.
3. Gentle agitation of the culture medium to assist diffusion of waste metabolites from the vicinity of the worm, with renewal of media at appropriate intervals to maintain the pH.
4. Compression of worms during culture within dialysis tubing to enable self-insemination to take place and fertilized eggs to result — The latter can be achieved by simply holding a loop of narrow-bore dialysis tubing in place with a bung or using a more sophisticated ground glass assembly (Figure 6) which can be prepared by a competent glass blower.

3. Results

As mentioned above, in the early experiments, eggs were infertile and cytological abnormalities were present in the testes. However, once the reasons for this were established and the culture conditions summarized above were applied, no difficulty was found in maturing worms *in vitro* with the production of fertile eggs. The highest level of fertility achieved was 88% which approaches the *in vivo* situation. Moreover, plerocercoids matured under these conditions in about the same time as in a bird (36 to 40 h). Highest fertility was achieved when two worms were cultured within the cellulose tubing (see Figure 6) so that the surface of the worms were pressed against each other. Sectioned worms showed the receptacultum seminis of cultured worms to be full of spermatozoa. It was not established whether or not cross-fertilization took place. Eggs from worms matured free in a culture tube rarely hatch, although occasionally a low level (>6%) of fertility occurred. A curious feature of these experiments was that a few (apparently) infertile eggs developed parthenogenetically, resulting in the formation of "miniature" oncospheres which failed to hatch.[20]

4. Evaluation

This is probably one of the easiest cestodes to mature *in vitro* once the basic techniques are mastered and the fact that fertile eggs can be obtained means that animal definitive hosts can be eliminated entirely. Although the techniques described above are recommended, it is likely that simpler techniques would suffice. Renewing the media frequently and the use of antibiotics may make absolute sterility unnecessary. The worm is an excellent model for biochemical experiments, as the fact that maturation can be "triggered" by simply raising the temperature to 40°C implies that a metabolic "switch" is operating and the biochemistry of this system has been much investigated.[18,19]

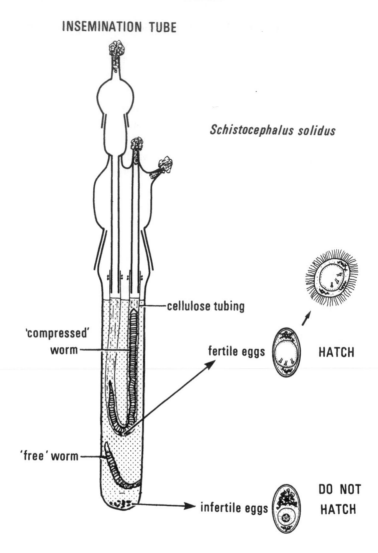

INSEMINATION TUBE

Schistocephalus solidus

cellulose tubing

'compressed' worm

fertile eggs   HATCH

'free' worm

infertile eggs   DO NOT HATCH

FIGURE 6.   Culture tube which enables *Schistocephalus solidus* to undergo self-fertilization during maturation *in vitro* at 40°C. Insemination only occurs when worms are compressed during maturation thus enabling the cirrus in each proglottid to enter the vagina; this results in fertile eggs. Worms cultured ''free'' in the medium produce only infertile eggs. (From Smyth, J. D. and McManus, D. P., *The Physiology and Biochemistry of Cestodes*, Cambridge University Press, London, 1989, 263. With permission.)

C. *IN VITRO* CULTURE TECHNIQUE: GROWTH OF PLEROCERCOID[29,30]

No serious attempt has been made to culture procercoids of this species to plerocercoids, but relatively small plerocercoids (2 to 200 mg fresh weight, F.W.) have been cultured to larger plerocercoids *in vitro*. This was achieved[29] using a medium of 25% horse serum, 0.5% yeast extract, and 0.65% glucose in Hanks's BSS at pH 7.1, at 21°C in a gas phase of 5% CO_2 in air, plus antibiotics. Larvae were cultured in 5 ml of medium, renewed every 48 h, in 125 × 20 mm roller tubes. Up to 500% growth in 8 d was reported. Somewhat better results were achieved in a slightly modified medium (0.6% glucose, 0.5% yeast extract, and 20% horse serum in Hanks's BSS).[30]

III. *LIGULA INTESTINALIS*

Cultured by: Smyth;[21,31,32] Flockart[33]
Definitive hosts: numerous species of fish-eating birds
Experimental hosts: birds, rabbits, cats, dogs[34]
Prepatent period: 48 to 72 h
First intermediate host: procercoid in fresh-water copepods
Second intermediate host: plerocercoid in numerous species of fresh-water fish
Distribution: Europe, Russia, North America, cosmopolitan?

A. GENERAL BIOLOGY AND MORPHOLOGY

1. Adult

The adults have been reported from numerous fish-eating birds (especially herons and ducks) which become infected by eating fish infected with the plerocercoid. The parasite is commonly found in reservoirs inhabited by fish-eating birds and up to 100% of fish may be infected. The strobila is unusual in being unsegmented, but transversely wrinkled. The genitalia (Figure 7) closely resemble those of *Schistocephalus solidus* (see Figure 3), but are more condensed. Because the plerocercoid is progenetic (see below), maturation takes place rapidly in the bird gut (within 2 to 3 d).

2. Plerocercoid

The plerocercoids are much larger than those of *Schistocephalus* and in large fish may reach a length of 50 cm. The progenesis of the genitalia is somewhat less than that in *Schistocephalus*, but the anlagen of the genitalia are clearly visible in stained sections of the plerocercoid (see Figure 7A). The plerocercoid is widely distributed having been reported from some 70 species of fish of the family Cyprinidae. Its presence frequently causes parasitic castration in fish (Figure 8) as evidenced by the marked reduction in the development of the gonads.[8] The ovary is often reduced to the level found in a "spent" fish and the testis is similarly affected. These effects are associated with the reduction in size and granulation of the basophil cells in the transitional lobe of the pituitary, but the mechanism of this effect is not understood.[8]

3. Life Cycle and Laboratory Maintenance

The life cycle is almost identical with that of *S. solidus* (see Figure 5) and will not be discussed further here. Details of the maintenance of the life cycle in the laboratory have been given by Orr and Hopkins.[35] After feeding plerocercoids to ducklings, eggs were found in the feces after 72 to 168 h of infection. Eggs were extracted by sieving and were washed by decantation and placed in water in crystallizing dishes. Flockart[33] found eggs in sections of adult worms as early as 48 h and reported that ducklings 2 to 7 d old were the most suitable host for laboratory infection.

Eggs can be embryonated at 25°C in a incubator in the dark and hatched by exposure to a bright light after 10 d. Egg fertility of *in vivo* eggs has been reported to be 84 to 90%.[33] When emergent coracidia were fed to the copepods, *Diaptomus fragilis* and *Mesocyclops leuckarti*, infective procercoids developed in the body cavity after 10 d at 25°C. *Cyclops strenuus* and *C. bicuspidatis* have also been reported as suitable intermediate hosts.[35] Plerocercoids have been found to develop satisfactorily in tropical fish fed on infected copepods, the following species being infected experimentally:[35] Schuberti barbs (*Barbus sachsi*), Giant danios (*Danio malabaricus*), and White Cloude Mountain minnows (*Tanichthys albonubes*). Of these, Giant danios, which reach a size of 10 to 12 cm, are ideal as laboratory hosts; in these, plerocercoids possess genitalia anlagen after 60 d at 25°C.

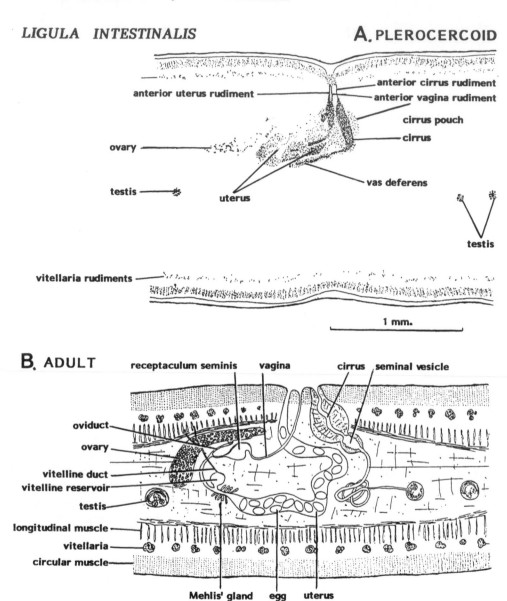

FIGURE 7.   *Ligula intestinalis*, comparison of the genitalia in (A) plerocercoid from fish; note the rudiments (anlagen) of the genitalia are already present; (B) sexually mature adult obtained by *in vitro* culture of a plerocercoid. (From Smyth, J. D., *Parasitology*, 38, 173, 1947. With permission.)

B. *IN VITRO* CULTURE: PLEROCERCOID TO ADULT[21,31-33]
1. Whole Plerocercoids

Because the culture techniques required for *Ligula* are so similar to those described for *Schistocephalus*, they will only be dealt with briefly here. The main difference in that the large size of the *Ligula* plerocercoids requires larger culture tubes and more frequent changes of media as the acid excretory products rapidly bring about a pH drop which inhibits development.

The original technique[31] used bacteriological peptone broth (ox heart broth + 1% peptone) — at that time, commercial media were not available. Later experiments[21,32] used meat extract media + 0.5% saline + 1% peptone and also undiluted horse serum. The most successful results were obtained using 50 ml of undiluted serum in tall culture tube

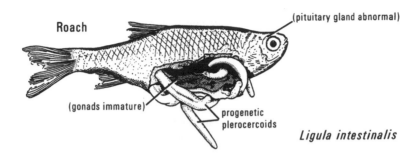

Roach

(pituitary gland abnormal)

(gonads immature)

progenetic
plerocercoids

Ligula intestinalis

FIGURE 8.   Roach infected with plerocercoids of *Ligula intestinalis*. The presence of the larvae causes "parasitic castration" of the fish.

(9 × 1 in.) (providing semi-anaerobic conditions), the medium being changed every 24 h with one larva per culture tube. Egg production took place at 70 h and about 6% of eggs proved to be fertile.

Some 30 years later, Flockart,[33] using the modern tissue culture media now available, was able to improve on these results. She used Parker 199 or NCTC 135 with 20% calf serum and cultured larvae within cellulose dialysis tubing (to induce insemination) as used for *Schistocephalus* (see Figure 6). As the plerocercoids of *Ligula* are larger than those of *Schistocephalus*, diffusion of the toxic, acidic waste products of metabolism was increased by making large number of minute perforations in the cellulose tubing with a sharp needle. This system proved to be very successful although the fertility of the eggs never exceeded 31%. It is likely that the buffering powers of the medium could be improved by the addition of HEPES, although this does not yet appear to have been attempted.

2. Plerocercoid Fragments

As indicated above, culturing whole plerocercoids creates difficulties due to the large size of a larva, especially in removal of the metabolic waste products and the large amounts of media required. An alternative method is to culture short lengths (about 6 to 12 mm) cut (aseptically) from the central region of the plerocercoid (Figure 9). Such fragments can be cultured in small tissue culture flasks or culture tubes in 20 ml of medium, changed every 1 to 2 d, depending on size, at 40°C. Eggs should appear on the third day of culture. Eggs produced by this method have been reported as having 11 to 19% fertility.[33] This technique is especially useful for teaching purposes as each student can culture his own fragment. It is also a valuable technique for biochemical studies.

Plerocercoid fragments can also be matured by implantation in the peritoneal cavity of mice, the eggs produced having a fertility of 11 to 21%.[33] Such fragments eventually become encapsulated.

C. EVALUATION

The large size of the plerocercoids of *L. intestinalis* and the ease with which whole larvae or fragments cut from them can be cultured to sexual maturity *in vitro*, with the production of eggs, makes them especially useful for research in comparative biochemistry and developmental biology and a number of valuable studies have been carried out.[36-38] The species also provides delightful experimental material for teaching purposes!

Compared with *Schistocephalus* — in which eggs produced *in vitro* may have a fertility as high as 88% — the fertility of eggs produced by *Ligula* is low (>31%) and clearly the technique needs improving to provide better conditions for insemination to take place. This is a research area which would reward further study.

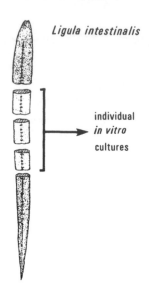

FIGURE 9.   *Ligula intestinalis*, plerocercoid cut in fragments each of which may be cultured individually to sexual maturity *in vitro*. (Modified from Smyth, J. D., *J. Exp. Biol.*, 26, 1, 1949. With permission.)

IV. *SPIROMETRA MANSONOIDES*

Cultured by: Berntzen and Mueller[16,17,39,40]
Definitive hosts: cat, dog, racoon[41]
Experimental hosts: cats[42]
Prepatent period: 12 to 13 d[42]
First intermediate host: procercoid in fresh-water copepods[39]
Second intermediate host: water snake, amphibia, alligators, birds, mammals including man
Distribution: Europe, Russia, North America, probably cosmopolitan

A. GENERAL COMMENT

This species is widely distributed and is best known for its ribbon-like plerocercoid or "*sparganum*" which occurs in the viscera and/or muscles of a wide range of cold-blooded and warm-blooded hosts. The pathological condition it induces in man is known as "*sparganosis*".[11] Although the most worked-on species is referred to as *Spirometra mansonoides*, the speciation is complex and there appears to be several strains in different countries (Table 1). Closely related species are *Spirometra erinacei* in cats and dogs, with plerocercoids in snakes and amphibia, and *Spirometra (Diphyllobothrium?) mansoni* with its adult in dogs and its plerocercoid in snakes.

The plerocercoid has been the subject of much biochemical research as it releases a growth factor which induces abnormal growth in experimental mice (see below). In a marathon effort spanning many years, Mueller and colleagues[16,17,39-42] have developed elegant techniques which enabled them to culture this species from the procercoid to the egg-producing adult and thus reproduce a substantial part of the life cycle *in vitro*.

B. GENERAL BIOLOGY AND MORPHOLOGY
1. Adult

The morphology of the adult,[43] which closely follows that of species of *Diphylloboth-*

TABLE 1
Development of Different Strains of *Spirometra* spp. *In Vitro*

Species	Gravid[a] (d)	Eggs in medium[b] (d)	Proglottids shed (d)	Length at termination (cm)	Total cultivation period (d)
Spirometra mansonoides	16	18	24	44.5	42
	18	22	28	31.0	30
Australian[c]	14	16	?[d]	38.0	20
	14	18	?[d]	36.0	20
	Terminated due to contamination				10
Malay[c]	15	18	?[d]	42.5	20
	Terminated due to contamination			—	11
	Terminated due to contamination			—	8

[a] Time required to become gravid. This was determined by periodic removal of the culture chamber and examination of the worms with an inverted microscope. As the uterus filled with eggs, it became dark and opaque.

[b] Time when eggs first appeared in culture medium.

[c] Australian and Malay worms are regarded as strains of a single species.

[d] Not observed; if shedding occurred, it was in very small pieces.

From Berntzen, A. K. and Mueller, J. F., *J. Parasitol.*, 58, 750, 1972. With permission.

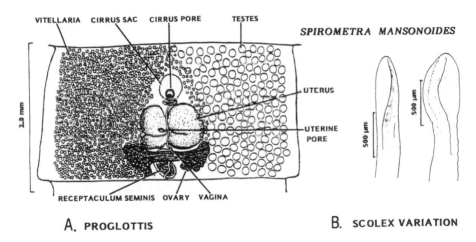

A. PROGLOTTIS   B. SCOLEX VARIATION

FIGURE 10. *Spirometra mansonoides*, morphology of adult proglottis. (From Mueller, J. F., *J. Parasitol.*, 21, 114, 1935. With permission.)

rium, is shown in Figure 10. It has been most studied in the cat in which it has been reported to survive for 3.5 years.[41] For details of the laboratory maintenance in cats, see Mueller.[42] Like the adult *Diphyllobothrium latum*, *Spirometra* absorbs large quantities of vitamin B_{12} which probably gives the scolex its pinkish color.

2. Sparganum (= Plerocercoid)

The sparganum is ribbon-like and ivory white in color, with the scolex poorly developed (if at all) and the strobila wrinkled, but unsegmented. Intermediate hosts can become infected with spargana in three ways: by direct ingestion of copepods infected with procercoids, by ingestion of an intermediate host already infected with a plerocercoid (e.g., a frog) and subsequent release of the spargana and penetration of the intestine, and (especially man) by local application of poultices (e.g., split frogs) to wounds or sore eyes; the spargana migrate into the human tissue under the influence of body heat.

3. Life Cycle and Laboratory Maintenance

The life cycle (Figure 11) generally follows the pseudophyllidean pattern,[41] and a valuable review of the techniques for maintaining the life cycle in the laboratory is given by Mueller.[44] Eggs are passed in the feces of infected cats and, on reaching water, embryonate within 10 d at 25 to 27°C and hatch when exposed to light. Eggs can be stored for a year at 4°C if fungus development is controlled by adding a few drops of alcoholic iodine. The eggs are resistant to iodine which eventually disappears by evaporation or organic combination.[42] When hatched coracidia are eaten by a susceptible copepod, such as *Cyclops vernalis*, procercoids develop to an infective stage in 10 to 14 d at 23°C.[41] Like many pseudophyllideans, the late nauplii and early copepodid stages are most susceptible; adults are difficult to infect. For maintenance of copepod cultures, see Mueller.[44]

The chief natural second intermediate host appears to be the water snake, but amphibia, reptiles, birds, and mammals, including man, are also infected naturally. The sparganum is highly paratenic and, if it enters the intestine of an unsuitable host, rapidly migrates through the gut wall and establishes itself as a tissue parasite. Mice make excellent laboratory intermediate hosts and can be infected with the spargana by feeding with infected copepods or procercoids or with a whole sparganum or by injecting sparganum scoleces subcutaneously or intraperitoneally. Mueller[44] recommends injecting 0.5 ml of saline + antibiotics into a lightly etherized mouse in the lumbar or sacral region, a procedure which raises a small blister. The scoleces are then injected directly into this blister using 1 ml of saline + antibiotics, a 1-ml disposable syringe, and a 19-gauge needle. Spargana migrate throughout the body and can occur almost anywhere in the tissues. Larvae become infective to the definitive host after as little as 4.5 d in the mouse.

In the natural life cycle, it is not known how cats are able to catch and consume water snakes and thus acquire this infection. In the laboratory, cats are readily infected by feeding with a sparganum, all but 3 to 4 mm of the anterior end being cast off before strobilar differentiation commences.[41] The North American strain of *S. mansonoides* releases a substance, with properties resembling those of a mammalian growth hormone,[45] which stimulates excessive growth in the mouse and has been the subject of much research.[8]

C. *IN VITRO* TECHNIQUE: PROCERCOID TO PLEROCERCOID

In a series of elegant techniques, Mueller[39-42,44] developed a system for obtaining procercoids of *S. mansonoides* in a sterile condition and culturing them to infective plerocercoids.

1. Basic Procedure

Large quantities of eggs were embryonated in 500-ml Erlenmeyer flasks, the whole being shaken continuously, except for a short break of 1 min every hour, the water being changed frequently to avoid bacterial growth. A few drops of iodine were added to control bacteria, if necessary. Mass hatching took place in strong sunlight, and the released coracidia were allowed to infect nauplii and early copepodid stages of *Cyclops vernalis*.

The basis of obtaining sterile procercoids from the copepods was the observation that under the influence of heat, the procercoids forced themselves out of the copepods and could be harvested in a modified Baermann apparatus (Figure 12) adapted for obtaining sterile cultures. If after collection, procercoids are found to be contaminated, a more sophisticated technique can be applied. This involves placing the copepods after harvesting in a large beaker of clean, sterile water with a little activated charcoal added and stirring. The copepods clean out their intestines and the charcoal settles. After 2 h, the copepods are again concentrated by the Baermann technique (above). A sufficient number of sterile procercoids are obtained by application of these methods.[44] Procercoids were cultured in a roller tube system, using a medium of Parker 199 + 10 calf serum + chick embryo extract in a proportion of 1.5 ml of CEE to 6 ml of 199-CS.

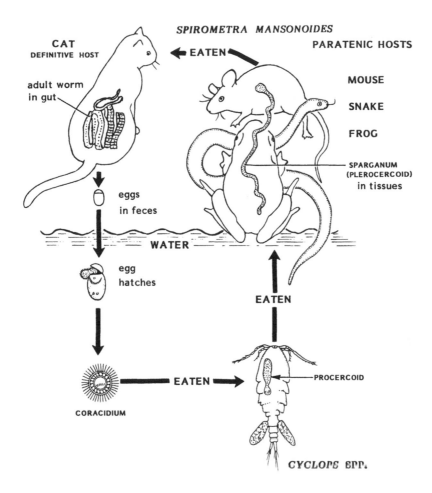

FIGURE 11. *Spirometra mansonoides*, life cycle. The snake is probably the natural intermediate (paratenic) host, but the spargana are also found in amphibia and mammals (including man). (Modified from Mueller, J. F., *J. Parasitol.*, 60, 3, 1974. With permission.)

2. Results

Although a few cultures were sometimes contaminated, the technique was very successful. A remarkable degree of growth was achieved, larvae doubling in size every 24 h during the first week but slowing later. Plerocercoids up to 20 mm were obtained in 3 weeks and up to 30 mm were obtained in 9 weeks. This rate of growth is comparable to that obtained in mice, and cultured plerocercoids were shown to be infective to cats in which they grew to mature worms. For further details regarding the above techniques, the original papers should be consulted.

D. *IN VITRO* TECHNIQUE: PLEROCERCOIDS TO ADULT WORMS[16,17]

1. Early Experiments

In a highly successful series of experiments, Berntzen and Mueller[16,17] have grown the plerocercoids of *S. mansonoides* to egg-producing adults *in vitro*. In the first series of experiments,[16] plerocercoids grown from procercoids obtained by the technique above were cultured in a continuous flow apparatus (as used for *Trichinella spiralis*) using a complex medium (Medium 115). The gas phase was 5 or 10% CO_2 in N_2. Plerocercoids were treated with an evaginating solution of bile salts and trypsin. Other details are given in the later experiments described below. Maximum development achieved by this technique was the

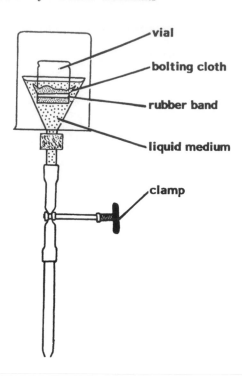

FIGURE 12.   Modified Baermann apparatus for harvesting procercoids of *Spirometra mansonoides* for *in vitro* culture. (Adapted from Mueller, J. F., *Trans. Am. Microsc. Soc.*, 78, 245, 1959. With permission.)

development of 194 to 301 morphologically normal proglottides, the posterior of which contained some genital anlagen, but sexual maturity was not achieved.

2. Later Experiments

Later experiments[17] modified the technique used in the early experiments (described above) and improved it to such an extent that sexually mature, egg-producing worms were obtained.

Source of material and pretreatment — In contrast to the early experiments, spargana from laboratory mice were used instead of those cultured from procercoids. They were cut to a length of 8 to 10 mm, with sterile precautions, and the "tail" was discarded. They were then placed in an evaginating solution for the first 72 h. This solution comprised sodium taurocholate, 2 g; trypsin (1:250), 1 g; glutathione (reduced), 400 mg; in 100 ml of Earle's BSS, used in a ratio of 3:8 with medium.[16]

Medium — The culture medium used was Medium 115 as used in the earlier experiments,[16] with the exception that 0.1 ml of cat bile was added to every liter of medium and after 120 h, the reducing component of the medium (Solution G) was omitted. For details of the composition and preparation of Medium 115, the original paper[17] should be consulted. Details of this are also given in Taylor and Baker.[3]

Culture conditions — Worms were cultured in a complex, continuous flow culture apparatus developed by Berntzen.[46] The gas phase was 10% CO_2 in N_2, the pH was 7.2 to 7.3, and the culture temperature was 38°C. The medium flow rate was regulated to replace 12.5 ml of metabolised medium with preconditioned new medium every 4 h. This system obviously produced a carefully stabilized environment. Over a period of 20 to 42 d, worms grew to a length of 31 to 44 cm and eggs appeared between 16 to 22 d of culture. Some 80% of the eggs were found to be fertile and when embryonated, the hatched coracida proved to be infective to copepods (species not stated). This clearly was a most successful system and similar results were obtained with the Australian and Malaysian "strains" of *Spirometra mansonoides* (see Table 1.)

3. Evaluation

This system represents a most successful culture technique for this species. It does not appear to have been repeated or extended by these or other workers for further studies on the developmental biology of *Spirometra*. If combined with the earlier culture systems for procercoids, it should also be possible to devise a system to culture procercoids from the oncospheres and thus complete the entire life cycle of this species *in vitro*, but so far this has not been attempted.

V. *HYMENOLEPIS DIMINUTA*

Cultured by: Berntzen;[14] Schiller;[47] Voge;[48] Roberts and Mong[49]
Definitive hosts: Rat and other rodents; occasional accidental parasite of man
Experimental hosts: laboratory rat; rejected in mouse after 11 d
Prepatent period: 12 to 13 d[50]
Intermediate host: numerous species of Orthoptera, especially *Tribolium* spp., occasionally in Lepidoptera or Coleoptera
Distribution: cosmopolitan

Four species of *Hymenolepis* have been extensively used as experimental models for *in vitro* culture, *H. diminuta*, *H. nana*, *H. microstoma*, and *H. citelli*. Much of the basic work was carried out on *H. diminuta* and the techniques developed for this species were then adapted for the others. *H. diminuta* was the first cestode to be cultured through its entire life cycle *in vitro* — a result which marked a milestone in the development of cestode *in vitro* culture.

A. GENERAL BIOLOGY

1. Adult

Hymenolepis diminuta is widely used as a laboratory cestode model and its morphology and life cycle are well known and will not be discussed in detail here. Most aspects of its biology have been comprehensively reviewed by Arai.[50] The adult is normally maintained in the laboratory in the rat in which the prepatent period is generally quoted as being 12 to 13 d. However, it can vary from 12 to 21 d, depending on the size of the infection.[51] The worm shows the two well-known migration patterns of an age-dependent forward migration and a diurnal (circadian) migration. The worm size depends not only on the species of definitive host, but also on the worm load; in light infections, worms are about 70 cm long.[50] It is now recognized that different isolates of *H. diminuta* may represent different "strains" which may have different nutritional or metabolic characteristics. These differences may be reflected in different requirements for *in vitro* culture, as is known to occur in different isolates of *Echinococcus granulosus*, although this phenomenon has not been demonstrated for any *Hymenolepis* sp. The life span of the species is not known, but it has been reported that the worm can live for at least 14 years.[50]

2. Life Cycle and Laboratory Maintenance

When proglottides become gravid, they are shed and disintegrate while still within the host intestine. The eggs, which are embryonated when laid, are immediately infective to a wide range of intermediate hosts. The most commonly used laboratory intermediate hosts are the beetles, *Tribolium confusum* or *T. molitor*, but the larvae can develop in numerous species of insects. The optimum temperature for development of the larva in *T. molitor* is 30°C, an infective cysticercoid developing in about 180 h.[2] In *T. confusum*, larvae develop as early as 5 d at 37°C, but require 24 d at 20°C.[52] If the insects are starved before feeding, they rapidly ingest the eggs and the hatched oncospheres penetrate the gut and develop into cysticercoids in the hemocele. The collection, sterilization, and storage of eggs have been described in detail by Hundley and Berntzen[53] as described below.

B. *IN VITRO* CULTURE: ONCOSPHERE TO CYSTICERCOID[48]
1. Eggs: Collection, Sterilization, and Storage

Although it is possible to obtain reasonable quantities of eggs in a sterile condition by dissecting surface-sterilized gravid proglottides,[48] larger quantities of eggs require a more elaborate treatment. The following technique of Hundley and Berntzen[53] allows substantial quantities of eggs to be collected, sterilized, and stored.

Feces were allowed to accumulate on trays beneath rat cages and were collected twice a week. The fecal pellets were then scooped out with a strainer, rinsed free of hair with tap water, and soaked for another hour in pans of water. The material was then liquified by stirring, and filtered through two layers of gauze and the filtrate was collected in a large beaker which was allowed to stand for a further 2.5 h. The supernatant was then siphoned off and the sediment was resuspended in water. This process was repeated until the supernatant was clear. After the supernatant was removed for a final time, the sludge remaining was resuspended in saturated NaCl (sp gr, 1.262), poured into 50-ml centrifuge tubes, and centrifuged at 100 × g. The eggs, which formed a dark ring at the top of the tube, were removed by a vacuum trap apparatus. Eggs were then rinsed free of salts by alternate centrifugation, siphoning, and resuspension in distilled water.

Eggs obtained in this way were transferred to 15 × 125 mm screw-top tubes and treated with sterilizing agents for 20 min, followed by washing five times in sterile distilled water. It was found that water dilutions of Osyl, 1:500; Zephiran, 1:5000; and iodine, 1:9, were equally effective in sterilizing the eggs. Sterile eggs were stored in 50-ml Erlenmeyer flasks or culture tubes — 6-ml suspensions of concentrated eggs in each container — with the addition of 4 ml of 1% streptomycin and were stored at 5 or 21°C. Egg viability was preserved better in flasks than in tubes.

2. Oncosphere Culture Technique

Oncospheres were first cultured to cysticercoids with a rather elaborate technique with monolayers of rat fibroblasts;[54] the resultant cysticercoids were shown to be infective to rats and were also grown to adults *in vitro*. The whole life cycle was thus completed *in vitro*. This monolayer technique is not described here because a much simpler system, based on that used for *H. citelli*,[55] was later developed by Voge[48] and this is given below.

Eggs were dissected from surface-sterilized proglottides and suspended in Earle's saline in screw-top vials with glass beads (3 mm) and shaken to break the shells. They were then transferred to a sterile hatching solution of 1% trypsin or 25,000 units per milliliter of Tryptar (Armour Pharmaceutical) and 1% bacterial amylase in Earle's BSS with biocarbonate. Most oncospheres hatched within 20 min and were transferred to the basic culture medium as used for *H. citelli* (see Table 7) and were then transferred to the final culture medium of basic medium plus a reducing medium of L-cysteine to give a final concentration of 8.93×10^{-4} M to 1.62×10^{-3} M. Approximately 150 larvae can be grown in a tube with 6 ml of medium.

Tubes were sealed with parafilm and incubated at 28°C in an inclined position; 2 ml of culture medium was replaced after 3 to 4 d and was replaced weekly thereafter or more frequently if it became excessively acid. The hollow-ball embryonic stages were present at 3 d, ovoid stages were present at 7 d, and tripartite body divisions were present at 9 d. A few withdrawn organisms were seen at 14 d, and most organisms had withdrawn the scolex at 18 d.

C. *IN VITRO* CULTURE: CYSTICERCOID TO MATURE ADULT

1. Continuous Flow Technique of Berntzen[14]

The first successful culture of *H. diminuta* (and later of *H. nana*) was achieved by Berntzen,[14,56] a result which represented a milestone in helminth culture. His method differed from those of earlier workers in that (1) it involved an ingenuous continuous flow apparatus (Figure 13) utilizing large quantities of an extremely complex medium which contained almost every major metabolite known to occur in cellular metabolism; (2) the flow system meant that the waste metabolites were almost immediately removed from the worms, and (3) the physicochemical conditions were apparently controlled, at least initially, but the actual pCO_2 or pO_2 or Eh levels were not stated.

With this system, Berntzen[14] reported that cysticercoids grew to sexually mature adults with gravid proglottides in 15 d — only 2 d longer than required *in vivo*. Unfortunately, this author appears to have failed to provide sufficient details of his system for others to be able to repeat these findings. For example, it was found to be impossible to prepare his culture medium according to the published formula, and, in addition, later workers (including Berntzen and colleagues[57]) failed to obtain consistent results with this method and it also proved to be unsuccessful for other cyclophyllidean species. Some of his original conclusions have also been questioned as being equivocal.[58]

On account of these difficulties with Berntzen's published technique, it is not given here, but details can be found in the original paper[14] or in the condensed account given in Taylor and Baker.[3] A later, modified technique, developed for *H. nana* proved to be somewhat better and is discussed later. Although this technique proved to be unsatisfactory, this result proved to be a major stimulus to other workers and a number of successful, but simpler techniques have since been developed. Some of these are described below.

2. Schiller's Diphasic Blood-Agar Technique[47]

Source of Material — Cysticercoids were dissected from flour beetles at 16 d post-infection and were excysted artificially by incubation in undiluted ox bile for 30 min at 37°C. Before culture they were washed three times in sterile Hanks' BSS + 100 units per milliliter of penicillin and 100 μg/ml of streptomycin.

Medium — The diphasic medium employed consisted of a blood-agar base overlaid with Hanks' BSS. This medium was prepared as follows.

Nutrient agar, 16 g, + 3.5 g of NaCl were dissolved in 700 ml of distilled water. To this was added 300 ml of sterile, defibrinated rabbit blood (inactivated at 56°C for 30 min) with thorough mixing. After autoclaving, the blood-agar mixture was dispensed in 10-ml units into cotton-stoppered Erlenmeyer flasks. After cooling to gelation, 10 ml of Hanks' BSS, adjusted to pH 7.5 with $NaHCO_3$ plus antibiotics, as above, was added to each flask. The medium was then preincubated at 32°C for 24 h to allow diffusion and to test for sterility. After gassing with 3% CO_2 in N_2, the medium was adjusted to a final pH 7.5 with NaOH.

Culture conditions — Cysticercoids (10 to 15) were cultured in each flask agitated on a Dubnoff metabolic shaking incubator (set at 30 c/min) at 37°C in a gas phase of 3% CO_2 in N_2.

Procedure — After 6 d, worms were transferred aseptically to fresh media and on day 8, powdered glucose (1 mg/ml) was added to the fluid overlay. On day 10, worms were transferred to individual flasks. Thereafter, transfers were made every 24 h. On day 20, the volume of BSS overlay was increased to 20 ml and glucose was added as above.

Results — These are summarized in Table 2. Worms began shedding proglottides with viable eggs on day 24. Although the worms produced were smaller than *in vivo* grown worms, the morphology appeared to be normal and eggs proved infective to beetles. When 6-d worms from rats were used, instead of cysticercoids, larger-sized worms were obtained

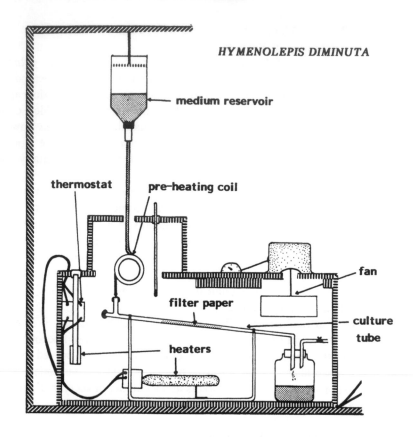

FIGURE 13.   Continuous flow culture apparatus used by Berntzen for the first successful *in vitro* culture of adult *Hymenolepis diminuta* from cysticercoids. (Modified from Berntzen, A. K., *J. Parasitol.*, 47, 351, 1961. With permission.)

TABLE 2
Hymenolepis diminuta: *In Vitro* Growth and Development From Artificially Excysted Cysticercoids in Schiller's Diphasic System

	18 d *in vitro* (4 worms)		29 d *in vitro* (9 worms)		44 d *in vitro* (5 worms)	
	Mean	**Range**	**Mean**	**Range**	**Mean**	**Range**
Total length (mm)	35	20—50	87	71—123	120	115—137
Max width (mm)	1.0	0.5—1.2	1.3	1.0—2.0	2.0	1.3—2.5
Total number of proglottids	365	185—534	480	418—523	529	458—600
Immature	235	133—286	167	140—223	161	126—185
Mature	55	52—61	45	36—60	47	42—76
Pregravid	64	48—68	68	42—77	69	52—91
Gravid	0	—	147	109—199	254	230—270
Percent fertile	0	—	36	24—57	38	20—61
Shedding events	0	—	2.6	1—3	4.7	2—6
Proglottids shed/event	0	—	41	17—77	45	20—122
Development of infective cysts	–	–	+	–	+	–

From Schiller, E. L., *J. Parasitol.*, 51, 516, 1965. With permission.

TABLE 3

Hymenolepis diminuta: *In Vitro* Growth and Development of Worms from Rats 6 d
Postinfection in Schiller's Diphasic System

	0 d *in vitro*[a] (12 worms)		10 d *in vitro* (23 worms)		14 d *in vitro* (14 worms)	
	Mean	Range	Mean	Range	Mean	Range
Total length (mm)	37	27—46	182	130—291	230	164—32
Max width (mm)	0.6	0.4—0.8	1.4	1.0—2.0	1.5	1.0—2.5
Total number of proglottids	436	339—606	1,054	732—1,322	1,144	971—1.4
Immature	436	339—606	443	251—710	465	162—70
Mature	0	—	225	81—379	270	134—54
Pregravid	0	—	274	130—656	182	69—43
Gravid	0	—	92[b]	0—278	189[c]	0—36
Percent fertile	0	—	20[b]	3—62	15[c]	0—36
Development of infective cysts	–	–	+	–	+	–

[a] 6-d-old controls.
[b] 14 of 23 worms were gravid.
[c] 11 of 14 worms were gravid.

From Schiller, E. L., *J. Parasitol.*, 51, 516, 1965. With permission.

and gravid proglottides developed earlier (Table 3). This latter method was further improved
by Roberts and Mong[49] and details of their technique is given below. Roberts[59] later showed
that sheep blood could replace rabbit blood, but horse blood gave poor results.

3. Roberts and Mong's Modification of Schiller's Technique[49]

Culture conditions — Roberts and Mong[49] modified Schiller's technique slightly and
in particular studied the development under different gas phases of 0, 1, 5, or 20% in N_2
and, in contrast to results of previous workers, found that the development was unaffected
by any of these gas phases. The medium was prepared according to Schiller[47] except that
Hanks' BSS was gassed with one of the above gas phases for 10 min and sealed before it
was added to the solidified blood agar in the culture flasks. Flasks were also sealed and
preincubated for 24 h at 34°C instead of 32°C.

Source of material — As starting material, these authors used 6-d-old worms from rats
previously infected with 40 cysticercoids from *Tribolium confusum*. Worms were rinsed in
Krebs-Ringer's solution adjusted to pH 7.4 with phosphate buffer, damaged worms being
rejected. After rinsing a further four times in Hanks' BSS, worms were transferred to
Erlenmeyer flasks containing 10 ml of Hanks' BSS with 5000 U of penicillin, 5 mg of
streptomycin, and 500 U of Mycostatin (Squibb nystatin) and were incubated for 1 h at
37°C using a Dubnoff water bath operating 30 excursions per minute. The worms were then
blotted on clean, sterile Whatman No. 50 filter paper, weighed individually, and cultured,
one worm per flask.

Procedure — Worms were cultured, using the shaking water bath system as above for
5 d in 25-ml Erlenmeyer flasks with 5 ml of blood agar and 5 ml of Hank's BSS overlay.
They were then transferred to 50-ml flasks with 10 ml of each media component for the
subsequent 7 d of the experiment. Worms were transferred to a flask with fresh medium
every 24 h. Cultures were evaluated after 12 d.

Results — This system was very successful, producing gravid worms in 12 d — a time
comparable to that obtained *in vivo*; results are summarized in Table 4. Although some
proglottides were abnormal, most produced large quantities of eggs. Moreover, these were

TABLE 4
Hymenolepis diminuta: In Vitro Growth and Development of Worms From Rats 6 d Postinfection in Roberts and Mong's Modification of Schiller's Diphasic System

Experiment number	Oxygen in gas phase (%)	Total proglottids at 0 time[a] (mean ± SE, n = 5)	Proglottids after 12 d *in vitro* (Mean ± SE, n = 10, or mean, range, n = 2)	
			Total	Gravid[b]
I		236 ± 35		
	0		788 ± 91	22 ± 6
	1		755 ± 88	14 ± 7
II		339 ± 26		
	0		1092 (1055—1129)	98 (90—106)
	5		1025 (1025—1026)	86 (56—116)
III		283 ± 52		
	0		876 (684—1069)	23 (21—26)
	20		901 (895—907)	14 (8—19)
IV		249 ± 29		
	0		896 (866—927)	73 (17—129)
	20		919 (864—975)	114 (113—116)

[a] Numbers of proglottids in samples of worms fixed at 0 time (6 d *in vivo*).
[b] Proglottids containing substantial numbers of eggs with apparently normal morphology (not total in "gravid" region, see text).

From Roberts, L. S. and Mong, F. N., *Exp. Parasitol.*, 26, 166, 1969. With permission.

embryonated and proved highly infective to beetles. A major result from this experiment was that the level of oxygen present, over the range 0 to 20%, appeared to make no difference to the growth or differentiation and substantial numbers of proglottides were produced over the entire range.

D. EVALUATION

Although Berntzen's original system appeared to produce results in his hands, published details of the system may have been incomplete because other workers, including the writer, have been unable to reproduce his results. It should be said, however, that his later modification of this method — as developed for *H. nana* (and discussed later in this text) — has proved to be more satisfactory.

Either Schiller's original technique[47] or Roberts and Mong's modification[49] of it appear to work well. The latter method, which avoids the difficulties of growing the early post-oncospheral stages, has been widely used for experimental purposes and appears to be the most satisfactory for general use. It is clear, however, that as a result of their experiments with the gas phase, this parameter may not need to be as strictly controlled as was formerly thought necessary. In a later experiment of *Hymenolepis* nutrition, Roberts and Mong[60] successfully used 5% CO_2 in N_2.

The original technique used rabbit blood in the solid phase and sheep blood has also been shown to be effective. It should be noted, however, that Turton[61] could only obtain preoncospheres using horse blood.

VI. *HYMENOLEPIS NANA*

Cultured by: Berntzen;[56] Sinha and Hopkins[62]
Definitive hosts: rat and other rodents; man

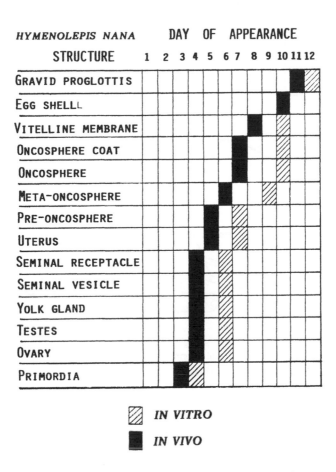

FIGURE 14. *Hymenolepis nana*, comparison of times of development of genitalia in worms grown *in vitro* and *in vivo*. (After Berntzen, A. K., *J. Parasitol.*, 48, 785, 1962. With permission.)

Experimental hosts: laboratory mouse and rat (different strains?)
Prepatent period: 11 to 16 d in the rat; 14 to 25 d in the mouse[11]
Intermediate host: *Tenebrio* spp., *Tribolium confusum*
Distribution: cosmopolitan

A. GENERAL BIOLOGY AND LIFE CYCLE

The general biology of this species is so well known that it will not be discussed in detail here. Morphologically it differs from *H. diminuta* chiefly by the presence of a well-developed rostellum with a crown of hooks. The life cycle differs markedly from that of *H. diminuta* in that in addition to utilizing insect intermediate hosts, eggs can also hatch in the rodent gut and penetrate the villi and develop into cysticercoids there. Hence there is both an indirect and a direct life cycle. From the point of view of assessing the success of a particular *in vitro* technique, it is important to know the pattern of development in the definitive host and the time of appearance of the various genitalia. These are shown in Figure 14.

B. *IN VITRO* CULTURE: ONCOSPHERE TO CYSTICERCOID[63]

The technique adopted was essentially that first used for the culture of the oncospheres of *H. citelli* to cysticercoids *in vitro*; this is described in detail under the latter species. Growth and development of the cysticercoid was found to occur only in the presence of 5%

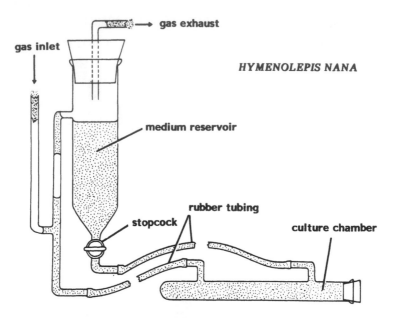

FIGURE 15. *Hymenolepis nana*, a single recirculating unit used for the successful cultivation of cysticercoids to adult worms. (Modified from Berntzen, A. K., *J. Parasitol.*, 48, 785, 1962. With permission.)

CO_2 in 95% N_2. Cysteine was found not to be an essential component of the medium, as cysticercoids developed with or without reducing agent, as long as the cultures were gassed with the above gas mixture. Completion of morphological development took place in 15 d.

The developmental sequence was as follows: 4 d, hollow ball stage; 6 d, early tripartite organisms; 11 d, fully developed tripartite organisms; 14 d, scolex withdrawn; 15 d, fully developed. After 22 d *in vitro*, cysticercoids were fed to mice and patent worms were recovered 10 d later.

C. *IN VITRO* CULTURE: CYSTICERCOID TO MATURE ADULT
1. Continuous Flow Technique of Berntzen[56]
Berntzen[56] followed up his culture of *H. diminuta*[14] by culturing *H. nana* from cysticercoids to sexual maturity using a modified continuous flow apparatus (Figure 15) and two culture media, Medium 101 and Medium 102. However, it must be pointed out that at least one experienced worker, Hopkins,[58] failed, over a period of 18 months, to reproduce Berntzen's results. The nature of the complex media used makes it unlikely that these will ever be useful for routine culture studies and, for that reason, data on their composition are not given here. For details, the original paper can be consulted. As Hopkins points out, it is likely that he may not have been exactly duplicating some essential step in Berntzen's complex procedures. Hopkins also analyzed many of Berntzen's conclusions regarding the importance of the gas phase and of various components of his media and other factors and concluded that few of these could be justified on the limited data provided. Nevertheless, there is no doubt that Berntzen's system worked, in his hands, and the rate of development achieved *in vitro* only lagged slightly behind that *in vivo* (see Figure 14).

2. Simplified System of Sinha and Hopkins[62]
As with *H. diminuta*, later workers found that *H. nana* could be grown to sexual maturity in a much simpler medium and in a standard roller tube culture system. Details of this are given below.

TABLE 5
Hymenolepis nana: Liver Extract Medium Which Supports Growth of Cysticercoids *In Vitro* to Egg-Producing Adults in 12 d

Component and concentration in final medium	Volume (ml) of stock solutions to prepare 100 ml of medium
30% Horse serum	30
0.3% Glucose	5
0.5% Yeast extract	10
40% Hanks' saline	40
10% Rat liver extract (20% aqueous)	10
Antibiotics (100 IU of sodium penicillin G + 100 µg of streptomycin sulfate (Crystamycin, Glaxo)/ml of medium $NaHCO_3$ (1.4%) + NaOH (0.2 *M*)	≃ 5
pH 7.2, gas 95% N_2 + 5% CO_2, roller tubes at 37°C	

From Hopkins, C. A., *Symp. Br. Soc. Parasitol.*, 5, 27, 1967. With permission.

Source of cysticercoids and excystment — Cysticercoids were dissected from infected *Tribolium confusum* into Hanks' BSS. The free cysticercoids were washed twice in sterile BSS and thereafter all procedures were carried out at 37°C in sterile solutions with antibiotics (100 units per milliliter of penicillin + 100 µg/ml of streptomycin).

Excystment was carried out by (1) treating larvae with 1% pepsin (1:2500) in BSS at pH 1.7 (pH adjusted with 0.2 *N* HCl) for 12 to 15 min followed by (2) washing three times in BSS and treating with a trypsin-bile solution (0.5% trypsin + 0.3% sodium taurogly-cocholate) at pH 7.2 in BSS. About 90% of worms excysted within 8 to 10 min. These were washed three times in BSS and collected in a petri dish in preparation for culturing.

Medium — The medium used is shown in Table 5. Particular attention is drawn to the fact that it was found[58] that the growth properties of the yeast extract varied substantially in different batches from different manufacturers. The sample used in the original experiments (Original Difco YE, Table 5) promoted excellent growth, yet when this was used up and replaced with other samples, much variation occurred (Figure 16). This is a well-recognized phenomenon in all *in vitro* culture fields and may sometimes be responsible for the differences obtained by different workers apparently using the same medium.

Rat liver extract was first used, but lamb liver (which is easier to obtain) extract was later used. This was prepared as follows.

The lamb liver was transported from the abbatoir in an iced container, cut into pieces of about 20 g, and stored at −15°C. When required, it was thawed at 4°C for 48 h. One part liver tissue (by weight) was homogenized with four parts (by volume) of deionized water at 4°C. The brei was adjusted to pH 4.0 with 1 *N* HCl (approximately 2.5 ml/100 ml of brei), squeezed through muslin, and centrifuged at 5420 × g for 1 h at 4°C. The supernatant was collected, filtered through a 0.45-µm membrane filter, and sterilized by passing through a 0.22-µm filter. The sterile extract was stored in aliquots of 5 and 10 ml at −15°C.

Culture conditions — Excysted larvae (16 to 20) were used per culture tube; the tubes were gassed twice with 5% CO_2 in N_2. The pH of the tubes was 7.2 + 0.2 some 2 to 3 h after setting up. The media was changed on days 3, 6, 9, 11, and 13. No details of volumes of media used were given.

Results — During 14 d of cultivation, cysticercoids grew to strobilated worms 7 to 30 mm long; 60 to 70% had mature genitalia and 20 to 30% produced eggs in some proglottides. Discharged eggs were found in the medium from the tenth day onward and these eggs proved to be infective to beetles and normal cysticercoids developed. Evans[6] reported similar results using a slightly modified medium, but full details were not reported.

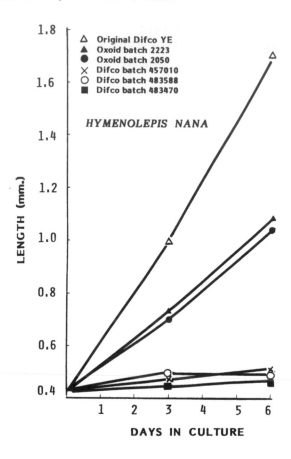

FIGURE 16. *Hymenolepis nana*, comparison of growth properties of different samples of yeast extract in the medium (see Table 5) used for *in vitro* culture of cysticercoids to adult worms. (Modified from Hopkins, C. A., *Symp. Br. Soc. Parasitol.*, 5, 27, 1967. With permission.)

D. EVALUATION

As pointed out above, although Berntzen's system worked in his hands, it has not been successfully used by other workers and cannot be considered to be a viable system. On the other hand, the system of Sinha and Hopkins[62] represents a simple system which has also been successful in the hands of other workers. It would appear to be a system which has much to recommend it for general use.

VII. *HYMENOLEPIS MICROSTOMA*

Cultured by: De Rycke and Berntzen;[64] Evans;[65] Seidel[66]
Definitive hosts: mouse and other rodents
Experimental hosts: laboratory mouse, rat, and hamster[67]
Prepatent period: 14 to 16 d in the mouse[68,69]
Intermediate host: *Tribolium confusum*
Distribution: cosmopolitan?

A. GENERAL BIOLOGY AND LIFE CYCLE

Unusual for a cestode, this species occupies an extraintestinal site, namely, the common bile duct and the extrahepatic ducts, although in some hosts (e.g., hamsters, rats) there is

TABLE 6

Hymenolepis microstoma: **Sequence of Development**
in the Mouse (Adult Male; Swiss Albino)

Days PI	Observations
1 2 }	No internal segmentation; or grossly visible anlagen
3	Some internal segmentation; appearance of anlagen
4 5 }	External segmentation; ♂ and ♀ anlagen discernible
6	Testes in few
7	Testes defined
8	Early mature to mature segments
9 10 }	Mature segments
11	Disappearance of ♀ glands; few preoncospheres
12	Preoncospheres: no hooks
13	Semigravid proglottids; oncospheres
14	Near gravid proglottids
15 16 }	Gravid proglottids

Note: PI, postinfection.

From De Rycke, P. H., *Z. Parasitenkd.*, 27, 350, 1966. With permission.

a tendency to attach in the duodenum.[67] The mouse appears to be the most satisfactory laboratory host. The worm lives in the intestine for the first 3 d postinfection and then migrates into the bile duct on the fourth day.[69] The prepatent period varies slightly with the maintenance temperature of the host, being 14 d at 21°C and 15 d at 35°C.[68]

The approximate sequence of segmentation and genital differentiation is given in Table 6 and the rate of growth is shown in Figure 17.[69] Cysticercoids develop readily in *Tribolium confusum*, becoming infective in 8 d in beetles kept at 30°C.[70]

B. *IN VITRO* CULTURE: ONCOSPHERE TO CYSTICERCOID

Seidel[66] cultured *H. microstoma* from eggs to oncospheres and on to adult worms.

1. Culture Technique

Eggs were dissected from surface-sterilized gravid segments in Earle's BSS and shells were broken by shaking with glass beads. The eggs were then transferred to 150×25-mm centrifuge tubes containing 5% $NaHCO_3$ in 0.85% NaCl and 50,000 units of Tryptar which had been gassed with 5% CO_2 in N_2 for 10 min and then again, with eggs present, for 20 min. After hatching, the oncospheres were centrifuged at 5000 rpm for 2 min and washed once in culture medium before placing them in culture vessels.[66]

Oncospheres were cultured in 130×10-mm screw-top tubes in 6 ml of medium, with 30 to 70 organisms per tube. The tubes were sealed with Parafilm and kept stationary in an inclined position, at 24 and 28°C, in a gas phase of air. The medium remained unchanged during the first week of culture, but thereafter 2 ml of medium was replaced three times a week.

2. Medium

This was a modification of that used by Voge and Green[55] for *H. citelli*, but organisms grew best in media without reducing agents — in contrast to the conditions required by *H. diminuta* and *H. citelli*.

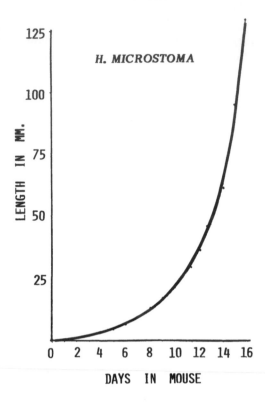

FIGURE 17. *Hymenolepis microstoma*, growth in mouse (adult male; Swiss albino). (Modified from De Rycke, P. H., *Z. Parasitenkd.*, 27, 350, 1966. With permission.)

3. Results

In media without the addition of reducing agents and containing 10% fetal calf serum, morphologically normal cysticercoids developed after 16 d in culture, but were cultured until 21 d when used later for culturing to adult worms (see below).

C. *IN VITRO* CULTURE: CYSTICERCOID TO MATURE ADULT

1. Early Work

De Rycke and Berntzen[64] were the first to grow cysticercoids of *H. microstoma* to near sexual maturity *in vitro*. They used a medium referred to as HM67, a medium derived from Medium 115 previously used by Berntzen and Mueller[16] for the cultivation of *Spirometra mansonoides*, plus glucose, serum, and hamster bile in a gas phase of air. Although the worms grew and strobilated, they remained small (Figure 18) and complete sexual maturity was not achieved.

2. Technique of Evans[65]

Evans[65] was the first to culture *H. microstoma* to egg-producing adults, using a modification of the method used by Sinha and Hopkins[62] for *H. nana*. A useful, abbreviated account of this technique is given by Evans.[6]

Medium — The medium consisted of 60 ml of (modified) Eagle's medium (plus antibiotics) with 10 ml of sheep or hamster liver extract and 30 ml of horse serum; 1 to 10 ml of ox bile was added and the pH was adjusted to 7.6 and dispensed into 5-ml volumes into roller tubes. The tubes were gassed for 30 s with 5% CO_2 in N_2, immediately sealed with rubber bungs, and rotated in an incubator at nine revolutions per hour at 37.5°C for 2.5 h.

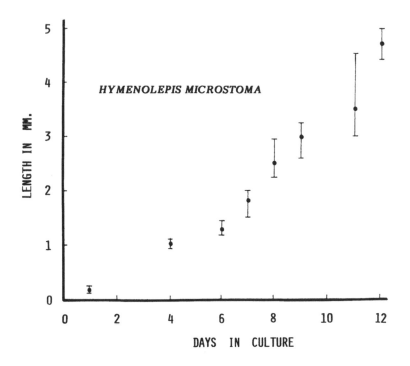

FIGURE 18. *Hymenolepis microstoma*, growth *in vitro* in Medium HM67 with serum and bile. (Modified from De Rycke, P. H. and Berntzen, A. K., *J. Parasitol.*, 53, 352, 1967. With permission.)

Culture procedure — Cysticercoids were obtained from *Tribolium confusum* and, after excystment, were washed three times in sterile Hanks' and 20 to 30 were placed in each culture tube. The tubes were regassed for 30 s before culturing in the roller tube system. The medium was changed on day 6, and on day 7 the larger worms were transferred to new medium (one to three per tube); the medium was renewed on days 9 and 11 and daily thereafter, the tubes being gassed and sealed on each occasion.

Results — Organisms (80 to 100%) had strobilated by day 7 and 38 to 100% showed gonad development. By day 9, both male and female systems were fully developed and sperm were present. Gravid proglottides with apparently normal eggs had developed by day 16, but these were not formed in cultures lacking ox bile. Gravid worms were obtained from cultures gassed with 5 and 10% O_2, but not in those gassed with 15, 20, or 30% O_2.

3. Technique of Seidel[66,71]

Using cysticercoids from *Tribolium confusum* and a diphasic medium (see below), Seidel[71] first showed that hemin was a fundamental requirement for strobilation in *H. microstoma*. The maximum development he achieved in these early experiments was the development of proglottides containing preoncospheres, but not fully formed oncospheres. He later improved[66] this technique and, by using cysticercoids grown from oncospheres, was able to obtain fully gravid worms, thus completing the entire life cycle of this cestode *in vitro* — a remarkable achievement! His later technique is given below.

Medium — The diphasic medium consisted of 20 ml of nutrient agar slants (with or without 5% whole human blood) overlaid with 20 ml of Triple Eagle's Medium with 30% inactivated horse serum, as used in the early experiments.[71] If nutrient agar was used without blood, it was necessary to incorporate hemin into the liquid overlay. NCTC 135 plus 20% horse serum was also satisfactory as an overlay. The final concentration of glucose in all fluid media was adjusted to 5 mg/ml. The pH was adjusted to 7.0 to 7.2 with $NaHCO_3$. A stock hemin solution is prepared by dissolving 100 mg of hemin in 1 to 2 ml of triethanolamine. This is diluted in water to give a 1 mg/100 ml solution and is sterilized by membrane filtration.

Culture procedures — These followed those used in early experiments.[71] Culture vessels (not stated) containing five to ten organisms were sealed with Parafilm and incubated in an inclined position at 37°C. During the first 10 d of culture, 5 ml of medium was replenished every second day. When worms grew to about 10 mm, cultures were subdivided so that each tube contained not more than three worms. The pH was checked periodically.

Results — During the first 4 d of culture, the pattern of development in all media closely followed that *in vivo* reported by De Rycke[69] (see Figure 17). Normal adult worms measuring 90 to 150 mm with gravid proglottides developed; the patency time was not stated, but was probably more than 14 d.

D. EVALUATION

The fact that this species has been grown through its entire life cycle *in vitro* makes it a valuable model for studies in developmental biology and metabolism, although its advantages have not been much exploited. The separate systems for growing the cysticercoid from the egg and the adult from the cysticercoid are relatively simple and appear to be readily reproducible.

VIII. *HYMENOLEPIS CITELLI*

Cultured by: Voge and Green[55]
Definitive hosts: *Citellus* spp. (squirrels)[72,73]
Experimental hosts: hamster;[55] mice[74]
Prepatent period: 18 to 19 d (ground squirrel)[73]
Intermediate host: *Tribolium confusum*
Distribution: cosmopolitan?

A. GENERAL BIOLOGY AND LIFE CYCLE

Although this species of *Hymenolepis* is not widely used as an experimental cestode model, it is important as being the first species whose oncosphere was grown to a viable cysticercoid.[55] The experience gained with this species was then applied to *H. diminuta*, *H. nana*, and *H. microstoma* with equal success. To date, the adult *H. citelli* has not been grown from a cysticercoid *in vitro*.

H. citelli was first reported in the genus *Citellus*,[72] and its general biology and life cycle in the ground squirrel, *C. beecheyi*, were elaborated further by Voge,[73] who also documented the morphological differences between it and *H. diminuta* (Figure 19). The life cycle is similar to that of *H. diminuta*, described earlier, and will not be dealt with in detail here. The hamster has been most used as a laboratory host;[73] however, the worm can be grown in mice, but it is more readily rejected, particularly in heavy infections.[74]

B. *IN VITRO* CULTURE: ONCOSPHERE TO CYSTICERCOID
1. Culture Technique[55]

Gravid proglottides (from hamsters) were washed in three changes of 0.85% NaCl containing penicillin (300 units per milliliter) and streptomycin (300 μg/ml), being kept in each wash for 10 to 15 min. Eggs were teased out in Earle's BSS and shaken with glass beads in sterile screw-cap tubes to break the shells. When most egg shells were broken, eggs were transferred to small petri dishes with trypsin solution (25,000 units of Tryptar) in BSS at room temperature and checked every 5 min for hatching. Having removed most

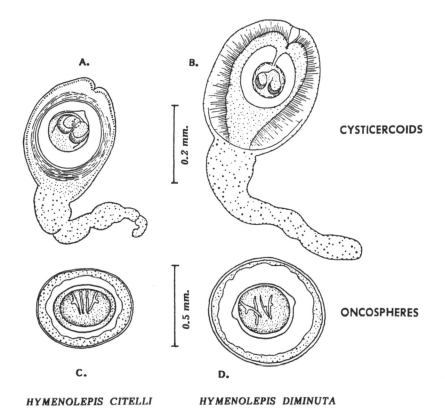

A.   B.   CYSTICERCOIDS

0.2 mm.

0.5 mm.

C.   D.   ONCOSPHERES

HYMENOLEPIS CITELLI   *HYMENOLEPIS DIMINUTA*

FIGURE 19. Cysticercoids and eggs of *Hymenolepis citelli* compared with those of *H. diminuta*. (A) and (B) 20-d-old cysticerci from *Tenebrio* spp.; (C) and (D) live embryonated eggs. Camera lucida drawings. (Modified from Voge, M., *J. Parasitol.*, 42, 485, 1956. With permission.)

of the broken shells, oncospheres were then concentrated in the bottom of the dish by swirling and transferred to 130 × 10-mm screw-cap culture tubes with 6 ml of medium. The exact number of organisms per tube was not stated, but >250 appeared to be too many, at the hollow ball stage, and some were transferred to other tubes.

2. Medium

The culture medium (Table 7) was a modification of that used by Landureau[75] for the culture of cockroach cells. The ingredients were dissolved in 900 mg of distilled water and 100 ml of fetal calf serum (noninactivated) was added and the pH was adjusted to 7.0 with $NaHCO_3$. Reducing agents were added before use, as it was established that reducing conditions were necessary to stimulate growth.

Reducing agents were added to individual cultures to give concentrations as follows: glutathione, $9.9 \times 10^{-4} M$; L-cysteine, $1.9 \times 10^{-3} M$; L-ascorbic acid, $4.9 \times 10^{-4} M$; dithiothreitol, 4.0 to $4.9 \times 10^{-4} M$. For details of preparation of stock solutions of these reducing reagents, the original paper should be consulted.

3. Results

Although the results showed some variation, with abnormal organisms appearing in all cultures, the evidence clearly indicated that the presence of reducing agents were necessary for normal growth and differentiation to take place. Best results were obtained in cultures with ascorbic acid or cysteine in which complete development to cysticercoids was achieved. The authors[55] point out, however, that both of these substances have other metabolic func-

TABLE 7
Hymenolepis citelli: **Modified Landureau's Medium**[75]
used for the Cultivation of Oncospheres to
Cysticercoids

L-Arginine HCl	2.0 g	L-Valine	0.152 g
L-Aspartic acid	0.200	$M_8SO_4 \cdot 7H_2O$	0.670
L-Glutamic acid	1.0	KCl	0.894
α-Alanine	0.120	$CaCl_2$	0.484
β-Alanine	0.0445	NaCl	7.42
L-Cysteine	0.101	$NaHCO_3$	0.424
L-Glutamine	0.559	$NaH_2PO_4 \cdot H_2O$	0.010
Glycine	0.200	Glucose	2.5
L-Histidine HCl	0.404	Trehalose	6.9
L-Leucine	0.249	α-Ketoglutaric acid	0.365
L-Lysine	0.0124	Citric acid	0.0153
L-Methionine	0.492	Fumaric acid	0.0058
L-Proline	0.748	L-Malic acid	0.058
L-Serine	0.083	Succinic acid	0.0059
L-Threonine	0.020	Yeast extract	1.0
L-Tyrosine	0.362	Lactalbumin hydrolysate	3.5

Note: The above are dissolved in 900 ml of double-distilled water, filtered through a Millipore filter, 0.22-μm pore size, and stored at −25°C until used.

From Voge, M. and Green, J., *J. Parasitol.*, 61, 291, 1975. With permission.

tions, in addition to reduction, so that other factors may be involved. Development *in vitro* was substantially longer than that *in vivo*; the time which elapsed between withdrawal and the formation of the external "protective" layers was about 2 weeks — twice as long as that required *in vivo*. The total time to develop infective cysticercoids *in vitro* was 45 d, compared with an (estimated) time of 20 d *in vivo*. The infectivity of the *in vitro*-grown cysticercoids was demonstrated by feeding 45-d-cultured organisms to hamsters, numerous eggs being passed 21 d later.

C. EVALUATION

The success of this technique is reflected in the fact that it, or minor modifications of it, have been used by other authors to successfully culture the oncospheres of other species, *H. diminuta*, *H. nana*, and *H. microstoma*, to infective cysticercoids; for details see individual species.

IX. *MESOCESTOIDES CORTI*

Cultured by: Barrett et al.;[76] Thompson et al.;[77] Ong and Smyth;[78] Ong[79]
Definitive hosts: carnivores, especially dogs and skunks[80]
Experimental hosts: cats, dogs[82]
Prepatent period: 12 d (in dogs)[77]
Intermediate hosts: first, unknown?; second, reptiles, birds, mammals
Distribution: cosmopolitan?

This species has proved to be an excellent model for the study of differentiation in cestodes on account of the ease with which the second larval stage — the *tetrathyridium* can be cultured *in vitro* to develop either sexually or asexually. An additional advantage is that the tetrathyridium can be maintained easily in the laboratory by intraperitoneal passage through mice.[83] Its life cycle is also unique in having an asexual phase in the intestine of the definitive host (see below).

A. GENERAL BIOLOGY

1. Morphology

Voge[80] regards *M. variabilis* and *M. manteri* as synonyms of *M. corti*. The genus *Mesocestoides* is characterized by the presence of a *paruterine organ* — a thickened sac (Figure 20) which appears to provide additional protection for the delicate oncospheres which lack a protective capsule. The morphology of *M. corti* is shown in Figure 20 and will not be discussed in detail here.

2. Life Cycle

The adult of *M. corti* occurs in a number of carnivores, such as the dog, fox, cat, skunk, etc., although other species infect other mammals (especially rodents) and reptiles. This particular species, however, is unique in having the property to undergo asexual development during the intestinal phase in the definitive host. This was first dramatically demonstrated by Eckert et al.,[81] who in one of several experiments fed 1000 tetrathyridia to a dog and obtained, on autopsy, some 40,600 worms 11 weeks later; a similar multiplication took place in other dogs and skunks. The other widely used experimental species, *M. lineatus*, does not divide asexually in this way. Although all the details of the life cycle are not clear, the pattern of development in the intestine (Figure 21) appears to be as follows.[8] When a tetrathyridium is taken into the gut, it first sheds it posterior tissue and its scolex divides somewhat unevenly into a large and a small form, each with two suckers. The smaller form regenerates two suckers and may divide again; the larger form also regenerates two suckers, but strobilates and forms an adult worm. Furthermore, the adult worm appears to be capable of further asexual divisions by either longitudinal splitting of the scolex (to produce a two sucker form which can regenerate, as above) or a bud can arise from the strobila. In dogs, gravid proglottides have been reported at 12 d,[77] but the prepatent period may be as early as 10 d, as in the related species, *M. lineatus*.[82]

The remainder of the life cycle is still incompletely known, but it is thought that two intermediate hosts are involved. The first is unknown, but may be a coprophagous arthropod. This hypothesis is based on an observation by Soldatova[84] of various developmental stages from oncospheres to a (procercoid-like)[85] cysticercoid in oribatid mites. It was assumed that this stage developed into a tetratyridium when eaten by the second intermediate host. In different species, this can be a species of amphibia, reptile, bird, or mammal. The second intermediate host appears to act as an (obligatory) paratenic host. In the laboratory, mice serve as excellent intermediate hosts,[83] the tetrathyridium localizing in the liver and (uniquely) dividing by splitting longitudinally from the scolex. The specimens of *M. corti* used in laboratories throughout the world probably all came from the original isolate obtained by Specht and Voge[83] from the fence lizard, *Scelporus occidentalis biseriatus*.

B. *IN VITRO* CULTURE: ONCOSPHERE TO TETRATHYRIDIUM[86]

Early experiments by Voge[85] succeeded in growing the oncosphere to procercoid-like stages and to young tetrathyridia containing an apical organ and outlines of suckers (Figure 22). This technique was improved on by Voge and Seidel,[86] who succeeded in completing the growth of the oncosphere to a fully developed tetrathyridium. This modified technique is given below.

1. Culture Technique

Gravid proglottides were obtained from the feces of skunks or dogs and washed for 30

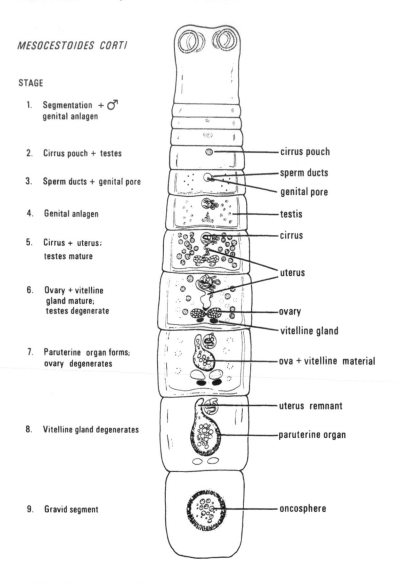

MESOCESTOIDES CORTI

STAGE

1. Segmentation + ♂ genital anlagen

2. Cirrus pouch + testes

3. Sperm ducts + genital pore

4. Genital anlagen

5. Cirrus + uterus; testes mature

6. Ovary + vitelline gland mature; testes degenerate

7. Paruterine organ forms; ovary degenerates

8. Vitelline gland degenerates

9. Gravid segment

cirrus pouch

sperm ducts

genital pore

testis

cirrus

uterus

ovary

vitelline gland

ova + vitelline material

uterus remnant

paruterine organ

oncosphere

FIGURE 20.   *Mesocestoides corti*, diagrammatic representation of the stages of sexual differentiation during development from a tetrathyridium to an adult worm *in vivo* and *in vitro*. (From Barrett, N. J., Smyth, J. D., and Ong, S. J., *Int. J. Parasitol.*, 12, 315, 1982. With permission.)

min in Earle's saline containing antibiotics (penicillin, 900 IU/ml, and dihydrostreptomycin, 900 μg/ml) before transferring them to depression slides. Eggs were freed from the paruterine organs by squeezing gently with fine forceps. Hatching was initiated by adding two drops of Tryptar (25,000 units per milliliter) in Earle's saline with biocarbonate at pH 7.0 and placing the slides in covered petri dishes at room temperature. Hatching was usually accomplished in 30 min and oncospheres were transferred to 10 × 1000 mm culture tubes by pipette in 3 ml of Medium A (see below). The tubes were sealed with a double layer of Parafilm and kept at room temperature (24 to 28°C) in a slanted position. Each tube received approximately 50 to 60 oncospheres. Medium (1 ml) was changed once a week for the first 2 weeks and three times a week thereafter.

Under these conditions, larva developed to early tetrathyridia with clacareous corpuscles and sucker primorida (see Figure 22G). To complete development, larvae which had attained a length of approximately 400 μm were transferred to the diphasic blood-agar Medium B (see below) and cultured at a higher temperature (37°C).

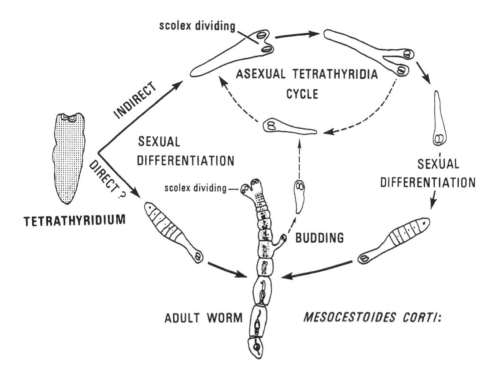

FIGURE 21. *Mesocestoides corti*, pattern of asexual/sexual development in the intestine of the definitive host (cat/dog).

2. Medium

For the first stage of development, a monophasic medium (called here, for convenience, Medium A) was used. This consisted of Triple Eagle's Medium plus 30% inactivated horse serum and antibiotics (penicillin, 300 IU/ml; dihydrostreptomycin, 300 µg/ml). For later development to fully developed tetrathyridia, a diphasic medium (Medium B) was used. This consisted of blood agar slants of 5% citrated blood in nutrient agar in 16 × 125 mm tubes overlaid with 8 ml of Triple Eagle's Medium (pH 7.0). Medium was changed once a week.

3. Results

At 64 d, the best cultures developed tetrathyridia 1.5 mm long and 1 mm wide; the sucker musculature was well developed and the apical organ had disappeared. These larvae appeared indistinguishable from those grown *in vivo*. There was, however, much variation of development within a culture tube. Transformed tetrathyridia which had been kept at 30°C on blood agar for 5 weeks, or longer, began to multiply asexually and appeared normal.

C. EVALUATION

This technique appears to have been reasonably successful, but clearly could be considerably improved to provide more uniform results. It has the advantages that the media and systems used are relatively simple and could be readily prepared in most biological laboratories. Theoretically, tetrathyridia grown in this way could be grown to adults by the technique described below and the whole life cycle would then be completed *in vitro*.

D. *IN VITRO* CULTURE: ASEXUAL MULTIPLICATION OF TETRATHYRIDIA

Voge and Coulombe[87] were the first to examine asexual multiplication of tetrathyridia *in vitro* in a number of different media. They found that asexual multiplication occurred in

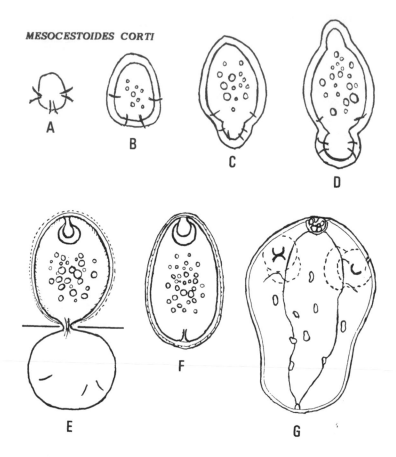

MESOCESTOIDES CORTI

FIGURE 22.   *Mesocestoides corti*, diagrammatic representation of the *in vitro* development of an oncosphere to a young tetrathyridium. (A) Hatched oncosphere; (B) early growth stage with external coat; (C) tail differentiation; (D) anterior differentiation; (E) constriction of tail, loss of hooks, and appearance of subtegumental muscles and tegument; broken line denotes partial disintegration of external coat; (F) loss of tail, appearance of excretory pore, tegumental spines, calcareous corpuscles, and primordia of suckers. The clear vesicles are probably lipid granules. (From Voge, M., *J. Parasitol.*, 53, 78, 1965. With permission.)

both monphasic and diphasic media provided that blood or plasma was present. Culturing was carried out in Erlenemeyer flasks containing 20 ml of fluid media with whole blood or solid agar bases. The greatest increase occurred in whole blood and an initial pH of 7.4 to 7.6 in NCTC 109 medium. This preliminary technique will not be described further here because it has been overtaken by the more standardized techniques of Ong and Smyth[78] in which much greater levels of asexual multiplication were obtained (see Table 11 and Section IX.E.3).

E. *IN VITRO* CULTURE: TETRATHYRIDIA TO MATURE ADULTS[76-78]

Two groups of workers, Barrett et al.[76] and Thompson et al.,[77] simultaneously published results of experiments which resulted in the sexual differentiation of tetrathyridia of *M. corti in vitro*. Both of these groups of workers used a medium based on that developed for the cultivation of the strobilar phase of *Echinococcus*[88,89] with minor modifications. Both of these results can be regarded as preliminary because although some sexually mature adults were obtained, they provided little unequivocal evidence as to the nature of the factors inaucing sexual or asexual differentiation. Based on these early results, a more critical

TABLE 8

Mesocestoides corti: **Liquid Medium Used by Thompson et al. for *In Vitro*
Culture of Tetrathyridia to Adult Worms in a Diphasic Medium Based on
S.10E.H[a] as used by Smyth[88,89] for Culture of *Echinococcus granulosus***

Liquid Phase[a]

Medium CMRL 1066 with glutamine (Flow Laboratories) or Medium 199 (modified) with Earle's salts and glutamine (Flow Laboratories)	325 ml
Fetal calf serum (Commonwealth Serum Laboratories/Grand Island Biological Company)	125 ml
30% D-Glucose (British Drug Houses)	7.25 ml
5% dog bile in Hanks' saline	2.1 ml
5% yeast extract in Hanks' saline (Difco)	45 ml
Sodium bicarbonate	2.1 g
HEPES	20 mM
Neomycin-penicillin (Commonwealth Serum Laboratories)	0.4 μg/ml
Gentamycin (Schering)	100 μg/ml
Amphotericin (Squibb)	2 μg/ml

Solid Phase[a]

Bovine serum coagulated at 76°C for 75 min

[a] See Table 13.

From Thompson, R. C. A., Jue Sue, L. P., and Buckley, S. J., *Int. J. Parasitol.*, 12, 303, 1982. With
permission.

examination of the factors controlling asexual/sexual differentiation was carried out later by
Ong and Smyth.[78] These various experiments are described separately below.

1. *In Vitro* Technique of Thompson et al.[77]

Culture procedures and media — Tetrathyridia were removed aseptically from
Quackenbush mice in which they had been maintained by serial passage. They were treated
with 0.05% pepsin at pH 2.0 for 5 min at 38°C and rinsed three times in Hanks' BSS plus
antibiotics (0.4 μg/ml of penicillin; 0.4 μg/ml of neomycin) before placing in culture vessels.
Both monophasic (liquid) and diphasic culture systems were used. The composition of the
liquid medium is shown in Table 8; the solid phase in the diphasic system was prepared by
coagulating bovine serum at 76°C for 75 min; 150 larvae in 10 ml of culture fluid were used
in 30-ml tissue culture flasks. Cultures were gassed every third day with 10% O_2/5%
CO_2/85% N_2 and the liquid medium was changed every 3 d. The pH of the medium was
not stated, but was probably 7.4, at least initially. The number of cultures used in these
experiments was not stated.

Results — It was reported that "consistent development" of segmented, mature, adult
worms was obtained, using Medium 1066 in the liquid phase, the percentage of strobilated
adults being higher in diphasic media (30 to 70%) than in monophasic media (1 to 2%).
Asexual multiplication of tetrathyridia also occurred in all cultures, either by scolex division
or the formation of buds, as described in the *in vivo* cycle (see Figure 21). It was particularly
interesting to note that self-insemination was observed in many worms, but sperm were not
seen in the copulation canal nor were fully developed oncospheres observed. The authors
speculated that the failure to produce viable oncospheres could have been due to some
abnormality in the reproductive system that precluded fertilization or that nutritional factors
may have been involved.

TABLE 9

Mesocestoides corti: **Medium Used by Barrett et al. for the *In Vitro* Culture of Tetrathyridia to Adult Worms in a Diphasic System Based on Medium S.10E.H[a] as Used by Smyth[88,89] for *Echinococcus granulosus***

Liquid Phase (Modifications of S10E.H)[a] (including 20 m*M* HEPES)

Fetal calf serum (Gibco)	100.0 ml
CMRL 1066 (Flow) including 100 IU/ml of penicillin and 100 μg/ml of streptomycin (Flow)	260.0 ml
5% Yeast extract (Gibco) in Hanks' saline	36.0 ml
30% Glucose (Hopkin & Williams) in distilled water	5.6 ml
5% Dog bile in Hanks' saline (or 0.2% sodium taurocholate (Sigma) in distilled water)	1.4 ml

Variations

Substituting CMRL 1066 with
  RPMI 1640 (Flow)
  McCoys 5A (Flow)
  TC 199 (Gibco)
Substituting fetal calf serum with
  Horse serum (Flow)

Solid Phase[a]

Newborn calf serum (Flow) coagulated at 76°C until it becomes opaque and ''sets''

[a]   See Table 13.

From Barrett, N. J., Smyth, J. D., and Ong, S. J., *Int. J. Parasitol.*, 12, 315, 1982. With permission.

2. *In Vitro* Technique of Barrett et al.[76]

Culture procedures and medium — Tetrathyridia were removed aseptically from TFI mice in which they had been maintained by intraperitoneal passage. In an initial experiment, after 50 d of asexual multiplication *in vitro*, one tetrathyridium spontaneously segmented and developed into a sexually mature adult. This observation stimulated further work on the possible factors inducing asexual/sexual differentiation. Monophasic and diphasic media were used, the liquid medium being S10E.H (Table 9) — as used for *Echinococcus* — and the solid phase being coagulated bovine serum. Other culture parameters examined included number of larvae per culture, type of culture vessel and volume of media, gas phase, enzyme pretreatment, shaking or stationary conditions, and frequency of media changes; for details, see Table 10.

Results — Although sporadic sexual maturation occurred in most cultures, it occurred most often in the following system: 5 or 10 ml of Medium S10E.H (see Table 9), in Leighton tubes, slowly rotated in an incubator at 38°C with 100 to 200 worms, changing the medium every 2 or 3 d. Segmentation occurred more frequently in cultures gassed with 20% CO_2 than with other gas mixtures, the earliest segmentation being reported at 15 d in liquid medium. There was, however, much inter- and intravariation in cultures. A few shelled eggs with hooked oncospheres were found in one mature worm, suggesting that insemination had occurred in at least one instance.

3. *In Vitro* Technique of Ong and Smyth[78]

In the light of the above results, Ong and Smyth[78] investigated, under more carefully controlled conditions, the effect of various physicochemical factors on asexual/sexual differentiation comparing results, where possible, using 95% confidence intervals. This resulted

TABLE 10

Mesocestoides corti: **Various Parameters Used by Barrett et al. for the *In Vitro* Culture of Tetrathyridia to Adult Worms**

Number of tetrathyridia/culture	20—400	
Amount of media/culture	(a) Liquid (ml)	(b) Solid (ml)
Sterilin t.c. flasks	20	—
Costar t.c. flasks	10	10
M.D. bottles	20	20
Leighton t.c. tubes	5 or 10	2 or 3
Gas phase	(a) Air	
	(b) 10% CO_2, 5% O_2, 85% N_2	
	(c) 20% CO_2, 80% N_2	
Pretreatment	(a) 0.2% pepsin in Hanks' saline (pH 2), rotated for 20 min at 38°C	
	(b) 0.5% trypsin in distilled water, treated as (a)	
Movement of cultures	(a) "Still" — in incubator	
	(b) "Activated" — either rotated at 1 revolution/5 min (Leighton t.c. tubes) or shaken in a waterbath, continuously or intermittently (all t.c. flasks and M.D. bottles)	
Intervals between changing media	2—5 d	

Note: Tissue cultures = t.c.

From Barrett, N. J., Smyth, J. D., and Ong, S. J., *Int. J. Parasitol.*, 12, 315, 1982. With permission.

in a greatly improved, standardized technique which enabled sexual development *in vitro* to proceed at the same rate as that *in vivo* (in dog), although only a low production of embryonated eggs was obtained.

Culture conditions and media — The monophasic medium (S.10E.H, Table 9) was a modification of that used by Barrett et al.[82] (see above), except the HEPES buffer was replaced with 20 mM HEPPSO (N-hydroxyethylpiperazine-N'-2-hydroxy propane sulfonic acid, Research Organics, Inc., Ohio), which gave a higher, stable pH of 7.73 at 37°C. Tetrathyridia removed aseptically from TFl mice were first screened to exclude division stages or individuals with only two suckers or those without suckers. Hence, only tetrathyridia with four suckers were used in the cultures; 25 to 100 larvae were used per culture tube, the latter being 19 × 105 mm Leighton tubes set at an angle of about 10° in a roller tube apparatus at a speed of one revolution per minute at 38°C in an incubator. To maintain the gas phase and pH (see below), culture tubes were sealed with two layers of Parafilm. The following gas phases were tested: (1) air; (2) 5% CO_2/95% N_2; (3) 5% CO_2/10% O_2/85% N_2; and (4) 20% CO_2/80% N_2. The effect of pH on sexual differentiation was tested using the following buffers: MOPS (pK_a 7.01); HEPES (pK_a 7.30); and HEPPSO (pK_a 7.73) all at 37°C.

Pretreatment — The earlier experiments, described above,[76,77] utilized pepsin apparently to stimulate evagination. This is, in fact, unnecessary as evagination takes place in saline.[78] However, in order to test the possible effect of pretreatment — before culturing — the following preliminary experiments were performed. All larvae were rinsed briefly in Hanks' BSS before the following pretreatments:

1. No pretreatment, apart from preliminary rinsing
2. Incubated in BSS at 38°C overnight

3.   Incubated in BSS adjusted to pH 2.0 with HCl and shaken for 10 min at 38°C
4.   Incubated in 0.2% pepsin in BSS at pH 2.0 and shaken for 10 min at 38°C
5.   Incubated in 0.5% trypsin in BSS at pH 7.4 and shaken for 20 min at 38°C
6.   Treated as (4) followed by (5)

Results — Rather unexpectedly, it was found that the only cultures which underwent segmentation were those which had been incubated overnight in Hanks' BSS. Little difference was observed between cultures containing from 25 to 100 tetrathyridia. When the effect of pH and O_2 was investigated (Figure 23), it was found that a pH >7.4 and an anaerobic gas phase gave significantly greater segmentation rates than a lower pH or an aerobic gas phase. Thus, a gas phase of 5% CO_2/95% N_2 combined with a pH higher than 7.4 provided the most suitable conditions for segmentation and sexual differentiation. In the best cultures, organisms began segmentating *in vitro* within 1 week, a period compatible with the rate of development in the dog.[77] The paruterine organ began developing 8 d after the onset of segmentation. Although sexually mature adults were grown consistently from tetrathyridia, fully developed oncospheres were only obtained sporadically. The size, morphology, and activity of these oncospheres appeared to be similar to those in oncospheres produced *in vivo*.[79] Due to the fact that the first intermediate host is unknown, it was not possible to test their viability. In any case, this would have been difficult due to the small numbers of oncospheres produced. Although self-insemination was sometimes observed, it is clear that conditions for this process to occur readily *in vitro* were not present in the culture conditions provided. Asexual multiplication readily occurred, but only in those diphasic cultures in which the base had been perforated with small holes (Table 11) into which the tetrathyridia could penetrate. This result agrees with the observation of Voge and Coulombe,[87] who found that in blood agar tubes, the numbers of organisms only increased if they became embedded in the agar. In diphasic media, without holes in the bases, large number of adults, but few asexual forms, were produced.

F. EVALUATION

M. corti must be regarded as one of the most valuable models for studying the phenomenon of asexual/sexual differentiation in cestodes or, indeed, any helminth parasite. It shares this position with the hydatid organism, *Echinococcus granulosus*, whose development *in vitro* can likewise be triggered to develop asexually or sexually by varying environmental conditions.[90] *M. corti* has the added advantage that it can be more readily maintained in the laboratory in mice in which large numbers of tetrathyridia can be produced. In addition, since all developmental stages have now been cultured *in vitro*, it should be possible to complete the entire life cycle *in vitro*. To date, this has not been achieved.

It is difficult to critically compare the results of the three groups of workers discussed above[76-78] because it is likely that minor (and probably unrecorded) differences in protocols may have been used. Thus, Ong and Smyth[78] found that sexual development only took place in organisms which had been "preconditioned" in BSS at 38°C overnight and sexual differentiation was stimulated by a pH >7.4 and anaerobic conditions. It is not known if the other groups stored or otherwise treated their larvae before culturing. It is clear that asexual development takes place best in diphasic media either with holes in the solid base or with a semisolid phase in which the larvae can burrow. The addition of blood may be an advantage.[87] The results of Ong and Smyth[78] were the only ones in which nondividing organisms were initially screened before culturing and statistical analysis applied to the results. Nevertheless, there may be other factors, such as Eh, which may be important and which were not apparently controlled by any group and this parameter may need further examination.

A major problem encountered by all of the above workers was the low level of self-insemination and fertilization achieved, resulting in only very small number of oncospheres

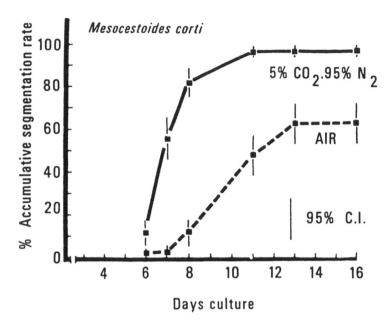

FIGURE 23. *Mesocestoides corti*, effect of the gas phase on the induction of sexual differentiation of tetrathyridia *in vitro*. The anaerobic phase induces significantly higher segmentation rates than air; C.I., confidence limits. (From Ong, S.-J. and Smyth, J. D., *Int. J. Parasitol.*, 16, 361, 1986. With permission.)

being produced. This is a common problem with other cestodes, such as *Schistocephalus solidus* and *Echinococcus granulosus*, and further work is clearly needed on this problem in *M. corti*.

X. *MESOCESTOIDES LINEATUS*

Cultured by: Kawamoto et al.[91]
Definitive hosts: carnivores, especially dogs and foxes[80]
Experimental hosts: cats, dogs,[82,91] hamsters[92]
Prepatent period: 10 d (in cats)[82]
Intermediate hosts: first, unknown?; second, tetrathyridia in reptiles (especially snakes), birds and mammals
Distribution: cosmopolitan

A. GENERAL BIOLOGY

Unlike *Mesocestoides corti*, this species does not multiply asexually in the intermediate host or in the intestine of the definitive host; otherwise its life cycle is similar. *M. lineatus* has the added advantage, however, that as well as developing in cats and dogs (which are difficult laboratory hosts), it will also develop in the hamster[92] — a much easier host to maintain. The first intermediate host is unknown, there being unsupported evidence that it may be coprophagous arthropod.[84] The morphology generally resembles that of *M. corti* and will not be discussed further here.

When a tetrathyridia is taken into the gut of the definitive host, it sheds its posterior body region within 24 h, and all viable worms become attached in the anterior third of the gut.[91] Gravid proglottides are formed by day 10.

B. *IN VITRO* CULTURE: TETRATHYRIDIA TO ADULTS[91]

In a series of elegant experiments, Kawamoto et al.[91] demonstrated that, unlike *M. corti*,

TABLE 11

Mesocestoides corti: **Distribution and Development of Asexual (Tetrathyridia) and Sexual (Adults) Organisms in Diphasic Media With and Without Holes in the Coagulated Serum Base**

Culture	Number free in liquid phase			Number situated in serum base			Total number of adults (a + c)	Total number of tetrathyridia (b + d)	Number of all organisms (a + b + c + d)	Increase (Y)
	Adults (a)	Tetrathyridia (b)	Total (a + b)	Adults (c)	Tetrathyridia (d)	Total (c + d = X)				
Serum base punched										
E	4	103	107	0	10	10	4 (3.4%)	113	117	17
F	21	135	156	2	118	120	23 (8.3%)	253	276	176
G	0	123	123	0	145	145	0	268	268	168
H	3	176	179	0	159	159	3 (0.89%)	335	338	238
Serum base not punched										
I	149	3	152	42	3	45	191 (97%)	6	197	97
J	106	7	113	11	9	20	117 (88%)	16	133	33
L	79	29	108	3	4	7	82 (71%)	33	115	15

Note: There were 100 tetrathyridia per initial culture.

From Ong, S.-J. and Smyth, J. D., *Int. J. Parasitol.*, 16, 361, 1986. With permission.

TABLE 12
Medium NCTC-135-Plus Used by Kawamoto et al. for the *In Vitro* Cultivation of Tetrathyridia of *Mesocestoides lineatus* to Adult Worms

Stock solutions[a]

    Stock solution A: 10 mM each of 8 kinds of amino acid: arginine HCl, aspartic acid, asparagine, alanine, cystine, glutamic acid, tryptophan, tyrosine; in distilled water

    Stock solution B: 10 mM each of 11 kinds of amino acid: glycine, histidine, isoleucine, leucine, lysine HCl, methionine, proline, phenylalanine, serine, threonine, valine; in distilled water

    Stock solution C: thyamine HCl, riboflavin, calcium pantothenate, pyridoxine HCl: *p*-aminobenzoic acid, niacin, folic acid, inositol, 25 mg each; biotin, 12.5 mg; choline chloride, 250 mg; vitamin B12, 50 mg; dissolved in 100 ml of distilled water

Components/liter of NCTC-135-plus[b]

Stock solution A	50 ml
Stock solution B	50 ml
Stock solution C	5 ml
NCTC-135 (Earle's salt with glutamine)	9.6 g
Sodium bicarbonate	2.1 g
HEPES	2.3 g
Streptomycin	100—200 mg
Kanamycin	100—200 mg

[a] Stock solutions were stored at $-20°C$ in adequate volumes.

[b] pH was adjusted to 7.4 with 1 N NaOH and then distilled water was added, bringing the final volume to a liter.

From Kawamoto, F., Fujioka, H., and Kumada, N., *Int. J. Parasitol.*, 16, 333, 1986. With permission.

sexual differentiation only occurred when organisms were pretreated with trypsin or other proteolytic enzymes. A minimum exposure of 24 h was necessary to induce segmentation which first occurred 5 d post-treatment.

1. Source and Laboratory Maintenance of Tetrathyridia

Tetrathyridia of *M. lineatus* were collected from snakes, mostly *Agkistrodon halys* and *Elaphe quadrivirgata*, in Japan.[82] Encapsulated tetrathyridia were freed from their cysts by digestive treatment with 0.5% pepsin for about 30 min at 37°C. After washing in phosphate-buffered saline (PBS), they were maintained in male ICR mice by intraperitoneal injection (20 larvae per mouse). When required for culture, the tetrathyridia were removed aseptically from the body cavity and washed at least three times in PBS containing 200 μg/ml each of streptomycin and kanamycin.

2. Culture Conditions and Medium

Tetrathyridia were cultured in monophasic liquid media in two types of plastic culture dishes: 60-mm culture dishes (Corning) or 90-mm petri dishes (Nissui Co.); 5 to 10 larvae were cultured in 8 ml of medium in the smaller dishes and in 15 ml of medium in the larger dishes. They were cultured at 37°C in a gas phase of 5% CO_2 in air or by a candle-jar method. Media were changed every 3 d.

Tetrathyridia were treated with enzymes (in NCTC-135 or M-199) for 3 h to 5 d. In the 5-d treatment, the medium was changed at 72 h and culture was continued for 2 d. After enzyme treatment, larvae were transferred to medium NCTC-135 plus, a medium supplemented with vitamins and amino acids (Table 12).

3. Results

After treatment with trypsin alone, tetrathyridia evaginated their scoleces, segmented, and grew to sexually mature adults. The induction of development was apparently related to the proteolytic activity of the trypsin because it was inhibited or decreased by the addition of soybean trypsin inhibitor. Pronase E was as effective as trypsin, but chymotrypsin showed less activity. In the best cultures, shelled eggs containing active oncospheres were observed.

C. EVALUATION

M. lineatus appears to require different "trigger(s)" to stimulate sexual differentiation than *M. corti*. Moreover, the physicochemical conditions successful with each species are different, *M. lineatus* strobilating under aerobic conditions, whereas *M. corti* favors anaerobic conditions. Sufficient data are not available to compare other parameters of the two culture systems.

The monophasic system for *M. lineatus* is easy to handle and its success is reflected in the fact that the average time for strobila formation (5 d) agreed with that required *in vivo*.[82]

XI. *ECHINOCOCCUS GRANULOSUS*

Cultured by: Smyth;[89,90,93-96] Smyth and Davies;[12,88] Smyth et al.[97]
Definitive hosts: wild, feral, and domestic carnivores, especially dogs and foxes[98]
Experimental hosts: dogs[97]
Prepatent period: 34 to 40 d[97,99]
Intermediate hosts: cystic larva (hydatid cyst) in man and numerous species of wild and domestic ungulates, especially sheep, goats, horses, pigs, and camels[100-102]
Distribution: cosmopolitan[100]

Recent work on the *in vitro* culture of *Echinococcus granulosus* and *E. multilocularis* has been reviewed by Smyth and McManus,[8] Smyth,[96] and Howell.[7] Earlier literature has been reviewed by Taylor and Baker.[3] In one early experiment, Webster and Cameron[103] obtained a few segmented forms of both these species, but with no genital development.

A. GENERAL BIOLOGY
1. Speciation

Echinococcus granulosus is well known to be the causative organism of hydatid disease (hydatidosis/echinococcosis), a zoonotic disease of man and domestic and wild animals caused by the larval stage or hydatid cyst. There is a vast literature on the biology, epidemiology, and pathology of this organism; the early work has been reviewed by Smyth[100,102] and more recent studies have been reviewed by Thompson.[101] The systematics of the species are now recognized to be complex due to the existence of intraspecific variants (strains?; isolates?) whose general biology, developmental biology, biochemistry, immunology, and physiology may vary substantially. Of particular relevance, it can be noted that although the isolate of *E. granulosus* from sheep can readily be grown to sexual maturity *in vitro*, the isolate from horse fails utterly to develop in the same culture system. Striking physiological differences have been shown to exist between these two isolates.[8,12,13] Recent work on the genetics of this species has been reviewed by Thompson and Lymbery.[104]

2. Life Cycle

The life cycle is sufficiently well known to need little elaboration. The adult worm, which is very small (2 to 8 mm),[11] inhabits the small intestine of carnivores, especially dogs. The typical taeniid-type eggs are passed in feces and if ingested by an ungulate such

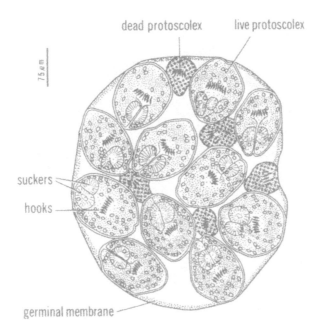

FIGURE 24. *Echinococcus granulosus*, brood capsule from sheep hydatid cyst. Note presence of some dead protoscoleces. (From Smyth, J. D. and Davies, Z., *Int. J. Parasitol.*, 4, 631, 1974. With permission.)

as sheep, they hatch in the duodenum, and the released oncosphere penetrates the mucosa and eventually reaches a tissue site, especially the liver or lungs, although any organ can be infected. The oncosphere develops into a characteristic hydatid cyst containing brood capsules (Figure 24), each of which contains protoscoleces. If — say after slaughter — the viscera containing the cysts are fed to a dog, each protoscolex evaginates in the duodenum and develops into an adult worm. However, the protoscoleces also have the unusual property of being able to dedifferentiate into a "secondary" hydatid cyst, should they leak out from a cyst in the intermediate host (Figure 25) — a situation which often arises in surgical operations in man for the removal of a cyst.

This ability of a protoscolex to differentiate sexually into an adult worm, if taken into the gut, or asexually into a secondary cyst, in a tissue site, has made this cestode one of the most fascinating organism to study. Elucidation of the factors which control asexual/sexual differentiation in *Echinococcus* and the reproduction of this phenomenon *in vitro* has represented one of the major challenges in parasitology and has been the subject of a large number of studies which have taken many years to resolve.[90,93-96] The situation now is that growth of a protoscolex in a cystic (i.e., asexual) direction or in a strobilar (i.e., sexual) direction can now be controlled *in vitro*. Before discussing these techniques, it is essential to have some understanding of the potential growth patterns of a protoscolex *in vitro*, and these are discussed below.

B. SOURCE AND HANDLING OF HYDATID CYST MATERIAL
1. Collection and Transport
Although it is often relatively easy to obtain hydatid cyst material from a local abbatoir or slaughter house, it is important to note that this material needs to be handled with great care, if it is to be used for setting up *in vitro* cultures. Some of the difficulties in handling such material have been reviewed by Smyth and Davies[88] and Smyth.[89]

Briefly, particular care should be exercised in transporting hydatid material to the lab-

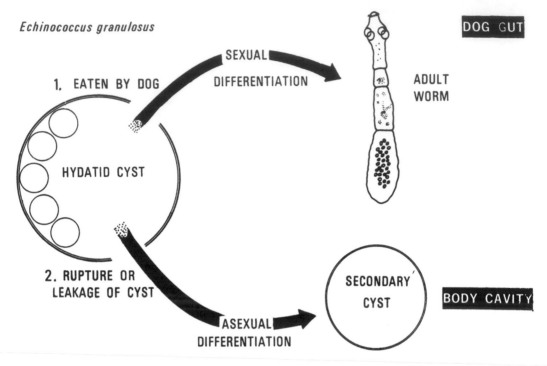

FIGURE 25.   *Echinococcus granulosus, in vivo* patterns of development of the protoscolex. (From Smyth, J. D., in *Molecular Paradigms for Eradicating Helmintic Parasites*, UCLA Symp. on Molecular and Cellular Biology, New Series, Vol. 60, MacInnis, A., Ed., Alan R. Liss, New York, 1987, 19. With permission.)

oratory. In particular, it should not be allowed to become overheated (as may happen in the summer months or in tropical countries) and transport in previously cooled, insulated containers is recommended. Ideally, hydatid cysts should be dissected on the same day, but they may be stored, for not more than 1 d, in a refrigerator at 4°C.

2. Dissection of Cysts for Protoscoleces

Although it is possible, with care, to obtain protoscoleces in a sterile condition from cysts dissected in a nonsterile room, the use of an open "laminar flow" (sterile) cabinet is strongly recommended. In the following procedures, sterile glassware and instruments are used throughout.

Warning — Protective glasses should be worn during dissection as hydatid cysts are under pressure and may squirt liquid containing protoscoleces into the eyes.

Procedure for dissecting cysts (Figure 26)[88,89,96] — Paint the surface of a cyst with 1% iodine in 95% ethanol, allow to dry, and repeat the process. Using a large-size needle (size 16 to 19) and a 10- to 50-ml hypodermic syringe, appropriate to the cyst size, draw off about half the fluid contents of the cyst and transfer it to a sterile container. Using scissors or a scalpel, remove the top of the cyst so that the inside can easily be seen for access. Remove the brood capsules, which form the bulk of the "hydatid sand" on the bottom of the cyst, with a Pasteur pipette and transfer them, in a little hydatid fluid, to a screw-top vial or other sealable container. With fresh cysts, it may be necessary to scrape off brood capsules adhering to the germinal membrane. Brood capsules are best stored at 4 to 6°C in 20-ml vials (universal containers) in about 10 ml of hydatid fluid. Many brood capsules burst during these procedures releasing protoscoleces, but treatment with pepsin (see below) is normally necessary to obtain a pure isolate of protoscoleces, free from germinal membrane fragments; 1 ml of settled brood capsules/protoscoleces provides about 10 to 12 cultures. This material can be stored in a refrigerator for 1 to 3 d, but is best used immediately.

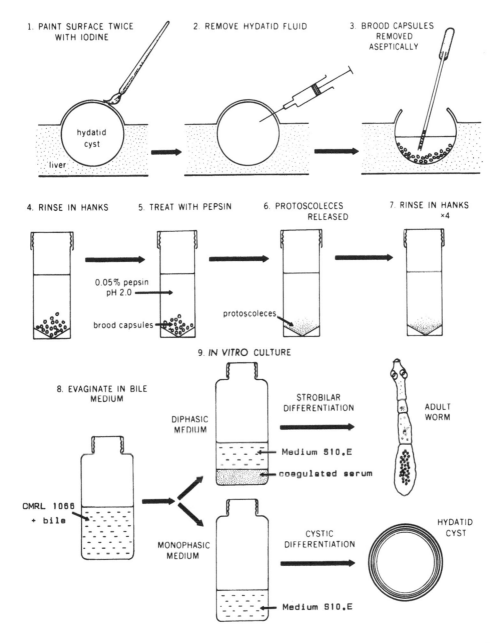

FIGURE 26. *Echinococcus granulosus*, technique for setting up *in vitro* cultures; all procedures are carried out under sterile conditions. (From Smyth, J. D., *Angew. Parasitol.*, 20, 137, 1979. With permission.)

Checking brood capsules/protoscoleces — Before setting up cultures, samples of brood capsules should be examined (microscopically) for presence of bacteria and viability. Due to indigenously infected cysts, bacteria are occasionally present and such material should be abandoned. Regarding viability, some cysts are either completely infertile or contain may dead protoscoleces, readily recognizable by their brown color (see Figure 24). Experience shows that brood capsules containing more than about 40% dead protoscoleces make poor material for culture. Viability is usually evident by observing slight movement (stimulated by a drop of bile, if necessary) or, if in doubt, by observing flame cell movement.

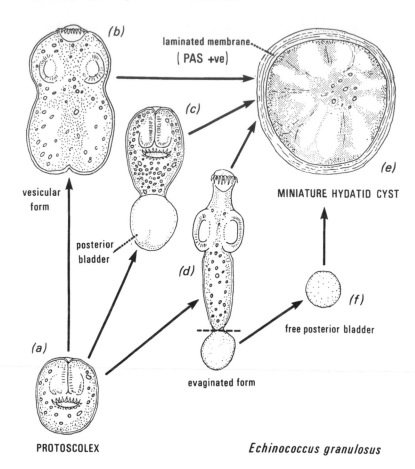

FIGURE 27. *Echinococcus granulosus*, the various forms developed from a protoscolex when cultured *in vitro* in monophasic liquid medium. (From Smyth, J. D. and McManus, D. P., *The Physiology and Biochemistry of Cestodes*, Cambridge University Press, Cambridge, 1989. With permission.)

Pepsin treatment — Before using for *in vitro* culture, samples of brood capsules must be treated with pepsin to release the protoscoleces and digest any dead protoscoleces present. Digestion is carried out at 38°C in 2 mg/ml of pepsin (1:10,000) in Hanks' BSS adjusted to pH 2.0 with 5 *N* HCl and sterilized by membrane filtration. Stirring with a magnetic stirrer assists the digestion process, but is not essential. The process should be examined at intervals, conveniently under an inverted microscope, the digestion being stopped when all the protoscoleces are freed from their brood capsules (15 to 45 min).

Washing protoscoleces — After pepsin treatment, rinse the protoscoleces four times in BSS, allowing them to settle thoroughly after each rinse. Each rinse may be conveniently carried out in a 20-ml vial on a roller tube, at 38°C, allowing 15 min for each rinse. After rinsing, protoscoleces should be used immediately for culturing.

C. *IN VITRO* CULTURE: GENERAL GROWTH PATTERNS (FIGURE 27)[8,93]

When cultured *in vitro*, a protoscolex can develop in a number of different ways, depending on the culture conditions, especially whether monophasic or diphasic media are used. It is important to understand this pattern before the more general problems of asexual/sexual differentiation are dealt with. In monophasic media, the following forms may appear in culture:

1. Unevaginated protoscolex (Figure 27a) — The protoscolex remains undifferentiated, as in a hydatid cyst.
2. Vesicular type (Figure 27b) — The protoscolex swells and becomes "vesicular" and rounded, eventually secreting a laminated membrane, and can now be regarded as a miniature hydatid cyst.
3. Posterior bladder type (Figure 27c) — A small bladder or vesicle develops in the posterior region of an unevaginated protoscolex, apparently arising from a few cells carried over from the germinal membrane of the brood capsule to which it was attached. This type secretes a laminated membrane and eventually also develops into a hydatid cyst.
4. Evaginated protoscolex with bladder (Figure 27d) — This type probably arises from the previous type, becoming evaginated when the posterior bladder is in an early stage of development. In monophasic medium, this, too, will develop into a small hydatid cyst.
5. Free posterior bladder (Figure 27f) — In the horse isolate (but not the sheep isolate), some posterior bladders may become separated from the protoscoleces, form independent bladders, secrete a laminated membrane, and become small hydatid cysts.[105]

In *diphasic* media, strobilar development (sexual differentiation) takes place under appropriate culture conditions (see Section XI.E).

D. *IN VITRO* CULTURE: CYSTIC (ASEXUAL) DIFFERENTIATION OF PROTOSCOLECES

The very early experiments on *in vitro* cultivation of protoscoleces of *E. granulosus*[93] used a variety of monophasic liquid media, including various combinations of bovine amniotic fluid, hydatid fluid, bovine serum, beef embryo extract, and Parker 199. In almost all combinations of media tried, the growth patterns followed those shown in Figure 27 and described above. Personal observations since then have confirmed that in almost any well-balanced liquid medium, such as NCTC 135, CMRL 1066, or Parker 199, plus 20% bovine or fetal calf serum, at a pH of about 7.4, in either an anaerobic or aerobic gas phase, cystic development takes place, resulting in miniature hydatid cysts with laminated membranes (see Figure 27e). In the longest experiment carried out,[96] over a period of 8 months of culture, the beginnings of brood capsules could just be observed through the cyst wall, but protoscoleces were not formed. To date, viable cysts with "fertile" brood capsules have not been grown *in vitro*. This is not surprising, since *in vivo* (in mice) cysts only become fertile after 8 months.[106]

The pattern in the *in vitro* development of protoscoleces from horse hydatid cysts is slightly different from that of sheep, in that the posterior bladders show a tendency to break off and form miniature hydatid cysts as in Figure 27; there are also other minor differences in development.[105]

E. *IN VITRO* CULTURE: PROTOSCOLECES TO ADULT WORMS
1. General Comment

Determination of the factors which induced a protoscolex to differentiate sexually, i.e., in a strobilar direction, has been one of the major challenges in cestode *in vitro* culture. All early attempts using the more standard culture procedures failed to induce strobilar growth.[93] This led to the conclusion that some unsuspected and unusual requirement was missing from the culture conditions provided. Further microscopic study of the adult worm *in situ* in the dog gut showed that the contact of the scolex was much more intimate than previously suspected, with the extended rostellum penetrating into the mouth of a crypt of Lieberkühn. This led to the conclusion that the missing requirement could be a solid substrate contact

TABLE 13
Medium S.10E.H.: Used for the *In Vitro* Culture of *Echinococcus granulosus* and *E. multilocularis* from Protoscoleces to Adult Worms[88,89,96]

Liquid Phase

Basic medium	
CMRL 1066	260 ml
Fetal calf serum	100 ml
5% Yeast extract (in CMRL 1066)	36 ml
30% glucose (in distilled H_2O)	5.6 ml
5% Dog bile or 0.2% Na taurocholate (in Hanks' BSS)	1.4 ml

Plus

Buffer
 20 mM HEPES + 10 mM NaHCO$_3$
Antibiotics
 100 μg/ml of gentamycin (and/or 100 IU/ml each of penicillin and streptomycin)

Solid Phase

Bovine serum[a] coagulated at 76°C for 30—60 min

[a]   Bovine serum is often sold commercially as "newborn calf serum". This should not be confused with "fetal calf serum" (as used in the liquid phase) which does not coagulate on heating.

with which, in some way, triggered strobilar differentiation, possibly via a neurosecretory mechanism.

Further experimentation showed that this hypothesis generally held for *E. granulosus* (but contrast *E. multilocularis*) and when protoscoleces were cultured in a *diphasic* medium with a suitable substrate (coagulated bovine serum), strobilar differentiation resulted and sexually mature adult worms developed (see Figure 26).[94] By varying the composition of the liquid phase (Table 13) and other parameters, a reasonably reliable technique has now been developed. This is shown in Figure 26 and described below.

2. Culture Technique (see Figure 26)
The basic technique has been described in some detail[88,89,96] and the account below has incorporated some minor variations.

Culture material and evagination — The initial culture material consisted of sterile, washed protoscoleces obtained from hydatid cysts by the protocol described above (Section XI.B.2). Before culturing, these are evaginated by treating in CMRL 1066 (or NCTC 135) containing sodium taurocholate (2 mg/100 ml) or sterile dog bile (1 ml of 5% dog bile in 100 ml of CMRL 1066). The sodium taurocholate should be as pure as possible. Crude mixtures of bile salts (often sold under the name of "sodium tauroglycocholate") may contain deoxycholate which has a powerful lytic effect on protoscoleces and may kill or damage them *in vitro*.[107]

Diphasic culture medium — The liquid phase consisted of Medium S.10E.H (see Table 13). The solid phase consisted of coagulated newborn calf or bovine serum, coagulation being obtained by heating at a precise temperature of 75 to 76°C in an oven for 30 to 60 min. The texture of the surface appears to be fairly critical and should be neither too hard nor too soft.

Culture conditions — A number of different types of glass or plastic containers have been used successfully as culture containers — milk dilution bottles (Kimax, U.S.), plastic flasks, or Leighton tubes have all been used. About 10,000 protoscoleces were used per

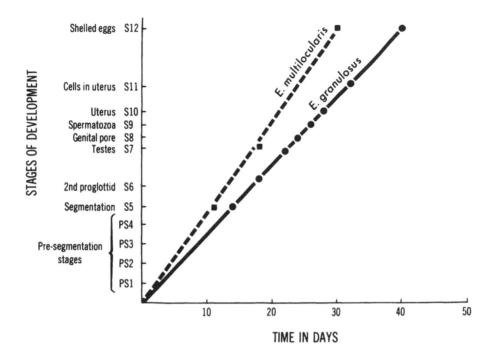

FIGURE 28. Notional diagram comparing the growth of *Echinococcus granulosus* (sheep strain) and *E. multilocularis* in the dog. (From Smyth, J. D., *Angew. Parasitol.*, 20, 137, 1979. With permission.)

culture vessel with 20 ml of liquid medium over the solid base. Cultures are incubated at 38.5°C (dog body temperature) preferably in a water bath with discontinuous shaking (2-min shaking three times per hour) or in a slow revolving roller tube system. The initial pH was 7.4 and the medium was gassed with 10% O_2/5% CO_2/85% N_2, but there is no substantial evidence that this gas mixture is critical or optimum for the system. Medium was renewed every 3 d.

3. Criteria for Assessing Development

In the dog (Figure 28), maturation normally requires 40 d (exceptionally 35 d) and segmentation takes place at 14 d. It is thus important to be able to recognize the small but gradual changes which take place up to segmentation, so that the success of a particular culture system under test can be evaluated as early as possible during cultivation. For convenience, development up to segmentation has been divided into stages as follows:

- Presegmentation Stage 1 (Figure 29, PS.1; see Figure 33A) — A protoscolex immediately after evagination with bile
- Presegmentation Stage 2 (Figure 29, PS.2; see Figure 33B) — Calcareous corpuscles become less dense and indistinct outlines of excretory canals appear
- Presegmentation Stage 3 (Figure 29, PS.3; see Figure 33C) — Excretory canals and posterior excretory bladder become clear; calcareous corpuscles now almost disappeared
- Presegmentation Stage 4 (Figure 29, PS.4 Figure see 33D) — "Banding" stage, in which the "pinching off" of a proglottis becomes evident, although no distinct partition is visible
- Segmentation Stage 5 (Figure 29, S.5) — The first proglottis forms
- Segmentation Stage 6 (Figure 29, S.6) — The second proglottis forms
- Segmentation Stage 7 (Figure 29, S.7) — Testes appear

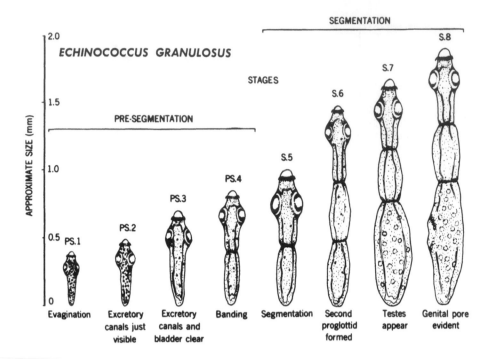

FIGURE 29.   *Echinococcus granulosus*, developmental stages up to genital pore development. (From Macpherson, C. N. L. and Smyth, J. D., *Int. J. Parasitol.*, 15, 137, 1985. With permission.)

- Segmentation Stage 8 (Figure 29, S.8; Figure see 33E) — Genital pore evident
- Segmentation Stage 9 — Spermatozoa present (not shown)
- Segmentation Stage 10 — Uterus evident
- Segmentation Stage 11 — Cells in uterus
- Segmentation Stage 12 — Shelled eggs in uterus

4. Results

Culture of hydatid isolates from sheep — In common with results with other helminths, much intra- and intervariation occurred. Differences were probably due to a variety of factors, such as unevenness in seeding cultures, contaminants, differences in culture vessels, mechanical failure of equipment, and (especially) variation in the hydatid samples and culture components, such as sera, yeast extract, bile salts, and basic chemicals. As is widely recognized, fetal calf serum proved to be the most variable component, its growth properties varying with each sample. Our experience was to test a number of samples and, if feasible, purchase and deep freeze a large quantity of a sample which gave good growth. The results of *in vitro* cultures over a period of some 3 years are summarized in Figure 30. In all these experiments, only one single culture segmented in 15 d (compared with 14 d in the dog); approximately 50% took between 16 to 22 d and the rest took more than 20 d.[88] Thus, most *in vitro* cultures lagged behind development *in vivo*, but in the best results, this lag was only 1 to 4 d, but in many, it was 50% behind that in the dog. The major differences between *in vivo* and *in vitro* worms was the failure of self-insemination to take place in the latter. It has been shown[108] that self-insemination takes place in dog worms (Figure 31) and although all sorts of experimental techniques have since been tried, it has not yet been possible to induce self-insemination *in vitro*.[109] This is in contrast with the situation in *Schistocephalus solidus* in which insemination was induced by cultivation within cellulose tubing (See Figure 6). The culture tube system, described above, can also be effectively replaced by the circulating "lift" system shown in Figure 32. This works very effectively and medium

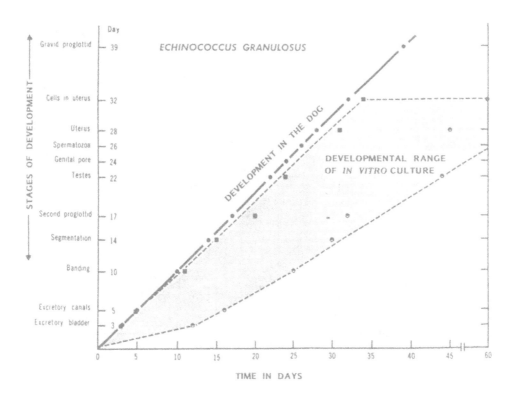

FIGURE 30. *Echinococcus granulosus*, range of development *in vitro* achieved over many years, compared with that in the dog; somewhat notional. (From Smyth, J. D. and Davies, Z., *Int. J. Parasitol.*, 4, 631, 1974. With permission.)

normally needs renewing once a week. A battery of these can be connected together using a single gas pipeline. It is partly based on a system developed by Berntzen[56] for *H. nana*. The above results were all obtained using hydatid material of sheep origin. It has since been demonstrated that similar results can be obtained with isolates from buffalo, goats, camels, cattle, and man,[110] but not horse which behaves entirely differently from any other isolate and fails to grow *in vitro*. This situation is discussed further below.

Culture of hydatid isolates from horse — The application of *in vitro* culture has revealed unexpected biological and biochemical differences between hydatid isolates of different origins. When the techniques used above were used with protoscoleces of horse origin, quite unexpectedly the organisms failed to strobilate *in vitro*. This was at first attributed to suspected faults in techniques or in media components. When, however, after 2 years of experiments involving some 200 cultures, horse material failed to strobilate, it was realized that isolates of *E. granulosus* from horse represented a different "strain" from that of sheep with probable nutritional requirements different from sheep and other isolates such as buffalo, goats, etc.[12] Further investigations have revealed that substantial physiological and biochemical differences exist between horse isolates and sheep.[8,13]

Anomolous monozoic development — In the course of culturing *E. granulosus* protoscoleces (of sheep origin) to sexual maturity, a proportion of organisms in one culture developed sexually mature genitalia without becoming segmented into proglottides, i.e., they were "monozoic" in form.[95] These monozoic forms were miniature in size compared with fully segmented specimens from dogs or from normal cultures, but they appeared to contain a full complement of genitalia. This unusual result points to independent control of somatic and sexual differentiation. Monozoic organisms similar to those developed in *E. granulosus* cultures developed much more readily in cultures of the related species *E. multilocularis* (see Figure 36).

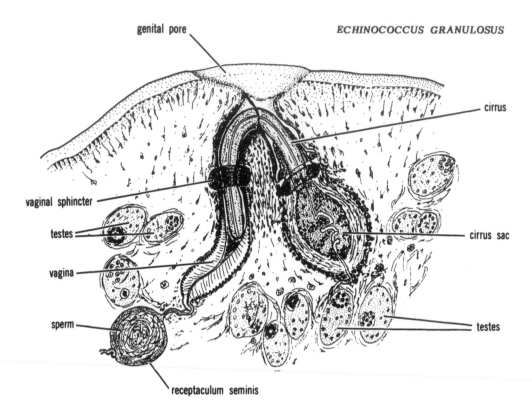

FIGURE 31.   *Echinococcus granulosus*, self-insemination as seen in whole mount of mature worm from dog. (After Smyth, J. D. and Smyth, M. M., *J. Helminthol.*, 43, 383, 1969. With permission.)

F. *IN VITRO* CULTURE OF PARTLY DEVELOPED *IN VIVO* WORMS[111]
1. General Comment

As emphasized above, adult worms grown in the above culture systems fail to undergo self-insemination and therefore do not produce fertile eggs. However, it is a relatively simple matter to obtain fertile eggs by commencing with partly matured (i.e., inseminated) *in vivo* worms and complete their maturation *in vitro*.[111] In the sheep isolate, insemination has been shown to take place at about 22 d,[97] but worms a few days later are recommended for use. As the shelled eggs produced by this technique are highly infective, due safety precautions should be adopted throughout.

2. Technique for Obtaining Gravid Worms

1.   Dogs are infected by feeding protoscoleces in the usual way and are autopsied about 5 d before shelled eggs are due to appear in the uterus. In the sheep isolate of *E. granulosus*, this is about 35 d, as the prepatent period in dogs is generally accepted to be about 39 to 40 d. However, great care must be taken in selecting this time, as it is known that the isolate from Swiss cattle has a prepatent period of 35 d[99] and other isolates may show similar differences.
2.   The infected gut is cut into 4- to 6-in. lengths, sliced open and placed in large glass containers of (preheated) BSS maintained in a water bath or incubator. Worms free themselves from the gut wall within about 10 to 15 min and can be picked up with a Pasteur pipette.
3.   The living worms are transferred to vials of sterile BSS with the usual antibiotics and washed repeatedly, by allowing to settle or on a roller tube system, with frequent changes of media. In this way, sterile, partly matured worms may be obtained.

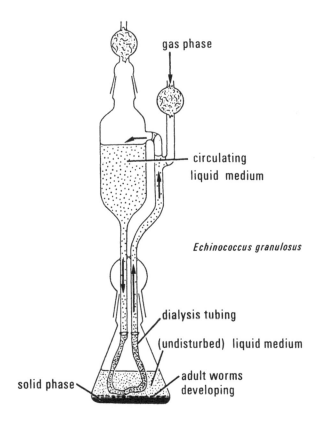

FIGURE 32. *Echinococcus granulosus*, a stationary, circulating "lift" for the cultivation of protoscoleces to adult worms. The dialysis tubing separates the circulating medium from that in the lower flask. This results in the interface between the worm and the solid phase remaining undisturbed, but nutrients and waste materials can be exchanged via the dialysis membrane. Up to six such systems can be connected to a single gas pipeline, providing a battery of cultures. (After Smyth, J. D. and McManus, D. P., *The Physiology and Biochemistry of Cestodes*, Cambridge University Press, Cambridge, 1989. With permission.)

4. These worms are then transferred into the *in vitro* culture systems described above and allowed to complete their development *in vitro* until egg production begins.

Thus by the use of this technique, worms gravid with infective eggs may be obtained and handled within the safety of a culture tube.[111] This method, or minor variations of it, has also been used successfully by Kumaratilake and Thompson[112] for obtaining fertile eggs on this species and it has also been used for obtaining fertile eggs of *E. multilocularis*.[113]

G. *IN VITRO* CULTURE: ONCOSPHERE TO EARLY CYSTIC LARVA[114]

The oncosphere of *E. granulosus* has been grown to an early cystic larva some six times the size of the original oncosphere (see Figure 37), but growth to a fully developed cyst has not yet been achieved.[114] The technique for this, which was more successful with other taeniid larva, is described in Section XIII.B.1, where cultivation of oncospheres of *Taeniidae* spp. is considered.

H. EVALUATION

Some of the techniques described above have now been developed to the stage where

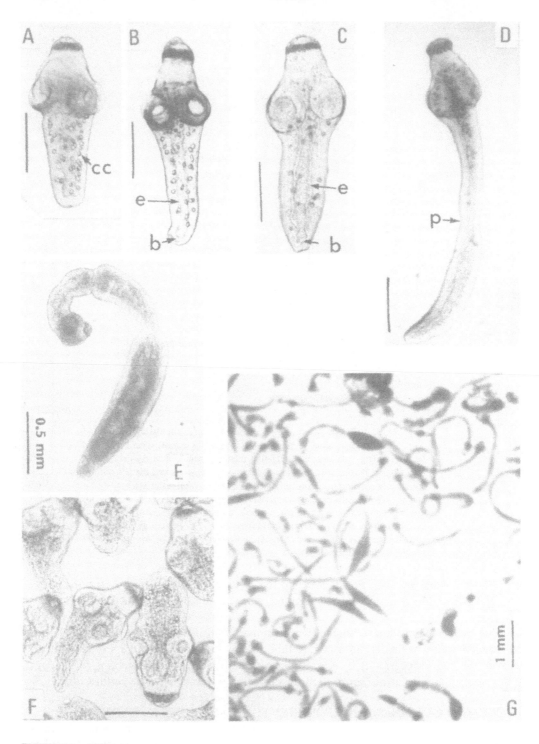

FIGURE 33.   *Echinococcus granulosus, in vitro* culture of adult worm from protoscoleces from sheep cysts (except F). Unlabeled scale bars = 100 μm. For stages see Figure 29. (A) Stage PS.1, freshly evaginated protoscolex; note calcareous corpuscles (cc); (B) intermediate stage between PS.2 and PS.3, calcareous corpuscles disappearing and excretory canals (e) and bladder (b) becoming visible; (C) Stage PS.3, excretory canals (e) and bladder (b) now very evident; (D) Stage PS.4, "banding" stage, with partition (p) of first proglottid appearing; (E) sexual mature adult worm with three proglottides; (F) horse isolate, Stage PS.3, 8-d culture; no further development occurs even after prolonged culture; (G) 66-d culture; note variation in individual development. (From Smyth, J. D. and Davies, Z., *Int. J. Parasitol.*, 4, 631, 1974. With permission.)

they can be reliably used routinely to develop normal (or near-normal) adult strobila of *E. granulosus* and *E. multilocularis*. It must be emphasized, however, that satisfactory results are very dependent on the quality of the constituents of the medium; and in this respect, variation in the growth-promoting properties of the fetal calf serum component is well documented and different samples are likely to produce different results. It must also be emphasized that the growth rate *in vitro* lags substantially behind that *in vivo* and the long culture periods required, often exceeding several months, is inclined to discourage workers.

The major outstanding problem in these culture systems remains the failure to induce self-insemination *in vitro* with the result that fertile eggs are not produced. This is likely to be a spatial relations problem (as in *Schistocephalus*), but other factors such as nutrition may also be involved. The solution to this problem remains a major challenge in this field.

XII. *ECHINOCOCCUS MULTILOCULARIS*

Cultured by: Smyth;[89] Smyth and Davies;[115] Smyth and Barrett;[116] Barrett[117]

Definitive hosts: wild, feral, and domestic carnivores, especially foxes of the genera *Vulpes* and *Alopex*[98]

Experimental hosts: cats,[118] dogs[113]

Prepatent period: 30 d[119]

Intermediate hosts: multilocular hydatid cysts in man and numerous species of wild rodents, especially those of the genera *Microtus, Lemmus,* and *Clethrionomys*[120]

Distribution: mainly Europe (especially France and Germany), Japan, subarctic Islands (especially St. Lawrence Island), North America, but distribution localized[120]

A. GENERAL BIOLOGY

Recent work on the *in vitro* culture of *Echinococcus multilocularis* has been reviewed by Smyth and McManus,[8] Smyth,[96] and Howell.[7]

1. Speciation

Echinococcus multilocularis is well known to be the causative organism of (multilocular) hydatid disease, a highly pathogenetic zoonotic disease of man and wild animals, caused by the larval stage or hydatid cyst. Unlike *E. granulosus*, which generally forms a well-defined unilocular cyst, which can often be removed surgically, the cysts of *E. multilocularis* are composed of numerous small vesicles, containing protoscoleces, which are extremely proliferative and invasive to host tissues. The cysts are inoperable in man and invariably result in death. There is a vast literature on the biology, epidemiology, and pathology of this organism; the early work has been reviewed by Smyth[100,102] and more recent studies have been reviewed by Thompson.[101,104] The systematics of the species do not appear to be as complex as those of *E. granulosus*, although evidence of different strains is beginning to emerge.[120]

2. Life Cycle

The life cycle generally resembles that of *E. granulosus* and will not be discussed in detail here, the major difference being that rodents, and not ungulates, serve as intermediate hosts. The adult worm is also much smaller (1.2 to 3.7 mm) than *E. granulosus* (1.5 to 6 mm), although their size range overlaps.[11] There are also minor differences in morphology.

Like *E. granulosus*, the cysts can differentiate into secondary cysts (in a tissue site) or into adult worms (in an intestinal site). Unlike *E. granulosus*, however, the cysts grow very rapidly in laboratory rodents which makes this species a useful — if somewhat dangerous — experimental organism.

B. SOURCE AND HANDLING MULTILOCULAR HYDATID CYST MATERIAL
1. Warning

E. multilocularis cystic material must be handled with the greatest of care and it is recommended that all those concerned with these operations (or even watching) should wear protective glasses and rubber gloves. It is strongly recommended that when animals are being dissected to remove cysts (see below) that the procedure should be carried out in a glass chamber with arm holes or, at the very least, under a glass or plastic shield held over the site of dissection between the animal and the operator. This is essential for safety, as the hydatid fluid in the cystic vesicles is often under considerable hydrostatic pressure and may squirt into the face or eyes of the operator. Care should also be taken when injecting experimental animals with protoscoleces, as an accidental scratch with a needle, or spillage into a cut in the skin, could result in the protoscoleces becoming established and proliferating in the tissue of the operator.

2. Maintenance in Laboratory Animals

E. multilocularis is readily maintained in cotton rats (*Sigmodon hispidus*) or gerbils (*Meriones unguiculatus*) by intraperitoneal injection.[121] Cotton rats are the better hosts in that the host tissues do not become so heavily calcified as in gerbils. Innocula for serial passage are easily prepared by chopping some cyst material into small pieces with a scissors (with precautions as outlined above!) in a little saline and pressing the fragments through a small fine-mesh sieve. This produces a suspension of protoscoleces plus much cellular and calcified debris. About a 0.2-ml injection per animal is adequate. The cysts take about 4 to 6 months to become fertile in cotton rats.

C. *IN VITRO* CULTURE: PROTOSCOLECES TO ADULT WORMS
1. Obtaining Protoscoleces from the Host[88,89]

Cystic material of *E. multilocularis* is dense and jelly-like and the surrounding host tissue often heavily calcified and therefore requires treatment with a much stronger pepsin solution (0.5%) than those of *E. granulosus* and digestion for a much longer time. A great deal of debris is also produced and the separation of this from the protoscoleces requires special care. The following procedure (Figure 34) is recommended:

1. Shave the abdomen of the animal and paint it with 1% iodine in 95% ethanol; allow to dry; and repeat.
2. Remove the surface skin, and then cut open the body wall to expose the body cavity.
3. Carefully dissect the cyst material into a weighed beaker or petri dish and transfer about 10 g of cysts into prewarmed 200 ml of sterile pepsin (5 mg/ml; 1:10,000) in Hanks' BSS, at a pH 2.0, in a sterile Waring blender. Alternatively, a tubular homogenizer (such as the Internationale Laboratoriums Apparate O GH 84) can be used.
4. Homogenize for 10 to 15 s only first, and then again, if necessary; too-long homogenizing may produce excess frothing. The solution immediately becomes alkaline due to the release of calcium corpuscles from the host tissue. Readjust the pH to 2.0 with 5 *N* HCl.
5. Transfer to a covered, sterile 500-ml beaker, stirring the mixture with a magnetic stirrer on a hot plate or in an incubator at 38.5°C, continually adjusting the pH to 2.0 with 5 *N* HCl. The pH may take some time (5 to 15 min) to stabilize. When the pH finally stabilizes at 2.0, allow digestion to proceed for a further 30 to 60 min, depending on the nature of the material.
6. After digestion, pour through a sterile plastic or metal sieve into another beaker. By frequent washings in Hanks' BSS, followed by repeated removal of supernatant debris, an almost pure suspension of protoscoleces can be obtained.

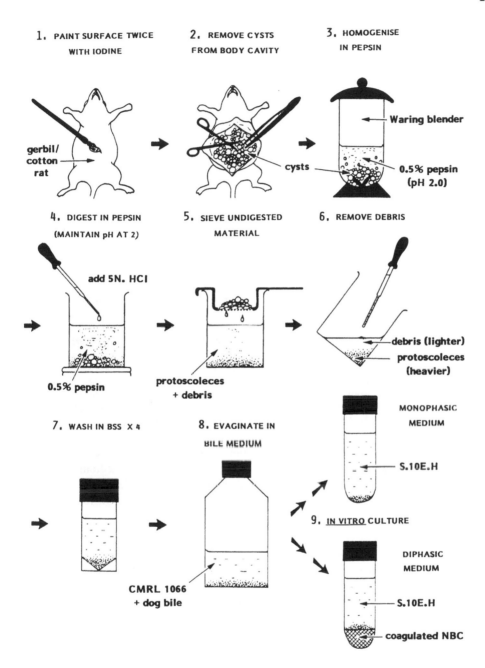

1. PAINT SURFACE TWICE WITH IODINE
2. REMOVE CYSTS FROM BODY CAVITY
3. HOMOGENISE IN PEPSIN

gerbil/cotton rat

Waring blender

cysts

0.5% pepsin (pH 2.0)

4. DIGEST IN PEPSIN (MAINTAIN pH AT 2)
5. SIEVE UNDIGESTED MATERIAL
6. REMOVE DEBRIS

add 5N. HCl

0.5% pepsin

protoscoleces + debris

debris (lighter)
protoscoleces (heavier)

7. WASH IN BSS X 4
8. EVAGINATE IN BILE MEDIUM

MONOPHASIC MEDIUM

S.10E.H

9. IN VITRO CULTURE

CMRL 1066 + dog bile

DIPHASIC MEDIUM

S.10E.H

coagulated NBC

FIGURE 34. *Echinococcus multilocularis*, protocol for setting up *in vitro* cultures based on the technique of Smyth.[89] (Diagram from Barrett, N. J., Ph.D. thesis, University of London, London, 1984. With permission.)

7. The pure suspension of protoscoleces should finally be transferred to a screw-top vial and washed four times, rotating the vial for 10 min on a rotator during each washing.

All apparatus and solutions used above should be sterile.

2. Setting Up Cultures[88,89]

The same medium as used for *E. granulosus* (see Table 13) suffices for *E. multilocularis* as do the same culture vessels and the general protocol. However, because (unexpectedly,

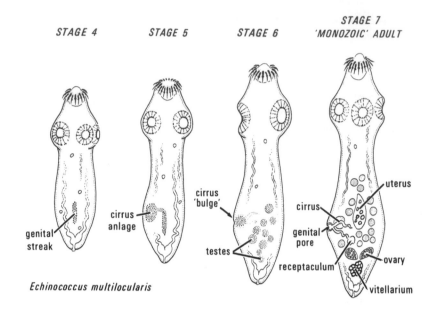

FIGURE 35.   *Echinococcus multilocularis*, development of "monozoic" forms *in vitro*; early stages (1 to 3) not shown. Such forms apparently develop only under abnormal conditions of culture; they fail to undergo strobilization, but develop a complete set of male and female genitalia. Note development of "cirrus bulge" due to development of genitalia, without corresponding somatic development (see also Figure 36C). Similar monozoic forms have occasionally been found in cultures of *E. granulosus*. (From Smyth, J. D. and Barrett, N. J., *Rev. Iber. Parasitol.*, 39, 39, 1979. With permission.)

see results) strobilization will take place in a monophasic medium (as well as a diphasic medium), a simple roller tube system, with 10 to 20 ml of liquid Medium S.10E.H in Leighton tubes or screw-top vials, with 0.05 ml of settled protoscoleces per culture, has eventually proved to be the most efficient system and is recommended for routine use.

3. Results

The first remarkable feature of the *in vitro* development was the fact that strobilization took place in monophasic (liquid) medium, in contrast to the culture of *E. granulosus* which required a diphasic system.

The second unusual feature was the fact that — again in contrast to *E. granulosus* — in a high percentage of cultures, early segmentation was suppressed and unsegmented and sexually mature "monozoic" forms developed (Figures 35 and 36C). This was particularly the case in the early experiments by Smyth and Davies[115] and Smyth and Barrett[116] in which monozoic development was the dominating pattern of development.

The monozoic forms have a characteristic triangular shape, with the cirrus region bulging out, as the somatic tissue does not appear to grow sufficiently quickly relative to the developing genitalia. During the early stages of culture, the development is similar to that in *E. granulosus*, but at about 21 d, a translucent, circular area appears about half way down the body. This has been called the "cirrus patch"[89] which eventually develops into a cirrus and genital pore. After about 25 d of culture, testes appear and by about 31 d the uterus cavity can be identified. In the best cultures, the uterus contains ova and vitelline cells, but like *E. granulosus*, fertile eggs are not produced, apparently due to the failure of insemination to take place.

In later *in vitro* experiments,[89,96] however, using slightly modified roller tube systems and media buffered with HEPES, development more nearly approached that *in vivo*, with

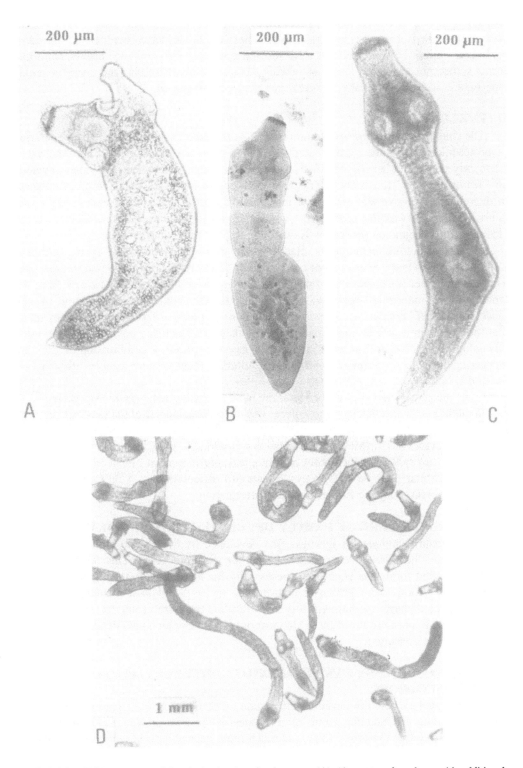

FIGURE 36. *Echinococcus multilocularis, in vitro* development. (A) Aberrant early culture with additional scolex bud forming; (B) normal adult with two proglottides, approaching sexual maturity; (C) "monozoic" form in 32-d-culture. Note triangular profile with "cirrus bulge" due to development of genitalia, without corresponding somatic growth; for details of morphology, see Figure 35; (D) 70-d culture showing a variety of forms developed, including some monozoic forms together with adult with worms 2 to 3 proglottides. (From Smyth, J. D., *Angew. Parasitol.*, 20, 137, 1979. With permission.)

two-segment and three-segment worms developing, although the interproglottid divisions were rarely clearly formed (see Figure 36D). These organisms have been called "pseudosegmented" forms.[116] When cultures were allowed to develop for very long periods, abnormal forms sometimes appeared, the most striking of these being a strobila with an extra scolex (see Figure 36A); occasionally, several extra scoleces developed.[116]

D. EVALUATION

The chief problem in preparing cultures of *E. multilocularis* is that of the initial separation of protoscoleces from the cystic material removed from the rodent host. This material is frequently dense and heavily calcified, as explained above, and when macerated develops an "ice cream-like" consistency. Treatment with pepsin frees the contained protoscoleces, in due course, but a certain amount of manipulative skill is required to separate the tissue debris during the washing procedures, without losing too much of the parasite material. However, this separation procedure is readily mastered with some practice.

The actual culture technique is relatively simple, and if a roller tube system utilizing Leighton tubes is used, development can be readily followed under an inverted microscope. That development achieved *in vitro* does not quite approach that *in vivo* is evident from the tendency for "monozoic" forms to develop and also the failure for insemination to take place, as with *E. granulosus*. However, the failure to produce fertilized eggs does mean that the organism is safe to handle *in vitro*, a result which has many advantages for routine physiological or biochemical work. If fertile eggs are required, a combination of growing worms in dogs up to insemination and then culturing them *in vitro* can be utilized (see Section E below).

From the physiological point of view, the most intriguing result from the culture of *E. multilocularis* is the fact that its protoscoleces will differentiate in a strobilar (sexual) direction in a monophasic medium, whereas *E. granulosus* will only do so in a diphasic medium. At present, no evidence is available to explain this remarkable phenomenon, which suggests that each species responds to different physiological, nutritional, or biochemical "triggers" to initiate strobilar (i.e., sexual) or cystic (asexual) differentiation. This clearly is a fundamental problem requiring further detailed investigation.

E. *IN VITRO* CULTURE OF PARTLY DEVELOPED *IN VIVO* WORMS

The original technique for this was first developed by Smyth and Howkins[111] for *E. granulosus* and has already been described earlier (Section XI.F). This technique, in a slightly modified form, has been used successfully for the production of eggs by *E. multilocularis*.[113] Worms were grown in dogs for 20 to 21 d and then matured *in vitro* in monophasic or diphasic medium for 8 d when fertile eggs were produced, although these were fewer than those produced by *in vivo* worms. For further technical details, the original paper should be consulted.

F. *IN VITRO* CULTURE: CYSTIC (ASEXUAL) DIFFERENTIATION OF PROTOSCOLECES

As indicated earlier, the germinal membrane of *E. multilocularis* is extremely proliferative, invading and branching out in the tissues of the host *in vivo*. It is therefore not surprising to find that the cystic stages are much more amenable to *in vitro* culture than the corresponding stage of *E. granulosus*.

There appears to be no recent work carried out on cystic development *in vitro*, although a number of interesting results were carried out by early workers. Thus, Yamashita et al.[122] cultured protoscoleces in a basic medium of lactalbumin hydrolysate in Hanks' BSS reinforced with various nutrients, such as bovine serum, bile and liver extracts. Only infertile cysts were formed, the developmental routes following closely that described for *E. gran-*

ulosus in Figure 27, i.e., they either became vesicular or formed posterior bladders before forming miniature hydatid cysts with laminated membranes. More successful results were achieved using fragments of germinal membrane. Thus, Rausch and Jentoft[123] cultured pieces of germinal membrane in a basic medium of 40% ascitic fluid in Hanks' BSS plus HeLa cells and nutrients such as vole embryo extract, but, unfortunately, full details of the technique used were not provided. In these cultures, tissues proliferated and produced vesicles after 29 d of culture and by 55 d some 22 protoscoleces were present in the vesicles; the latter proved infective to voles on intraperitoneal injection.

Somewhat similar experiments were carried out by Lukashenko,[124] who utilized protoscoleces, separate vesicles, and minced tissues. In a medium of Parker 199, supplemented with cotton rat embryo extract, bovine serum, and lactalbumin hydrolysate, vesicles developed after 38 d, a laminated membrane developed in 54 d, and protoscoleces developed after 99 d. The latter proved infective to cotton rats on intraperitoneal injection.

XIII. VARIOUS TAENIIDAE

A. GENERAL REVIEW

With the exception of *Echinococcus* spp., dealt with in the previous section, the taeniid cestodes have proved to be unusually difficult to culture *in vitro*, and to date no gravid adult of any *Taenia* species has been grown from a cysticercus. The nearest to achieving this goal has been the results of Osuna-Carrillo and Mascaro-Lazcano,[125] who grew sexually mature *T. pisiformis* from a cysticercus, but failed to obtain normal development or fertile eggs. Most progress has been made with the development of oncospheres of several *Taenia* species and *Echinococcus granulosus* to nearly fully developed cysticerci or early cystic stages. The most successful of these techniques are discussed below. Results of less-successful experiments, which were not followed up later, are not reviewed here; most of these are referenced in Taylor and Baker[3] and Voge.[4]

B. *IN VITRO* CULTURE: ONCOSPHERES TO CYSTIC LARVAE

1. General Comments

Techniques for the *in vitro* culture of taeniid oncospheres to cysticerci were first described by Heath and Smyth,[114] the species involved being *E. granulosus*, *Taenia hydatigena*, *T. ovis*, *T. pisiformis*, and *T. serialis*. The most successful results were obtained with *T. pisiformis*, oncospheres of which were grown to the cystic stage with development of hooks and suckers. Some early cystic development was obtained with the other species studied. Growth of these species is shown in Figure 37. A remarkable feature of the development of *T. pisiformis* was the fact that some larvae of *T. pisiformis* underwent transverse or longitudinal fission, a phenomenon probably due to the cellular "organizer" region becoming divided by the action of the centrifugation process involved in the early cultivation procedures and one worthy of further investigation. Later improvement of the initial technique by Heath[126] resulted in the development of almost normal cysticerci of *T. hydatigena*, *T. ovis*, *T. serialis*, and *T. taeniaeformis*; these had fully developed suckers, but hook formation was incomplete in that blades were formed but not hardened and shafts were absent. The application of these or similar techniques to *T. saginata* resulted in only limited success.[127,128] The early differentiation of the oncosphere of *E. granulosus in vitro* has also been examined in some detail.[129]

2. Technique of Heath[126]

The technique is essentially a modification of that of Heath and Smyth.[114] It basically involves:

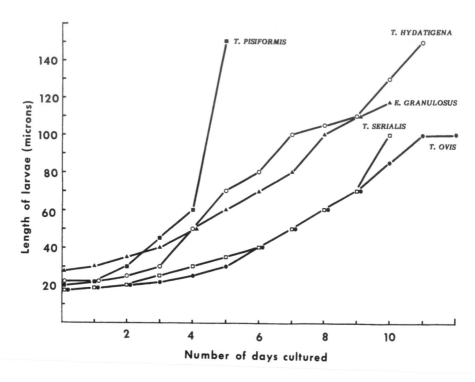

FIGURE 37.   A comparison of the early *in vitro* development of cystic larvae of *Taenia pisiformis*, *T. hydatigena, T. serialis, T. ovis*, and *Echinococcus granulosus* from oncospheres. (From Heath, D. D. and Smyth, J. D., *Parasitology*, 61, 329, 1970. With permission.)

1.   Teasing the eggs from a gravid proglottis
2.   Washing with BSS + antibiotics
3.   Treatment with pepsin
4.   Hatching and activating in pancreatin plus bile
5.   Washing to sterility
6.   Culturing in a tissue culture medium (Parker 858, NCTC 135, etc.) with the addition of fetal calf or other serum and red cells

Obtaining hatched oncospheres from eggs — The following technique is based on Heath:[126]

1.   Obtain gravid proglottides by purging or autopsying infected dogs.
2.   Wash worms with BSS until visually clean and chop worms in BSS and sieve worm debris from eggs.
3.   Allow eggs to settle overnight at 4°C and then treat with Hibitane (I.C.I. Ltd., London, active constituent chlorhexidine gluconate) for 20 min followed by washing three times in distilled water and storing at 4°C if required.
4.   Hatch and activate eggs by concentrating the requisite number of eggs by centrifugation in screw-capped culture tubes (Pyrex No. 9826) and pretreat with the artificial gastric solution (Solution A, Table 14) for 1 h at 37 to 39°C followed by centrifugation at 3000 revolutions per minute for 3 min.
5.   Draw off the supernatant leaving 2 to 3 mm of liquid above the eggs and add artificial intestinal solution (Solution B, Table 14), screwing the cap on quickly to retain CO_2; hatching and activation should take place, as in *E. granulosus* (Figure 38A).
6.   Centrifuge the solution and wash three times in the culture medium with added penicillin G (1000 U/ml), streptomycin (100 μg/ml), and nystatin (Squibb) 1000 U/ml.

TABLE 14
Artificial Gastric and Intestinal Solutions for Hatching of Taeniid Eggs

Solution A — Gastric solution
    1% pepsin (1:10,000 Sigma) + 1% concentrated HCl + 0.85% NaCl in distilled water
Solution B — Intestinal solution
    1% pancreatin (porcine 3 × N.F. Sigma) + 1% NaHCO$_3$ + 5% of the appropriate bile (sheep
    or rabbit — collected sterile) in distilled water (Bile was stored at $-10°C$ until required.)

From Heath, D. D. and Smyth, J. D., *Parasitology*, 61, 329, 1970. With permission.

7.   Transfer organisms to culture medium; in addition to activated oncospheres, the culture will now contain unactivated oncospheres (i.e., those still within their oncosphere membranes) and numerous keratin blocks from the disintegrated embryophores of the eggs (see Figure 38A).

Medium and culture procedure — The culture medium consisted of NCTC 135, with the dextrose level raised to 400 mg/100 ml with antibiotics as above, plus 1% washed, packed rabbit blood cells plus fetal calf serum. The concentration of fetal calf serum was 50% for the first 7 d, 20% for the next 14 d, and 10% thereafter. The gas phase was air and the temperature was 37 to 39°C and media were changed twice weekly. The culture vessels were 30- or 250-ml disposable plastic flasks (Falcon). Generally only about 10% of eggs yielded developing oncospheres, but the number varied with the source and age of the eggs. Optimum development was achieved when 0.1 ml of culture medium was provided for each developing oncosphere. The volume of fluid per larva was progressively doubled at each change by removing 50% of the larvae, until each larva ultimately received 10 ml of culture fluid at each change.

3. Results

The account below summarizes the best results obtained with this technique, or the earlier one of Heath and Smyth[114] on which it is based, or subsequent modifications developed for specific species.

T. pisiformis, T. ovis, T. taeniaeformis, T. hydatigena, and *T. serialis* — Oncospheres of all these species developed into immature cysticerci at a rate approaching that *in vivo* (see Figure 37), although there was a wide range of development within any one culture. As mentioned above, all *T. pisiformis* larvae divided into two organisms, many divisions taking place after 6 d *in vitro*. A few larvae of *T. ovis, T. hydatigena,* and *T. taeniaeformis* developed two or three scoleces, but did not divide in half, and in some *T. serialis* coenuri, scolex anlagen were observed to radiate from more than one site. Although hooks never developed fully, the larvae appeared to be normal in other ways and evaginated under the influence of bile and were resistant to digestion by artificial intestinal fluid plus bile. Pseudostrobilation of the neck region was observed in *T. pisiformis* and *T. taeniaeformis*. Oncospheres of the latter species were also cultured in a series of experiments, independent of those described above, but were generally less successful.

T. saginata — Attempts to culture oncospheres of *T. saginata* have not, in general, been very successful.[127,128] A major difficulty in culturing this species is shortage of suitable material, as only a low percentage of eggs appear to be viable. Thus, Heath and Elsdon-Dew[127] reported that in some worms " . . . although up to 50 terminal proglottides contained shelled eggs, most embryos were not capable of being activated . . . fully mature oncospheres were only obtained in large number after anthelmintic treatment of patients passing gravid segments.'' Eggs also needed careful washing and sterilizing as they were frequently

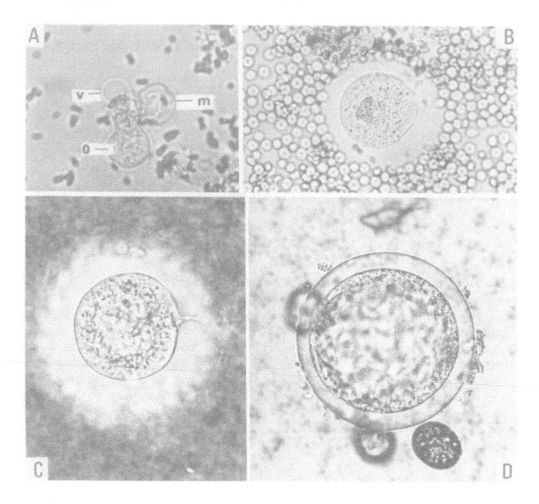

FIGURE 38. *Echinococcus granulosus*, early development *in vitro*.[129] (A) Hatched and activated oncosphere (o); note shed oncospheral membrane (m) and clear vesicle (v) from penetration gland; (B) 24 h *in vitro*; an apparent amorphous layer has formed around the organism; (C) 4 d *in vitro*; refractive bodies move to the periphery of the cyst and begin to differentiate. A central cavity begins to form; (D) 10-d culture in rabbit serum; laminations begin to appear in the amorphous layer. (Micrographs courtesy of Dr. D. D. Heath, Wallaceville Research Centre, MAFTech, Upper Hutt, New Zealand.)

contaminated with Gram-negative bacteria. Only early development *in vitro* was achieved up to cavity formation with some muscle development, the growth rate approximating that occurring *in vivo*.

Somewhat better results were obtained by Machnicka and Smyth,[128] using a culture medium consisting of Medium 858 (Difco, Detroit, MI) plus 20% inactivated fetal calf serum plus antibiotics. Although some 100% of eggs hatched, only about 10% became activated. Growth of larvae took place only for the first 10 d, although many survived for 17 d. The largest larvae measured about 0.08 mm in diameter and showed cells lining the central cavity and developed protrusions which may have represented scolex anlagen. In a parallel experiment, it was demonstrated that activated oncospheres, when injected into mice, which had been immunosuppressed with cyclophosphamide, could develop into early cystic larvae.[128]

Echinococcus granulosus — It must be emphasized that the eggs of this species are highly infective and therefore extremely dangerous to handle. It is strongly recommended that they only be handled in an isolation room specially prepared for such infective material.

Heath and Lawrence[129] provide practical details of the layout and use of such an isolation room.

As mentioned earlier, *E. granulosus* oncospheres were first cultured by a technique similar to that used for the *Taenia* spp. described above. A slightly modified technique was later used by Heath and Lawrence.[129] The media used consisted of NCTC 135, RPMI 1640, or McCoy's 5A (modified) plus 20% serum, either fetal calf serum or serum from 10-week-old rabbits. Rabbit blood cells (1%) were used in some cultures.

The best results were obtained in NCTC 135, fetal calf serum and blood cells, cysts growing to a mean diameter of 16 mm in 120 d, with a maximum of 20 mm. The early development *in vitro* showed some interesting features. Within 24 h, proliferation of oncospheral cells was apparent and a transparent halo of amorphous material appeared around the organism (Figure 38B). At 3 d, refractive bodies began to accumulate in the center of the organisms, and at 4 d (Figure 38C), the refractive bodies move to the outer cyst area and begin to differentiate. The transparent halo became more opaque in rabbit serum by 8 d, and laminations in it became evident in 10 d (Figure 38D). Proliferation of asteroidal cells in the germinal layer appeared later and these cells gradually formed a complex which became fully developed germinal membrane. Cultures were not continued beyond this stage.

C. *IN VITRO* CULTURE: CYSTICERCI TO ADULT WORMS

As mentioned earlier, no worker has succeeded in growing a species of the genus *Taenia* from a cysticercus or comparable larval stage to an adult worm. Serious attempts have only been made with two species, *T. crassiceps*[130] and *T. pisiformis*, with some preliminary experiments reported for *T. saginata*.[131] Because only preliminary techniques are available for these species, they are only described briefly here.

1. *Taenia crassiceps*[130]

Material and culture conditions — The material consisted of cysticerci of KBS and Toi strains of *T. crassiceps* maintained in the laboratory in rodents by intraperitoneal passage. The techniques used were essentially those developed for *Echinococcus granulosus* by Smyth[94] which have since been modified.[88,96] The medium consisted of the diphasic system described in Table 13, with the liquid phase made up of 100 parts of Medium 858 (Difco), 25 parts of inactivated fetal calf serum, and 12.5 parts of 5% yeast extract. To a liter of this was added 22 ml of 30% glucose, 5.0 ml of 2.0% KCl, plus antibiotics (100 IU/ml penicillin G and 100 μg/ml of streptomycin sulfate). The gas phase was either 10% O_2/5% CO_2/85% N_2 or 10% CO_2 in N_2; both systems gave similar results. Worms were cultured in 30-ml plastic culture flasks (Falcon).

Setting up cultures and evagination — Larvae dissected from mice were washed twice in warm Hanks' BSS and then digested in 50 ml of pepsin (10 mg/ml) at pH 1.9 for 20 min with agitation in a water bath. After washing further in three changes of BSS, they were evaginated in BSS plus, 3 mg/ml pancreatin, 0.1 mg/ml sodium taurocholate, and 0.15 mg/ml trypsin, the flask being kept in a shaking water bath at 38.9°C for 3 to 20 h. After washing in warm BSS, larvae were examined in a petri dish and only well-developed larvae used for culture. Culture flasks were maintained at 38.9°C (dog body temperature) in a shaking water bath. Medium was renewed every 48 h.

Results — In the best results, segmentation began within 5 to 7 d and by 18 d, testes and genital pores were formed. Due to tangling of the elongating strobila, abnormal proglottides were formed in some regions; in many the morphology of the proglottis appeared to be comparable with those grown *in vivo*. Figure 39B, C, and D show stained preparations of proglottides from 25-d-old cultured worms, in which the genital pore, genital ducts, ovary, vitellaria, ootype, and early uterus and testes can clearly be seen. Smyth and Rickard[132] repeated some of the above experiments and in several cases, strobila failed to lose their posterior cystic region which continued to bud so that remarkable organisms with an anterior strobilating region and a posterior budding region developed (Figure 39A).

2. *Taenia pisiformis*

Using essentially the same technique, with minor modifications, but using CMRL 1066 instead of Medium 858, Osuna-Carrilo and Mascaro-Lazcano[125] obtained similar but somewhat better results with *T. pisiformis*. Segmentation took place after 8 d with genital anlagen appearing at 11 to 12 d and the first sign of vagina and uterus formation clearly visible after 16 d. Testes were well developed after 29 d and the genital pore, which appeared about day 17, was open by day 22. Some proglottides showed abnormal development with the development of two to three genital pores. By day 32 some worms were fully sexually mature with the cirrus emerging from the pouch, but not ejaculating. Fertile eggs were not formed, presumably due to insemination not taking place.

3. *Taenia hydatigena*

Somewhat comparable results were obtained by Osuna-Carrilo et al.[133] In the best developed forms of this species, the genital pore and early, immature male and female genitalia were formed.

4. *Taenia saginata*

Some preliminary results were obtained with this species by Brandt and Sewell.[131] They used a diphasic system based on those already described above, but with a liquid phase of RPMI-1640, with L-glutamine and 25 mM HEPES buffer plus 10% inactivated fetal calf serum. Best results were obtained with a "disrupted" coagulated serum base (i.e., one broken up with a hypodermic needle). Larvae were evaginated with 10% ox bile and well washed in BSS before culturing.

Although these experiments may be regarded as only preliminary, they were encouraging in that in the best cultures, segmentation took place and the anlagen of the genitalia began to appear. The maximum length of a strobila was 70 mm with about 150 segments formed.

D. *IN VITRO* CULTURE: LARVAL MULTIPLICATION, *T. CRASSICEPS*

A useful technique for maintaining cysticerci of *T. crassiceps* in a budding condition was developed by Schiller,[134] although never formally published. The strain of *T. crassiceps* was not stated.

1. Culture Technique

The medium consisted of a diphasic medium of 70% neutral (Difco) agar plus 30% fresh defibrinated rabbit blood in Hanks' BSS plus penicillin (200 IU/ml) and streptomycin (200 µg/ml) at pH 7.8. Cultures were carried out in 50-ml Ehrenmeyer flasks, 25 ml of agar phase per flask. After gelation, 5 ml of the BSS were added as the liquid phase. Glucose (1.0 mg/ml) was added just before culture, and each flask was inoculated with five subspherical buds about 0.5 mm in diameter. Flasks were plugged with cotton wool and incubated in a Dubnoff shaking incubator (approximately 100 c/min) at 37°C in a gas phase of 5% CO_2 in N_2. Medium was renewed every 72 h.

2. Results

Buds rapidly increased in size, reaching a diameter of 3 to 5 mm in approximately 3 weeks. At this time, scolex development was complete and minute, exogenous buds were evident at the abscolex pole. During the next 3 to 4 weeks in culture, the parent bud enlarged and differentiated into a fully developed cysticercus. By the eighth week, F_1 buds began to

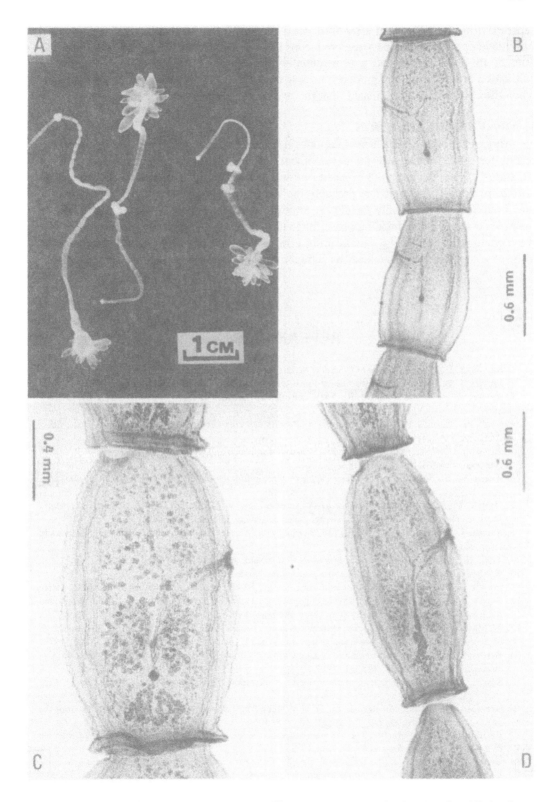

FIGURE 39.   *Taenia pisiformis, in vitro* development.[132] (A) Unusual *in vitro* development of strobilating forms showing budding of posterior bladder; 35-d culture; (B), (C), and (D) maturing proglottides showing early (B) and advanced development (C and D) of genitalia; 25-d culture.[132]

separate from the parent and after 54 d, these F_1 buds progeny had themselves produced fully developed scoleces and were producing F_2 buds. Bud counts indicated that after the 50th d, the reproductive rate was essentially logarithmic. During the longest period of cultivation (162 d) each bud produced no less than 157.3 progeny, nor was there any sign, when the cultures were terminated, that the reproductive rate was declining.

E. GENERAL EVALUATION

Although all stages of a taeniid cestodes have not yet been grown *in vitro*, encouraging results have been made with the development of oncospheres to well-developed (but not "mature") cysticerci and with cysticerci to near-mature adults in a number of species. *T. crassiceps* and *T. pisiformis* are probably the models of choice for further experiments as they both develop reasonably rapidly. *T. crassiceps* has the added advantage that it can be maintained *in vitro* in a budding phase for long periods — and possibly indefinitely — by the technique of Schiller,[134] thus assuring a regular supply of experimental material. It is also, of course, easily maintained, by intraperitoneal transmission, in laboratory rodents.

REFERENCES

1. **Smyth, J. D.,** *The Physiology of Cestodes,* W. H. Freeman, San Francisco, 1969.
2. **Smyth, J. D.,** *An Introduction to Animal Parasitology,* 2nd ed., Edward Arnold, London, 1976.
3. **Taylor, A. E. R. and Baker, J. R.,** *The Cultivation of Parasites in Vitro,* Blackwell Scientific, Oxford, 1968.
4. **Voge, M.,** Cestoda, in *Methods of Cultivating Parasites In Vitro,* Taylor, A. E. R. and Baker, J. R., Eds., Academic Press, London, 1978, 193.
5. **Arme, C.,** Cestoda, in *In Vitro Methods of Parasite Cultivation,* Taylor, A. E. R. and Baker, J. R., Eds., Academic Press, London, 1987, 282.
6. **Evans, W. S.,** The cultivation of *Hymenolepis in vitro,* in *Biology of the Tapeworm Hymenolepis diminuta,* Arai, H., Ed., Academic Press, New York, 1980, 425.
7. **Howell, M. J.,** Cultivation of *Echinococcus* species *in vitro,* in *The Biology of Echinococcus and Hydatid Disease,* Thompson, R. C. A., Ed., George Allen & Unwin, London, 1986, 143.
8. **Smyth, J. D. and McManus, D. P.,** *The Physiology and Biochemistry of Cestodes,* Cambridge University Press, Cambridge, 1989.
9. **Arai, H.,** *Biology of the Tapeworm Hymenolepis diminuta,* Academic Press, London, 1980.
10. **Kennedy, C. R.,** *Ecological Aspects of Parasitology,* North-Holland, Amsterdam, 1976.
11. **Smyth, J. D.,** *An Introduction to Animal Parasitology,* 1st ed., English University Press, London, 1962.
12. **Smyth, J. D. and Davies, Z.,** Occurrence of physiological strains of *Echinococcus granulosus* demonstrated by *in vitro* culture of protoscoleces from sheep and horse hydatid cysts, *Int. J. Parasitol.,* 4, 443, 1974.
13. **McManus, D. P. and Smyth, J. D.,** Differences in the chemical composition and carbohydrate metabolism of *Echinococcus granulosus* (horse and sheep strains) and *E. multilocularis, Parasitology,* 77, 103, 1978.
14. **Berntzen, A. K.,** The *in vitro* culture of tapeworms. I. Growth of *Hymenolepis diminuta* (Cestoda: Cyclophyllidea), *J. Parasitol.,* 47, 351, 1961.
15. **Smyth, J. D.,** Studies on tapeworm physiology. I. Cultivation of *Schistocephalus solidus in vitro, J. Exp. Biol.,* 23, 47, 1946.
16. **Berntzen, A. K. and Mueller, J. F.,** *In vitro* culture of *Spirometra mansonoides* (Cestoda) from the procercoid to the early adult, *J. Parasitol.,* 50, 705, 1964.
17. **Berntzen, A. K. and Mueller, J. F.,** *In vitro* cultivation of *Spirometra* spp. (Cestoda) from the plerocercoid to the gravid adult, *J. Parasitol.,* 58, 750, 1972.
18. **Beis, I. and Barrett, J.,** The contents of adenine nucleotides and glycolytic and tricarboxylic acid cycle intermediates in activated and non-activated plerocercoids of *Schistocephalus solidus* (Cestoda: Pseudophyllidea), *Int. J. Parasitol.,* 9, 465, 1979.
19. **Beis, I. and Barrett, J.,** Oxidative enzymes in the plerocercoids of *Schistocephalus solidus* (Cestoda: Pseudophyllidea), *Int. J. Parasitol.,* 10, 151, 1980.
20. **Smyth, J. D.,** Studies on tapeworm physiology. VII. Fertilization of *Schistocephalus solidus in vitro, Exp. Parasitol.,* 3, 64, 1954.

21. **Smyth, J. D.**, Maturation of larval pseudophyllidean cestodes and strigeid trematodes under axenic conditions; the significance of nutritional levels in platyhelminth development, *Ann. N.Y. Acad. Sci.*, 77, 102, 1959.

22. **Hopkins, C. A. and Smyth, J. D.**, Notes on the morphology and life history of *Schistocephalus solidus* (Cestoda: Diphyllobothriidae), *Parasitology*, 41, 283, 1951.

23. **McCaig, M. L. O. and Hopkins, C. A.**, Studies on *Schistocephalus solidus*. II. Establishment and longevity in the definitive host, *Exp. Parasitol.*, 13, 273, 1963.

24. **Braten, T.**, Host specificity in *Schistocephalus solidus*, *Parasitology*, 56, 567, 1966.

25. **Orr, T. S. C. and Hopkins, C. A.**, Maintenance of *Schistocephalus solidus* in the laboratory with observations on the rate of, and proglottid formation in, the plerocercoid, *J. Fish. Res. Bd. Can.*, 26, 741, 1969.

26. **Clarke, A. S.**, Maturation of the plerocercoid of the pseudophyllidean cestode *Schistocephalus solidus* in alien hosts, *Exp. Parasitol.*, 2, 223, 1952.

27. **Clarke, A. S.**, Studies on the life cycle of the pseudophyllidean cestode, *Schistocephalus solidus*, *Proc. Zool. Soc. London*, 124, 257, 1954.

28. **Smyth, J. D.**, Studies on tapeworm physiology. V. Further observations on the maturation of *Schistocephalus solidus* (Diphyllobothriidae) under sterile conditions *in vitro*, *J. Parasitol.*, 36, 371, 1950.

29. **McCaig, M. L. O. and Hopkins, C. A.**, Studies on *Schistocephalus solidus*. III. The *in vitro* cultivation of the plerocercoid, *Parasitology*, 55, 257, 1965.

30. **Hopkins, C. A. and Sinha, D. P.**, Growth of the fish tapeworm *Schistocephalus solidus* in vitro, *Parasitology*, 55, 19P, 1965.

31. **Smyth, J. D.**, Studies on tapeworm physiology. II. Cultivation and development of *Ligula intestinalis in vitro*, *Parasitology*, 38, 173, 1947.

32. **Smyth, J. D.**, Studies on tapeworm physiology. IV. Further observations on the development of *Ligula intestinalis in vitro*, *J. Exp. Biol.*, 26, 1, 1949.

33. **Flockart, H. A.**, *Ligula intestinalis* (L. 1758) — *In Vitro* and *In Vivo* Studies, Ph.D. thesis, No. D.28674/79, University of London, London, 1979.

34. **Yamaguti, S. Y.**, *Systema Helminthum*, Vol. II, *The Cestodes of Vertebrates*, Interscience, New York, 1959, 187.

35. **Orr, T. S. C. and Hopkins, C. A.**, Maintenance of the life cycle of *Ligula intestinalis* in the laboratory, *J. Fish. Res. Bd. Can.*, 26, 2250, 1969.

36. **Arme, C., Griffiths, D. V., and Sumpter, J. P.**, Evidence against the hypothesis that the plerocercoid larvae of *Ligula intestinalis* (Cestoda: Pseudophyllidea) produces a sex steroid that interferes with host reproduction, *J. Parasitol.*, 68, 169, 1982.

37. **McManus, D. P. and Sterry, P. R.**, *Ligula intestinalis*: intermediary carbohydrate metabolism in plerocercoids and adults, *Z. Parasitkd.*, 67, 73, 1982.

38. **Matskási, I. and Németh, I.**, *Ligula intestinalis* (Cestoda: Pseudophyllidea): studies on the properties of proteolytic and protease inhibitor activities of plerocercoid larvae, *Int. J. Parasitol.*, 9, 221, 1979.

39. **Mueller, J. F.**, The laboratory propagation of *Spirometra mansonoides* (Mueller, 1935) as an experimental tool. II. Culture and infection of the copepod host, and harvesting the procercoid, *Trans. Am. Microsc. Soc.*, 78, 245, 1959.

40. **Mueller, J. F.**, The laboratory propagation of *Spirometra mansonoides* as an experimental tool. III. *In vitro* cultivation of the plerocercoid larva in a cell-free medium, *J. Parasitol.*, 45, 561, 1959.

41. **Mueller, J. F.**, The biology of *Spirometra*, *J. Parasitol.*, 60, 3, 1974.

42. **Mueller, J. F.**, The laboratory propagation of *Spirometra mansonoides* as an experimental tool. I. Collecting, incubation and hatching of the eggs, *J. Parasitol.*, 45, 353, 1959.

43. **Mueller, J. F.**, A *Diphyllobothrium* from cats and dogs in the Syracuse district, *J. Parasitol.*, 21, 114, 1935.

44. **Mueller, J. F.**, The laboratory propagation of *Spirometra mansonoides* (Mueller, 1935) as an experimental tool. VII. Improved techniques and additional notes on the biology of the cestode, *J. Parasitol.*, 52, 437, 1966.

45. **Phares, C. K.**, Plerocercoid growth factor: a homologue of human growth hormone, *Parasitol. Today*, 3, 346, 1987.

46. **Berntzen, A. K.**, A controlled culture environment for axenic growth of parasites, *Ann. N.Y. Acad. Sci.*, 139, 176, 1966.

47. **Schiller, E. L.**, A simplified method for the *in vitro* cultivation of the rat tapeworm, *Hymenolepis diminuta*, *J. Parasitol.*, 51, 516, 1965.

48. **Voge, M.**, Axenic development of cysticercoids of *Hymenolepis diminuta*, *J. Parasitol.*, 61, 563, 1975.

49. **Roberts, L. S. and Mong, F. N.**, Developmental biology of cestodes. IV. *In vitro* development of *Hymenolepis diminuta* in presence and absence of oxygen, *Exp. Parasitol.*, 26, 166, 1969.

50. **Arai, H. P.**, Migratory activity and related phenomena in *Hymenolepis diminuta*, in *Biology of the Tapeworm Hymenolepis diminuta*, Arai, H. P., Ed., Academic Press, New York, 1980, 615.

51. **Roberts, L. S.,** The influence of population density on patterns and physiology of growth in *Hymenolepis diminuta* (Cestoda: Cyclophyllidea) in the definitive host, *Exp. Parasitol.,* 11, 332, 1961.
52. **Voge, M. and Turner, J. A.,** Effect of temperature on larval development of the cestode, *Hymenolepis diminuta, Exp. Parasitol.,* 5, 580, 1956.
53. **Hundley, D. F. and Berntzen, A. K.,** Collection, sterilization, and storage of *Hymenolepis diminuta* eggs, *J. Parasitol.,* 55, 1095, 1969.
54. **Graham, J. J. and Berntzen, A. K.,** The monoxenic cultivation of *Hymenolepis diminuta* cysticercoids with rat fibroblasts, *J. Parasitol.,* 56, 1184, 1970.
55. **Voge, M. and Green, J.,** Axenic growth of oncospheres of *Hymenolepis citelli* (Cestoda) to fully developed cysticercoids, *J. Parasitol.,* 61, 291, 1975.
56. **Berntzen, A. K.,** *In vitro* cultivation of tapeworms. II. Growth and maintenance of *Hymenolepis nana* (Cestoda: Cyclophyllidea), *J. Parasitol.,* 48, 785, 1962.
57. **Thorson, R. E., Digenis, G. A., Berntzen, A., and Konyalian, A.,** Biological activities of various lipid fractions from *Echinococcus scoleces* on *in vitro* cultures of *Hymenolepis diminuta, J. Parasitol.,* 54, 970, 1968.
58. **Hopkins, C. A.,** The *in vitro* cultivation of cestodes with particular reference to *Hymenolepis nana, Symp. Br. Soc. Parasitol.,* 5, 27, 1967.
59. **Roberts, L. S.,** Modifications in media and surface sterilization methods for *in vitro* cultivation of *Hymenolepis diminuta, J. Parasitol.,* 59, 474, 1973.
60. **Roberts, L. S. and Mong, F. N.,** Developmental physiology of cestodes. XIII. Vitamin B_6 requirement of *Hymenolepis diminuta* during *in vitro* culture, *J. Parasitol.,* 59, 101, 1973.
61. **Turton, J. A.,** The *in vitro* cultivation of *Hymenolepis diminuta*: the culture of 6-day-old worms removed from the rat, *Z. Parasitkd.,* 40, 333, 1972.
62. **Sinha, D. P. and Hopkins, C. A.,** The *in vitro* cultivation of the tapeworm *Hymenolepis nana* from larva to adult, *Nature,* 215, 1275, 1967.
63. **Seidel, J. S. and Voge, M.,** Axenic development of cysticercoids of *Hymenolepis nana, J. Parasitol.,* 61, 861, 1975.
64. **De Rycke, P. H. and Berntzen, A. K.,** Maintenance and growth of *Hymenolepis microstoma* (Cestoda: Cyclophyllidea) *in vitro, J. Parasitol.,* 53, 352, 1967.
65. **Evans, W. S.,** The *in vitro* cultivation of *Hymenolepis microstoma* from cysticercoid to egg producing adult, *Can. J. Zool.,* 48, 1135, 1970.
66. **Seidel, J. S.,** The life cycle *in vitro* of *Hymenolepis microstoma* (Cestoda), *J. Parasitol.,* 61, 677, 1975.
67. **Litchford, R. G.,** Observations of *Hymenolepis microstoma* in three laboratory hosts: *Mesocricetus auratus, Mus musculus* and *Rattus norvegicus, J. Parasitol.,* 49, 403, 1963.
68. **Evans, W. S. and Novak, M.,** Growth and development of *Hymenolepis microstoma* in mice acclimated to different environmental temperatures, *Can. J. Zool.,* 61, 2899, 1983.
69. **De Rycke, P. H.,** Development of the cestode, *Hymenolepis microstoma* in *Mus musculus, Z. Parasitkd.,* 27, 350, 1966.
70. **Voge, M.,** Development of *Hymenolepis microstoma* (Cestoda: Cyclophyllidea) in the intermediate host *Tribolium confusum, J. Parasitol.,* 50, 77, 1964.
71. **Seidel, J. S.,** Hemin as a requirement in the development *in vitro* of *Hymenolepis microstoma* (Cestoda: Cyclophyllidea), *J. Parasitol.,* 57, 566, 1971.
72. **McLeod, J. A.,** A parasitological survey of the genus *Citellus* in Manitoba, *Can. J. Res.,* 9, 109, 1933.
73. **Voge, M.,** Studies on the life history of *Hymenolepis citelli* (McLeod, 1933) (Cestoda: Cyclophyllidea), *J. Parasitol.,* 42, 485, 1956.
74. **Hopkins, C. A. and Stallard, H. E.,** Immunity to intestinal tapeworms: the rejection of *Hymenolepis citelli* by mice, *Parasitology,* 69, 63, 1974.
75. **Landureau, J. C.,** Cultures *in vitro* de cellules embryonnaires de blattes (insectes dictyopterès). II. Obtention de lignées cellulaires a multiplication continue, *Exp. Cell Res.,* 50, 323, 1968.
76. **Barrett, N. J., Smyth, J. D., and Ong, S. J.,** Spontaneous sexual differentiation of *Mesocestoides corti* tetrathyridia *in vitro, Int. J. Parasitol.,* 12, 315, 1982.
77. **Thompson, R. C. A., Jue Sue, L. P., and Buckley, S. P.,** *In vitro* development of the strobilar stage of *Mesocestoides corti, Int. J. Parasitol.,* 12, 303, 1982.
78. **Ong, S.-J. and Smyth, J. D.,** Effects of some culture factors on sexual differentiation of *Mesocestoides corti* grown from tetrathyridia *in vitro, Int. J. Parasitol.,* 16, 361, 1986.
79. **Ong, S.-J.,** *In vitro* Culture of *Mesocestoides corti* (Cestoda), Ph.D. thesis, University of London (Imperial College), London, 1984.
80. **Voge, M.,** North American cestodes of the genus *Mesocestoides, Univ. Calif. Publ. Zool.,* 59, 125, 1955.
81. **Eckert, J., Von Brand, T., and Voge, M.,** Asexual multiplication of *Mesocestoides corti* (Cestoda) in the intestine of dogs, *J. Parasitol.,* 55, 241, 1969.

82. **Kawamoto, F., Fujioka, H., Mizuno, S., Kumada, N., and Voge, M.**, Studies on post-larval development of cestodes of the genus *Mesocestoides*: shedding and further development of *M. lineatus* and *M. corti* tetrathyridia *in vitro*, *Int. J. Parasitol.*, 16, 323, 1986.

83. **Specht, D. and Voge, M.**, Asexual multiplication of *Mesocestoides* tetrathyridia in laboratory animals, *J. Parasitol.*, 51, 268, 1965.

84. **Soldatova, P.**, A contribution to the study of the development cycle in the cestode *Mesocestoides lineatus* (Goeze, 1782), parasitic in carnivorous mammals, *C. R. Dokl. Acad. Sci. U.S.S.R.*, 45, 310, 1944.

85. **Voge, M.**, Development *in vitro* of *Mesocestoides* (Cestoda) from oncosphere to young tetrathyridium, *J. Parasitol.*, 53, 78, 1967.

86. **Voge, M. and Seidel, J. S.**, Continuous growth *in vitro* of *Mesocestoides* (Cestoda) from oncosphere to fully developed tetrathyridium, *J. Parasitol.*, 54, 269, 1968.

87. **Voge, M. and Coulombe, L. S.**, Growth and asexual multiplication *in vitro* of *Mesocestoides* tetrathyridia, *Am. J. Trop. Med. Hyg.*, 15, 902, 1966.

88. **Smyth, J. D. and Davies, Z.**, *In vitro* culture of the strobilar stage of *Echinococcus granulosus* (sheep strain): a review of basic problems and results, *Int. J. Parasitol.*, 4, 631, 1974.

89. **Smyth, J. D.**, *Echinococcus granulosus* and *E. multilocularis*: *in vitro* culture of the strobilar stages from protoscoleces, *Angew. Parasitol.*, 20, 137, 1979.

90. **Smyth, J. D.**, Asexual and sexual differentiation in cestodes: especially *Mesocestoides* and *Echinococcus*, in *Molecular Paradigms for Eradicating Parasites*, New Series, Vol. 60, MacInnis, A., Ed., Alan R. Liss, New York, 1987, 19.

91. **Kawamoto, F., Fujioka, H., and Kumada, N.**, Studies on the post-larval development of cestodes of the genus *Mesocestoides*: trypsin-induced development of *M. lineatus in vitro*, *Int. J. Parasitol.*, 16, 333, 1986.

92. **Conn, D. B.**, The role of cellular parenchyma and extracellular matrix in the histogenesis of the paruterine organ of *Mesocestoides lineatus* (Platyhelminthes: Cestoda), *J. Morphol.*, 197, 303, 1988.

93. **Smyth, J. D.**, Studies on tapeworm physiology. X. Axenic cultivation of the hydatid organism, *Echinococcus granulosus*; establishment of a basic technique, *Parasitology*, 52, 441, 1962.

94. **Smyth, J. D.**, Studies on tapeworm physiology. XI. *In vitro* cultivation of *Echinococcus granulosus* from the protoscolex to the strobilate stage, *Parasitology*, 57, 111, 1967.

95. **Smyth, J. D.**, Development of monozoic forms of *Echinococcus granulosus* during *in vitro* culture, *Int. J. Parasitol.*, 1, 121, 1971.

96. **Smyth, J. D.**, *In vitro* culture of *Echinococcus* spp., in *Proc. 13th Congr. Hydatidology*, Asociacion International de Hydatidologia, Communidad de Madrid, Madrid, 1955, 84.

97. **Smyth, J. D., Miller, H. J., and Howkins, A. B.**, Further analysis of the factors controlling strobilization, differentiation and maturation of *Echinococcus granulosus in vitro*, *Exp. Parasitol.*, 21, 31, 1967.

98. **Smyth, J. D. and Smyth, M. M.**, Natural and experimental hosts of *Echinococcus granulosus* and *E. multilocularis*, with comments on the genetics of speciation in the genus *Echinococcus*, *Parasitology*, 54, 493, 1964.

99. **Thompson, R. C. A., Kumaratilake, L. M., and Eckert, J.**, Observations on *Echinococcus granulosus* of cattle origin in Switzerland, *Int. J. Parasitol.*, 14, 283, 1984.

100. **Smyth, J. D.**, The biology of the hydatid organisms, *Adv. Parasitol.*, 2, 169, 1964.

101. **Thompson, R. C. A.**, Ed., *The Biology of Echinococcus and Hydatid Disease*, George Allen & Unwin, London, 1986.

102. **Smyth, J. D.**, The biology of the hydatid organisms, *Adv. Parasitol.*, 7, 327, 1969.

103. **Webster, G. A. and Cameron, T. W. M.**, Some preliminary observations on the development of *Echinococcus in vitro*, *Can. J. Zool.*, 41, 185, 1963.

104. **Thompson, R. C. A. and Lymbery, A. J.**, The nature, extent and significance of variation within the genus *Echinococcus*, *Adv. Parasitol.*, 27, 209, 1988.

105. **Rogan, M. T. and Richards, K. S.**, *In vitro* development of hydatid cysts from posterior bladders and ruptured brood capsules of equine *Echinococcus granulosus*, *Parasitology*, 92, 379, 1986.

106. **Heath, D. D.**, The development of *Echinococcus granulosus* in laboratory animals, *Parasitology*, 60, 449, 1970.

107. **Smyth, J. D.**, Lysis of *Echinococcus granulosus* by surface-active agents in bile and the role of this phenomenon in determining host specificity in helminths, *Proc. R. Soc. London Ser. B*, 156, 553, 1962.

108. **Smyth, J. D. and Smyth, M. M.**, Self-insemination in *Echinococcus granulosus in vitro*, *J. Helminthol.*, 43, 383, 1969.

109. **Smyth, J. D.**, The insemination-fertilization problem in cestodes cultured *in vitro*, in *Aspects of Parasitology*, Meervitch, E., Ed., McGill University, Montreal, 1982, 393.

110. **Macpherson, C. N. L. and Smyth, J. D.**, *In vitro* culture of the strobilar stage of *Echinococcus granulosus* from protoscoleces of human, camel, cattle, sheep and goat origin from Kenya and buffalo origin from India, *Int. J. Parasitol.*, 15, 137, 1985.

111. **Smyth, J. D. and Howkins, A. B.,** An *in vitro* technique for the production of eggs of *Echinococcus granulosus* by maturation of partly developed strobila, *Parasitology,* 56, 763, 1966.

112. **Kumaratilake, L. M. and Thompson, R. C. A.,** Maintenance of the life cycle of *Echinococcus granulosus* in the laboratory and *in vitro* development, *Z. Parasitkd.,* 65, 103, 1981.

113. **Thompson, R. C. A. and Eckert, J.,** The production of eggs by *Echinococcus multilocularis* in the laboratory following *in vivo* and *in vitro* development, *Z. Parasitkd.,* 68, 227, 1982.

114. **Heath, D. D. and Smyth, J. D.,** *In vitro* cultivation of *Echinococcus granulosus, Taenia hydatigena, T. ovis, T. pisiformis,* and *T. serialis* from oncosphere to cystic larva, *Parasitology,* 61, 329, 1970.

115. **Smyth, J. D. and Davies, Z.,** *In vitro* suppression of segmentation in *Echinococcus multilocularis* with morphological transformation of protoscoleces into monozoic adults, *Parasitology,* 71, 125, 1975.

116. **Smyth, J. D. and Barrett, N. J.,** *Echinococcus multilocularis*: further observations on strobilar differentiation *in vitro, Rev. Iber. Parasitol.,* 39, 39, 1979.

117. **Barrett, N. J.,** Developmental Biology of *Echinococcus multilocularis In Vivo* and *In Vitro,* Ph.D. thesis, University of London, London, 1984.

118. **Kamiya, M., Ooi, H.-K., Oku, Y., Yagi, K., and Ohbayashi, M.,** Growth and development of *Echinococcus multilocularis* in experimentally infected cats, *Jpn. J. Vet. Res.,* 33, 135, 1985.

119. **Yamashita, J., Ohbayashi, M., and Kitamura, Y.,** Differences in the development of the tapeworm stage between *Echinococcus granulosus* (Batsch, 1786) and *E. multilocularis* (Leuckart, 1863), *Jpn. J. Vet. Res.,* 6, 226, 1958.

120. **Rausch, R. L.,** Life cycle patterns and geographic distribution of *Echinococcus* species in *The Biology of Echinococcus and Hydatid Disease,* Thompson, R. C. A., Ed., George Allen & Unwin, London, 1986, 44.

121. **Norman, L. and Kagan, I. G.,** The maintenance of *Echinococcus multilocularis* in gerbils (*Meriones unguiculatus*) by intraperitoneal inoculation, *J. Parasitol.,* 47, 870, 1961.

122. **Yamashita, J., Obhayashi, M., Sakamoto, T., and Orihara, M. E.,** Studies on echinococcosis. XIII. Observation on the vesicular development of the scolex of *Echinococcus multilocularis in vitro, Jpn. J. Vet. Res.,* 10, 85, 1962.

123. **Rausch, R. L. and Jentoft, V. L.,** Studies on the helminth fauna of Alaska. XXXI. Observations on the propagation of the larval *Echinococcus multilocularis* Leuckart, 1863, *in vitro, J. Parasitol.,* 43, 1, 1957.

124. **Lukashenko, N. P.,** Study on the development of *Alveococcus multilocularis* (Leuckart, 1863) *in vitro, Medskaya Parasitol.,* 33, 271, 1964.

125. **Osuna-Carrillo, A. and Mascaró-Lazcano, M. C.,** The *in vitro* cultivation of *Taenia pisiformis* to sexually mature adults, *Z. Parasitenkd.,* 67, 67, 1982.

126. **Heath, D. D.,** An improved technique for the *in vitro* culture of taeniid larvae, *Int. J. Parasitol.,* 3, 481, 1973.

127. **Heath, D. D. and Elsdon-Dew, R.,** The *in vitro* culture of *Taenia saginata* and *Taenia taeniaeformis* larvae from the oncosphere with observations on the role of serum for *in vitro* culture of larval cestodes, *Int. J. Parasitol.,* 2, 119, 1972.

128. **Machnicka, B. and Smyth, J. D.,** The early development of larval *Taenia saginata in vitro* and *in vivo, Acta Parasitol. Pol.,* 30, 47, 1985.

129. **Heath, D. D. and Lawrence, S. B.,** *Echinococcus granulosus*: development *in vitro* from oncosphere to immature hydatid cyst, *Parasitology,* 73, 417, 1976.

130. **Esch, G. W. and Smyth, J. D.,** Studies on the *in vitro* culture of *Taenia crassiceps, Int. J. Parasitol.,* 6, 143, 1976.

131. **Brandt, J. R. A. and Sewell, M. M. H.,** Preliminary observations on the *in vitro* culture of metacestodes of *Taenia saginata, Vet. Sci. Commun.,* 3, 317, 1979/80.

132. **Smyth, J. D. and Rickard, M. D.,** unpublished data, 1977.

133. **Osuna-Carrillo, A., Mascaro-Lazcanó, M. C., Guevara-Pozo, D., and Guevara-Benitez, D. C.,** Cultivo *in vitro* de *Taenia hydatigena, Rev. Iber. Parasitol.,* 38, 289, 1978.

134. **Schiller, E. L.,** Growth and development of larval *Taenia crassiceps in vitro,* paper presented to The American Society of Tropical Medicine and Hygiene, Houston, TX, November 1973, unpublished data.

Chapter 6

NEMATODA: FILARIOIDEA

J. Mössinger

TABLE OF CONTENTS

I. INTRODUCTION

Research on *in vitro* cultivation of filarial worms is receiving increasing attention. During the last decade, approximately 40 papers have been published, which is almost half of the total number of publications on the subject up to the present date. While until 1970 about three quarters of the articles dealt with the cultivation of the stages developing in the arthropod, during the last years more studies were performed on the stages parasitizing the vertebrate host. Many disciplines, e.g., immunology, developmental biology, and biochemistry, could highly profit from potent *in vitro* systems, especially as almost no suitable animal hosts are available for the filarial species parasitizing man.

In previous articles, various aspects of filarial cultivation have been reviewed. Taylor and Baker[1] in 1968 discussed about 20 papers, and in 1970, Weinstein[2] presented a summary of literature comprising 47 references, including the early studies. Pudney and Varma[3] in 1980 discussed the state of knowledge with emphasis on the arthropod phase in the filarial life cycle, and in 1987, Douvres and Urban[4] gave an up-to-date table including information on studies published since 1978.

The objectives of this chapter are to provide comprehensive information on the results obtained with the different filarial stages and species and to focus on the general problems arising with the cultivation of these parasites. As a baseline for the understanding and evaluation of the *in vitro* experiments, first the natural development of filarial worms is outlined, including a brief section on filarial nutrition.

II. LIFE CYCLE AND DEVELOPMENT OF FILARIAL WORMS

In total, over 500 filarial species infecting mammals, birds, reptiles, and amphibia are described,[5] and in wild animals new species are still being found. Information on the life cycles of 93 species has been reviewed by Schacher.[6] Research has predominantly concentrated on less than ten species, the life cycles of which either can be maintained in the laboratory or which parasitize man or domestic animals and are thus of medical or economic interest. The most important species infecting man are *Onchocerca volvulus* and *Wuchereria bancrofti*, the causative organisms of river blindness (onchocerciasis) and lymphatic filariasis (elephantiasis), respectively. The high degree of host specificity has largely impeded these species to be grown in surrogate hosts.

Filariae have a biphasic life cycle comprising development in an arthropod intermediate and a vertebrate definite host (Figure 1). No free-living stages intervene and transmission exclusively occurs via blood sucking vectors. The adult worms live in tissues, the lymph or blood system, or in the body cavities, but not in the alimentary canal. The viviparous female worms produce microfilariae which dwell either in the blood or the skin of the host. With some species, these microfilariae are still in the eggshell when released by the females and retain it as a sheath throughout their stay in the vertebrate hosts. With the other species, the microfilariae ecdyse while still *in utero* and are therefore unsheathed (Table 1). Being actively motile and able to survive for months up to years in the vertebrate host, the microfilariae are morphologically relatively undifferentiated.[7-9]

A. DEVELOPMENT IN THE VECTOR

Feeding on an infected vertebrate, the intermediate host ingests microfilariae either directly with the blood or with the lymph from the subcutaneous tissue. Sheathed microfilariae exsheath in the midgut soon after ingestion.[10,11] The microfilariae then penetrate the gut wall and migrate through the hemolymph to specific organs (see Table 1). *Brugia pahangi* microfilariae can be found in the flight muscles of the vector mosquitoes as early as one hour after feeding.[8] The subsequent development in the typical organ occurs *intracellularly*. The microfilariae become shorter and thicker, described then as "sausage form". This term merely refers to the shape of the larva, but not to a specific morphological stage.[12,13] The sausage form is still a first-stage larva. The typical shape is retained for several days during which characteristic development occurs. In the beginning, predominantly rearrangement of preexisting microfilarial structures takes place after which numerous mitotic divisions of primordial cells and increase in body length are observed (Figure 2). The gut is the first organ to become differentiated and is composed of the esophagus, intestine, and rectum.[8,13] The first molt occurs after about half of the total time of development in the vector or even later (see Table 1). The first stage cuticle is not necessarily cast off immediately, for in several filarial species it has been observed that the cuticle is retained throughout the development of the second stage until the second molt occurs.[13,21-23]

The second stage larva grows considerably and further differentiation of the gut occurs, resulting, among other developments, in an opening of the stoma to the exterior.[13] An anal plug, which is already visible in the sausage form, is usually prominent during the second stage. Furthermore, the genital primordium is formed.[8,13] After the second molt, the elongate wormlike third stage larva leaves the site of development and migrates to the head of the vector. The most prominent interior morphological features of the infective third stage larvae are the gut, with a short cuticular tube connecting the stoma with the esophagus, and the

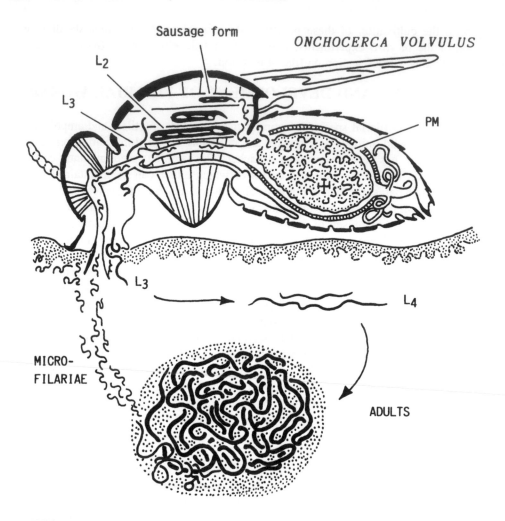

FIGURE 1. Life cycle of *Onchocerca volvulus*. Microfilariae taken up by a simuliid vector penetrate the peritrophic membrane and migrate to the thoracic muscles, where they develop intracellularly to the infective, third stage larvae. During a further blood meal of the vector, the infective larvae are transmitted to a human host. The infective larvae develop to the fourth and the adult stages which mate and produce the skin-dwelling microfilariae. Abbreviations: L_2, second stage larva; L_3, third stage larva; L_4, fourth stage larva; PM, peritrophic membrane. (Courtesy of P. Wenk, Tübingen.)

genital primordium, a compact group of cells the location of which is indicative of the sex of the larva.[15,16]

Growth during development to the infective stage varies considerably between different species. For example, the microfilariae of *Litomosoides carinii* measure approximately 85 μm in length and the third stage larvae are 850 μm long (tenfold increase).[14] In contrast, microfilariae of *O. volvulus* are much longer (300 μm), but the infective larvae are comparatively short (650 μm, twofold increase).[22] Generally, infective larvae of different species measure between 400 to 700 μm (*O. volvulus*, *O. lienalis*)[22,125,126] to 1400 to 1600 μm (*W. bancrofti*)[122,123] in length.

B. DEVELOPMENT IN THE DEFINITE HOST

After being transmitted to the definite host during a blood meal of the vector, the larvae grow relatively slowly within the first days, if at all. The head becomes more evenly rounded with several species, and the genital primordium increases slightly in size.[15,16] After a few days or after about 1 week, the molt to the fourth stage larva takes place with the cuticle

TABLE 1
Development of Filarial Worms In the Vector

Species	Vector	Microfilariae, sheathed (+) and unsheathed (−)	Site of development	Molts		Infective stage (L$_3$)[b]	Ref.
				L$_1$—L$_2$[b]	L$_2$—L$_3$[b]		
Dirofilaria immitis	Mosquitoes	−	Malpighian tubules	9—10	13	15—17	13
Brugia pahangi	Mosquitoes	+	Thoracic muscles	4—5	8	9	8
Acanthocheilonema viteae (syn. *Dipetalonema viteae*)	Ticks (*Ornithodorus moubata, O. tartakowski*)	−	Muscles	11—13	18—19	20—25	21
Litomosoides carinii	Mite (*Ornithonyssus bacoti*)	+	Fat body	6—7	10—12	13	14
Onchocerca volvulus	Simuliids	−	Thoracic muscles	4	6	7—8	22
Onchocerca lienalis	Simuliids	−	Thoracic muscles	4—5	5—6	7[c]	24
Wuchereria bancrofti	Mosquitoes	+	Thoracic muscles	8	12	12—13	87

Duration of development[a] (days after infection)

[a] The duration may vary considerably depending on ambient temperature. Periods reported here were observed under laboratory conditions at 25 to 29°C.

[b] L$_1$: first stage larva, L$_2$: second stage larva, L$_3$: third stage larva.

[c] Vector infected by intrathoracic inoculation of microfilariae.

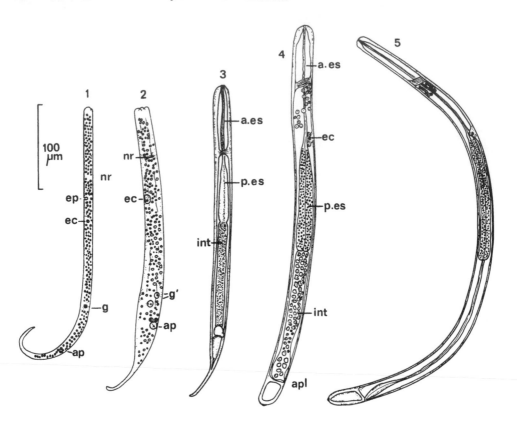

FIGURE 2.   Development of filarial worms (*Onchocerca* spp.) in the vector. From the late first stage (sausage form, 2) onward, the gut is differentiating, which is the most prominent internal structure of the larvae. (1) Microfilaria; (2) sausage form; (3) second stage larva; (4) preinfective, third stage larva; (5) infective, third stage larva. Abbreviations: ap, anal pore; apl, anal plug; a.es, anterior (muscular) esophagus; ec, excretory cell; ep, excretory pore; g, g cell; g', daughter cells of g cell; int, intestine; nr, nerve ring; p.es, posterior (glandular) esophagus. (Courtesy of H. Schulz-Key, Tübingen.)

of the third stage being cast off completely (Table 2). Subsequently, the fourth stage larva grows considerably; major changes occur in the genital system; other organs change little except in size. In both sexes the genital tract develops completely, being functional immediately after the final molt to the adult stage.[18] Male worms are distinctly smaller than females, and usually molt a few days earlier.[15-17] The stoma usually has a characteristic form in each stage being a useful feature, besides others, for the differentiation between third and fourth stage larvae and adults.

After mating, during a period of approximately 3 weeks, the first microfilariae develop in the uteri of the females.[16,18-20] Also during this time and later, the adult female worms, including their genital system, may grow further thus increasing the reproductive potential of the individual worm.[19,20] The adults live for several months up to many years, depending on the filarial species and the host-parasite system.

III. NUTRITION OF FILARIAL WORMS

Comparatively little is known about the nutritional requirements of filarial worms and although various physiological[35-37] and ultrastructural studies[38-41,97,98] have been performed, there is still some controversy about the role of the gut in nutrient uptake of larval stages. In all species and stages investigated so far, the cuticle is permeable to certain low molecular weight substances and plays an important part in filarial nutrition.[34]

TABLE 2
Development of Filarial Worms in the Definite Host

Species	Definite hosts (predominant)	Habitats of adults	Duration of development				
			Occurrence of molts (d after infection)		Intrauterine microfilarial development (d)	Prepatent period (weeks) (host investigated)	Ref.
			L₃→L₄[a]	L₄→AW[a]			
Dirofilaria immitis	Dog, cat	Heart, pulmonary arteries	2—3	50—70	<30	27—28 (dog)	15,25,103
Brugia pahangi	Cat, gerbils	Lymphatic vessels, heart, lung, testes, peritoneum	8—9 6—9	23—33 18—24	≈20	8 (cat) 8—12 (jird)	16 17
Brugia malayi	Man, monkeys, cat, gerbils	Lymphatic vessels, connective tissue, peritoneum, testes	5—10 7—8	35—40 29—35		11—14 (cat) 13 (jird)	26 27
Acanthocheilonema viteae (syn. *Dipetalonema viteae*)	Gerbils	Subcutaneous tissue, muscles, body cavities	6—7	22—23	20	7—8 (jird)	18,20
Litomosoides carinii	Cotton rat	Pleural cavity	8—9	22—24	18	7—9 (cotton rat)	19,28
Onchocerca volvulus	Man	Subcutaneous tissue	3—5[b]		<30[c]	60[d] (man)	29,30
Onchocerca lienalis	Cattle	Connective tissue	2—5	45—75		42—62[d] (cattle)	31
Wuchereria bancrofti	Man	Lymphatic system	8—11 (jird)	42 (monkey)		12—16[e] (man[d])	32,33

[a] L₃, third stage larva; L₄, fourth stage larva; AW, adult worm.

[b] Results from *in vitro* experiments.[125,129,131]

[c] Concluded from therapeutic studies.[29]

[d] For *Onchocerca* spp. and *W. bancrofti*, only limited information is available. In textbooks, usually prepatent periods of 0.75 to 1.5 years are reported. Generally, prepatency may be considerably longer than the time necessary for maturation of worms as it may take months until the sexes meet for mating and microfilarial densities reach a *detectable* level.

[e] Estimation.[32]

As the microfilariae lack a functional alimentary canal,[9,13] it is most likely that the uptake of all nutrients occurs across the cuticle only. *In vitro*, microfilariae of *Dirofilaria immitis* and *W. bancrofti* have been shown to utilize exogenous glucose, amino acids, and nucleic acid precursors.[35-37] In the vector, in parasitized muscle fibers, a dramatic decrease of glycogen levels probably indicates the use of exogenous carbohydrate by the developing larvae.[39,40] Additionally, during early development, the inner body, a cell group visible in microfilariae which probably is a food reserve, is resorbed.[38,96]

Second-stage larvae were shown to ingest host muscle mitochondria which were partly digested in the midgut, indicating that the gut was functional even though it was still blocked at the posterior end.[41] On the other hand, similar signs of degeneration were also found in the mitochondria in parasitized flight muscle cells, and it was suggested that solid particles pass into the gut by chance.[97] Mitochondria were also observed in the gut of third stage larvae.[39] In other studies, however, the gut of third-stage larvae was found to be tightly occluded at the esophageal-intestinal junction[42] and it was not possible to demonstrate oral uptake of vital stains.[43,109] It was concluded that the intestine of third stage larvae may be nonfunctional or at least play a minor role in nutrient uptake.[34,97]

The nutrition of the vector may distinctly influence the developing parasites. Repeated blood meals of mosquitoes after their infection enhanced the development of *D. immitis* and *B. malayi* larvae,[45] and in a defined diet-reared *Aedes aegypti* strain, the development of *B. pahangi* was only poor, compared to crude culture-reared controls.[46] *B. malayi* infective larval counts were higher when *A. aegypti* mosquitoes were fed with folic acid or its component *p*-aminobenzoic acid, and it was concluded that the vitamin directly might be utilized by the larvae.[44] Filarial development in the vector seems to be independent upon host hormones.[47,48]

The intestine of fourth stage larvae and of adult worms was found to be functional;[49,51,60] however, based on physiological and ultrastructural investigations, the rate of ingestion is supposed to be low.[50,60] Transcuticular uptake of adenosin, D-glucose, and L-amino acids has been demonstrated.[43,49,51] With glucose and amino acids as well, both saturable carrier transport systems and diffusion are involved.[53,54]

Except glucose, to a minor extent, filariae may also utilize other carbohydrates and carbolic acids as energy or carbon sources.[37,54-56] Besides the vitamin folic acid (see above), vitamin A might also play an important role in filarial nutrition. In *O. volvulus* adults, retinol concentrations were much higher than in the surrounding host tissue[57] and the development of microfilariae in adult *L. carinii* was retarded when the host animals were fed with a vitamin A-deficient diet.[58] Iron deposits found in the gut, intestinal cells, and uterine muscle cells of adult *O. volvulus*[60] might indicate the utilization of hemin.

Filarial worms, as possibly all parasitic helminths, probably require an exogenous source of certain sterols, fatty acids, and porphyrins.[59] However, it is not known which group of substances in kind and quantities is essential to support the development of the larvae and the considerable longevity of adults and microfilariae *in vivo*.

IV. SYSTEMS FOR CULTIVATION

A. COMPOSITION OF CULTURE SYSTEMS

The complex environments encountered by filarial worms in their hosts can be reproduced *in vitro* only to a limited extent. Additionally, culture systems have to be designed based on the very fragmentary knowledge on the *essential* physiological and nutritional requirements of these worms. Physical, chemical, and biological factors have to be taken into account: temperature, gas phase, substrate, pH, osmolarity, and frequency of medium changes, as well as nutritives provided by basic media and supplements. If cells or organs are cocultivated, their requirements also have to be considered. Numerous basic culture media have been tested, varying from simple salt solutions to complex media originally developed for insect of mammalian cell culture. Such media frequency consist of 40 to 70 defined inorganic and organic components. Nevertheless, semi- or undefined supplements derived from an arthropod or vertebrate have to be added to the basic medium. The different types of components are shown in Table 3. The favorable qualitative and quantitative combination of these components is largely determined on an empirical basis and depends on the parasite stage which is being cultivated.

With only very few exceptions, stationary culture systems with glass tubes or multiwell plastic culture plates as vessels have been used. Media were usually buffered with sodium bicarbonate in combination with a gas phase of 5% CO_2 in air. To increase the buffer capacity, occasionally HEPES was added to the media. Most workers used antibiotics, mainly penicillin/streptomycin or gentamycin and occasionally amphotericin B. Incubation temperatures varied with the parasite stages cultivated.

B. INITIATION OF CULTURES
1. Microfilariae

Except with *Onchocerca* spp., the microfilariae of which are obtained from skin or excised nodules, cultures are usually initiated with microfilariae from peripheral blood. An alternative source could be microfilariae directly from the uteri of female worms or those released *in vitro* or *in vivo*, e.g., into a body cavity. The separation of microfilariae from blood cells, if used at all, has been achieved by various methods such as agglutination of blood cells with phytohemagglutinin or specific serum, centrifugation, filtration, lysis of cells with saponin or enzymes, or a combination of those methods.[61-63] Saponin, however, proved to be toxic against microfilariae of *L. carinii* and *B. pahangi*.[64,85] With those microfilariae possessing a sheath (see Table 1), exsheathment, as observed in the midgut of the vector, probably is a prerequisite for further development.[8,64] Therefore, prior to culture, exsheathment has been induced artificially by exposure of microfilariae to 20 mM $CaCl_2$ or to proteolytic enzymes.[64,85] Microfilariae also exsheathed when placed on agar plates and maintained at 15 to 20°C[65,66] or in the presence of midgut tissue,[67] but these methods were found inappropriate for the initiation of cultures due to high variability.[64]

2. Developing First to Third Stage Larvae

These stages are obtained by preparation of vectors at appropriate times after the infective blood meal. Separation of third stage larvae from debris may be achieved by using a Baermann apparatus, other filter techniques, or by repeated transfer of larvae to fresh medium. Before preparation, the arthropod is usually surface sterilized by 70% ethanol and antibiotics are used at least for the initial washings of larvae. Cultures were also initiated with third stage larvae triggered *in vivo*. These larvae were removed from a mammalian host a few days after infection. After an appropriate period, fourth stage larvae may be obtained. To facilitate recovery, micropore chambers loaded with infective larvae have been implanted into host animals.[109-116]

3. Adult Worms

With most filarial species, the only possibility of obtaining adult worms is the autopsy of the host animal. *Onchocerca* spp. of cattle are isolated from skin or connective tissue. This material is usually available from abattoirs. *O. volvulus* are obtained by surgical removal of nodules from infected humans. To isolate the adults morphologically undamaged, the host tissue is digested with collagenase.[68]

TABLE 3
Various Components of Complete Culture Media

Undefined components of vector origin	Defined components	Undefined components of definite host origin
Hemolymph	Basic culture medium	Serum/plasma
Arthropod extracts	+ defined supple-	Serum fractions
Cells	ments	Organ extracts
Organs		Blood cells
		Cells

V. CULTIVATION OF MICROFILARIAE

Microfilariae are the most frequently cultured filarial stages. The two earliest studies published in 1912 were on *W. bancrofti* and *D. immitis*.[2] The latter has been studied repeatedly and a variety of culture systems have been applied. *D. immitis*, therefore, is a suitable species to demonstrate in detail the various approaches to microfilarial cultivation.

A. *DIROFILARIA IMMITIS*

Under minimal conditions in inorganic salt solutions, *D. immitis* microfilariae survived for only 4 to 8 h at 37°C and 6 to 13 h at 27°C. When glucose was added, survival time was increased to 20 to 45 h at 27°C.[71] Incubated in defibrinated blood, serum, or plasma, at temperatures varying from 4 to 40°C, the survival of microfilariae was 3 to 27 d.[2]

In tissue culture medium 199 plus 30% heat-inactivated dog serum, microfilariae survived 40 to 61 d (37°C). In medium 199 alone, survival was only 4 d.[69] This difference clearly points to the part complex host origin supplements may play. Cocultivation of dog kidney cells and microfilariae in medium containing 10% serum did not further improve survival time.[69] In another study, extracts of chick embryo, raw liver, or mosquitoes as supplements to pure or diluted serum were tested, but none of them had been beneficial. The longest survival was 14 d at 22°C in diluted Tyrode's solution with 33% horse serum and, as a vector component, inclusion of a whole mosquito gut in each culture flask.[70] Addition of red blood cells from the microfilariae donor to medium 199[70] did not extend the short survival period of only 4 d in medium alone.[69]

1. Whole Blood Culture

Partial development of microfilariae *in vitro* was first reported by Sawyer and Weinstein.[72,73] Based on length and width measurements, the authors distinguished four types of first stage larvae observed during cultivation. The most advanced type was the fully developed sausage form, corresponding to 4 d of development in the arthropod host. This was achieved in heparinized whole blood cultures (dog, rabbit, human) incubated at 27°C. Considerable variation (0 to 35%) of microfilariae reaching the sausage form was observed and differences were noted in blood cultures from different dogs. In medium NCTC 109, supplemented with different amounts of horse serum, comparable development was observed. The most favorable serum concentrations were 5 or 10% depending on the serum sample; higher concentrations of 20 and 40% inhibited growth. Heat inactivation of serum considerably improved the yield of sausage forms, which was 28% at most. Molting was not observed; however, the larvae were "in premolt as evidenced by the separation of the cuticle."[72] Development of microfilariae only occurred at 27°C or below, but not at 37°C. In a later study, Grace's insect tissue culture medium was found superior to medium NCTC 109.[74] The optimum concentration of heat-inactivated pony serum was 20% in both media (temperature, 26°C). Some developing larvae showed "internal development far beyond those seen in the first larval stage, yet no ecdysis had been observed."[74]

2. Insect Cell Culture

In 1966, Wood and Suitor[92] introduced the method of cocultivation of insect cell lines and microfilariae which was taken up by other workers and applied to several filarial species. *D. immitis* developed to the sausage form in four of six cell lines tested; addition of hemolymph seemed to enhance development.[75] In *Aedes vexans* or *Culicoides inornata* suspended cell culture in Grace's medium, the major part of the microfilariae developed to sausage forms with distinct anal plugs after 7 d; after 9 d, larvae showed a "loosened cuticle". However, subsequently over 90% of the original microfilariae population degenerated and with the remainder no complete molt was noted although they survived up to 51 d.[75] In another study, four mosquito cell lines were tested, but microfilariae survived only up to 10 to 15 d.[76] In one cell line of *Aedes malayensis*, maintained in medium MM/MK plus 20% inactivated fetal bovine serum (iFBS), occasionally sausage forms were observed. Interestingly, *without* insect cells in medium NCTC 135 plus 20% iFBS, considerably more microfilariae developed to the sausage form (15 to 20%).[76] These results demonstrate that the use of insect cell lines does not necessarily improve culture conditions. The basic medium and serum supplement are at least of similar importance.

3. Cell-Free Culture

In the search for an appropriate basic culture medium, a variety of commercially available media for mammalian or insect cell culture have been tested repeatedly, mostly in combination with different concentrations of iFBS (media MEM, L15, NCTC 135, RPMI 1640, Ham's F12, F12K, Waymouth's 750/1, mosquito culture medium — MCM, Schneider's *Drosophila* medium, Grace's medium and *Caenorhabditis briggsae* maintenance medium).[76-79] With eight of these media compared,[78] only very few microfilariae developed to late sausage forms (0 to 2.4%). Best results were obtained in *serum-free* Ham's F12 medium containing cells of *Aedes albopictus*. Also in two subsequent studies, Ham's F12 medium proved beneficial to maintenance or early development of microfilariae.[79,80] Without any further supplements, 50% of microfilariae survived for over 4 weeks in medium F12K, a modification of Ham's F12. After 2 weeks in culture, they were still capable to develop in mosquitoes. Development *in vitro*, however, was not induced, even if 5% iFBS was added (37°C).[79] In another study, with the same medium and serum concentration, but at 27°C, up to 22% of microfilariae developed to the sausage form.[80]

In confirmation with previous results,[72,73] this points to temperature as an important factor for early microfilarial development, which was significantly enhanced below 30°C.[77] If not stated otherwise, with the culture systems for vector specific stages described in the following, incubation temperature was 27 or 28°C. The optimum pH supporting survival of microfilariae was 7.2 to 7.6 (range investigated, 6.0 to 8.8).[77] The most favorable osmotic pressure for development to the sausage form was 330 to 390 mOsm/kg (range tested, 270 to 420 mOsm/kg).[81] A gas phase of 5% CO_2 in N_2 was found superior to oxygen containing gas mixtures.[78]

Summary: Early development of *D. immitis* microfilariae has been achieved in rather different culture systems. Whole blood cultures, insect cell cultures, and cell-free cultures with defined media plus mammalian serum yielded qualitatively comparable results. In none of these cultures, however, distinct development beyond the first stage occurred, which is also true for most other filarial species.

B. *BRUGIA* SPP.

Only a few studies have been performed on *Brugia* spp., the microfilariae of which are sheathed (see Table 1). Microfilariae of *B. pahangi* isolated from blood and placed in culture without further treatment failed to undergo any development in four mosquito cell lines tested, although they survived up to 60 d (medium MM/MK or MM/VP$_{12}$, 20% iFBS,

28°C).[76] When microfilariae were artificially exsheathed prior to culture, up to 14% developed to the sausage form in the same culture systems.[82] The removal of the sheath therefore is regarded a prerequisite for the initiation of development *in vitro*.[76,82,83] With exsheathed *B. malayi*, microfilariae up to 30% developed to the sausage form, but not beyond (*A. malayensis* cells, medium MM/MK).[82]

C. *ACANTHOCHEILONEMA VITEAE* (SYN. *DIPETALONEMA VITEAE*)

The natural development of *A. viteae* in the vector is longer than that of other filarial species (see Table 1). Possibly that is the reason why only in one paper maintenance of microfilariae, besides other stages, is reported.[84] With five combinations of media and different supplements, including Grace's medium or medium RPMI 1640 and iFBS, microfilariae survived only 4 to 11 d; no observations on growth or development were reported.

D. *LITOMOSOIDES CARINII*

Artificially exsheathed microfilariae were cultured in medium NCTC 109 with 20% iFBS plus amino acids, sugars, and organic acids of Grace's medium. The larvae grew progressively and reached a size corresponding to 4 to 5 d of development in the vector; a small percentage corresponded to 6 d of development *in vivo* which was still the first stage.[85]

E. *WUCHERERIA BANCROFTI*

With *W. bancrofti* spontaneous exsheathment of microfilariae, i.e., without special treatment with $CaCl_2$ or proteolytic enzymes, was observed in different studies.[2,70,86] In Tyrode's solution plus 50% horse serum or in a glucose-supplemented mixture of horse serum, Ringer's and medium 199 (3:3:1) exsheathed microfilariae survived only 8 d and no development occurred (22°C).[70] In medium 199 plus 10% inactivated human serum, spontaneous exsheathment occurred after 3 d or later and up to 10% of microfilariae developed to the sausage form within 14 to 20 d.[86] When the same medium was additionally supplemented with organic acids (malic, succinic, fumaric, α-ketoglutaric acid) and sugars (glucose, sucrose, fructose) of Grace's insect medium, development of microfilariae was enhanced (30% sausage forms after 8 d).[86] In a subsequent study with medium 199 plus 20% human serum, marginal improvement was achieved (2% late sausage forms resembling stages grown for 5 to 6 d in mosquitoes). As the authors suggest, this was possibly due to the addition of glutamic acid and trehalose among other supplements. These substances were found in high concentrations in the hemolymph of many insects.[87] In the same culture system as applied for *L. carinii*, 4 to 7% of previously artificially exsheathed microfilariae of *W. bancrofti* developed to the sausage form, mostly showing an anal plug.[85]

F. *ONCHOCERCA* SPP.

In a biphasic roller culture system originally used for the cultivation of the tapeworm *Hymenolepis diminuta* and of *Schistosoma* spp., *O. volvulus* microfilariae were maintained for considerable periods of time.[91] A mixture of agar (70%) and rabbit blood (30%) was overlaid with glucose-supplemented Hank's balanced salt solution and incubated at 37°C under a gas phase of 90% N_2/5% CO_2/5% O_2. Microfilariae from skin or nodule tissue inoculated to the liquid phase survived often as long as 60 d in "a strongly viable condition". In some experiments, less than 1% developed to stages "resembling those found in naturally infected black flies."[91] In a similar culture system with blood-agar slopes, *O. lienalis* microfilariae failed to develop.[89]

With one of six arthropod cell lines tested (tick cells, RA-243), microfilariae of *O. volvulus* and *O. gutturosa* developed to the late sausage form within 4 to 5 d (medium RPMI 1640 + 20% iFBS). Development of the gut was clearly visible and few larvae appeared to attempt molting. The most advanced stages obtained corresponded to 2 d of development *in vivo*, although nondeveloping microfilariae of *O. gutturosa* remained active for at least 40 d.[88] Primary cell cultures from larval tissue of *Simulium* spp. have been set up, but were not used for cultivation of microfilariae. The growth rate of *Simulium* cells was extremely slow, and the first cell monolayers were obtained not earlier than several months or up to half a year of culture.[3,88]

With *Aedes malayensis* cells in medium ML-15 or RPMI 1640 with 20% iFBS, *O. lienalis* microfilariae developed to an "intermediate stage" (30 to 35%) and the early sausage form (5 to 10%) by day 7 of culture.[89] In the absence of cells, microfilariae developed only to the "intermediate stage". Addition of reducing agents such as glutathione or L-cysteine hydrochloride enhanced development to the sausage form.[89] With reduced glutathione, this effect was confirmed in a cell-free culture system (medium L15, 10% FBS). Degree and frequency of development were dose related with a peak of activity at 15 mM glutathione.[90]

G. OTHER SPECIES
1. *Macacanema formosana*
This species parasitizes the Taiwan monkey, *Macaca cyclopis*. The adult worms live in the peritracheal tissue and the diaphragm; the unsheathed microfilariae occur in the blood.

Microfilariae were cocultivated with an insect cell line of *Aedes aegypti* by Wood and Suitor.[92] Grace's insect cell culture medium supplemented with 10% FBS and 0.5% hemolymph of a butterfly (*Philosamia cynthia*) served as growth medium.[1] In complete medium without cells, microfilariae survived only 6 to 7 d without development. In the presence of cells, microfilariae developed to "second and possibly to third stage larvae". The furthest stage of development was reached after 14 to 17 d with the larvae measuring about double the size of the microfilariae and an internal organization which was considered to be typical of the second stage larva. The external cuticle of the larvae had separated; however, throughout the experiment, no shedding of a cuticle was observed. It was therefore argued that probably the larvae "were actually only late first stage organisms that had molted but had been unable to complete ecdysis."[2] In any event, microfilariae developed distinctly beyond the sausage form and, interestingly, this did not occur when cultures were incubated at 28°C, but at 22°C. Unfortunately, with no other filarial species could this degree of development be achieved.

2. *Loa loa*
Division of a few cells was the only sign of development of microfilariae surviving for 10 d in medium 199 with red blood cells or in Tyrode's solution with 33% horse serum (22°C). No exsheathment was observed.[70]

3. *Dirofilaria repens*
Microfilariae developed to the sausage form in a 1:1 mixture of infected dog blood and Alsever's solution supplemented with glucose (25°C).[93]

4. *Dipetalonema* spp.
Not definitely determined microfilariae from dogs developed to the sausage form in heparinized whole blood cultures after a minimum of 12 d at 27°C.[73]

VI. CULTIVATION OF DEVELOPING FIRST AND SECOND STAGE LARVAE

In this approach, infested organs were removed from the arthropod vector at different times after the infective feed and the whole organ or isolated larvae was placed in culture.

A. *DIROFILARIA IMMITIS*

Infested whole Malphigian tubules of mosquitoes were cultured in different media. With Tyrode's solution plus horse serum (30%, 50%), larvae only molted from the second to the third stage in tubules recovered from the vector at 12 d after the infective feed or later. Younger larvae (5 to 11 d) remained alive for only 9 d and did not develop further.[70] With medium NCTC 109 plus serum (concentration not reported), a "small percentage" of the first stage larvae (tubules recovered 2 to 4 d after the infective feed) developed to the "late second stage and as far as the second molt."[2] With Schneider's *Drosophila* medium, which was one of seven media tested with 20% iFBS, up to 15% of the microfilariae developed to the second stage, but not beyond.[94] The Malphigian tubules were excised after 20 to 24 h. Cocultivation of *Aedes albopictus* cells with the tubules did not alter the results significantly.[94] Only in one case it was reported that third stage larvae were obtained from whole organ cultures initiated from one day old infections (medium NCTC 109, 40% serum).[99]

B. *BRUGIA PAHANGI*

Second stage larvae isolated from thoracic muscles were cultured cell free in Grace's medium plus 20% iFBS.[95] Interestingly, larvae of *different sizes* were recovered at 7 or 8 d from mosquitoes which were maintained at 25, 26, or 27°C. Measurements revealed that only second stage larvae which had a minimum length of 1164 μm when placed *in vitro* were able to molt to the third stage within a 5-day cultivation period. Smaller larvae failed to molt, although they grew beyond this critical size *in vitro*. The authors suggest that *in vivo* the "trigger of larval molt to the third stage" is related to the growth pattern of the larvae. *In vitro*, this trigger obviously was not present, although the larvae grew further. Addition of juvenile hormone analogue, β-ecdysone or extract of neurohemal organ to such cultures had no effect on molting.[95]

C. *ACANTHOCHEILONEMA VITEAE* (SYN. *DIPETALONEMA VITEAE*)

Second stage larvae were isolated from vector ticks 15 to 18 d after infection and maintained in Grace's medium with iFBS. Survival was only 6 d; no reports on growth or development were made.[84]

VII. CULTIVATION OF INFECTIVE, THIRD STAGE LARVAE

Concerning the early development in the definite host, some of the filarial species differ especially in the time after which the third larval molt occurs. With *D. immitis* and *Onchocerca* spp., the molt to the fourth larval stage takes place within 2 to 3 d after infection, but with the other species, the molt does not occur before 6 to 9 d (see Table 2). This difference is reflected in the results obtained with the cultivation of the third larval stage. If not stated otherwise, with all culture systems referred to in the following, the incubation temperature used was 35 to 37°C and the gas phase was 5% CO_2 in air or pure air.

A. *DIROFILARIA IMMITIS*

In various culture systems, vector-derived *D. immitis* infective larvae molted to the fourth larval stage. Complete molting, i.e., synthesis of the new cuticle and emergence of the larva from the old cuticle, occurred beginning after 2 d *in vitro*.

1. Cell-Free Culture

In pure dog serum and in 50% horse serum diluted with Tyrode's solution plus chick embryo extract, larvae molted although maximum survival in both media was only 3 d.[70] Survival of over 30 d was obtained in a culture medium consisting of 10% inactivated dog plasma, 50% Ringer's, and 40% medium 1066 plus heparinized dog blood (2%).[100] Molting was observed in larvae kept for 72 h or longer. The average length of fourth stage larvae cultured for 30 d was 1824 μm, compared to 1059 μm of third stage larvae removed from mosquitoes. In addition to growth, which has been the most marked reported for *D. immitis in vitro* so far, blunting of the anterior and changes of the posterior shape of the larvae have been noted.[100]

With culture medium NCTC 109 plus 10% human or horse serum, 100% of larvae molted to the fourth stage and survived up to 15 d.[102] Maximum development achieved appeared to be identical to that obtained after 5 d in host animals.[103] At 37°C, first shedding of the cuticle occurred after 54 to 65 h *in vitro*.[101] At lower temperatures of 21 to 25°C, survival was extended to up to 22 d, but larvae did not emerge from the cuticle. After 7 to 10 d, only "a clear space formed between the body and the cuticle" was observed with some larvae, but molting remained incomplete.[102] In a later study, it has been confirmed that a temperature of approximately 37°C is a prerequisite for complete molting.[104] At 37°C, even in *serum-free* NCTC 109, a few larvae molted completely, but survived only for 3 d.[102]

In recent studies, altogether about 15 cell culture media, supplemented with 10 or 20% serum or defined substances, have been tested for their ability to support molting and survival of third stage *D. immitis*.[104-106] Comparing RPMI 1640, medium 199, and Grace's insect medium, maximum molting rate (80%) and survival (26 d) was achieved in cultures with medium RPMI 1640 plus 10% human AB serum.[105] No significant differences in motility or molting response (always >50%) were detected between larvae cultured in media NCTC 135, F12K, CMRL 1066, or DMEM, all with 10% iFBS.[106] The frequency of molting increased, corresponding to the concentrations of FBS up to 20%; higher concentrations up to 40% had no further enhancing effect. Of three commercially available serum substitutes tested, only one (Nu serum, Collaborative Research, Waltham, MA, U.S.) had a similar effect as FBS. It is noteworthy, that even in serum-free medium F12K, the molting rate was between 5 to 40%, compared to 60% with FBS or Nu serum added.[106] In a comparable study, molting rates of 70 to 95% have been achieved with various media supplemented with 20% FBS.[104] Serum could be replaced by bovine albumin at concentrations of 10 to 30 mg/ml,[104] but not at lower concentrations of 1.25 to 5.00 mg/ml.[106] Addition of numerous defined supplements including transferrin, insulin, T_3, α-tocopherol, various amino acids, vitamins, and growth factors (GHL, platelet growth factor) did not improve larval molting or survival in serum free media.[104,106] Two gas phases tested (5% CO_2 in N_2 or 5% CO_2 in air) yielded identical results.[104]

2. Mammalian Cell Culture

When mammalian cells have been cocultivated with *D. immitis* third stage larvae, results have not been distinctly improved compared to cell free systems.[107,108] Four out of ten cell lines tested by Wong et al. supported molting and average survival of larvae of 12 to 18 d.[107] First molting was observed after 2 to 3 days in culture, the separation of and emergence from the cuticle of an individual larva was completed within 4 to 10 hours.[107] Using a dog sarcoma cell line cultured in seven different media plus 20% FBS, best results (80% molting by day 4 of culture) have been achieved with medium L-15.[108] Interestingly, under the same conditions, infective larvae of *B. pahangi* and *W. bancrofti* failed to molt. Molting rate and viability of *D. immitis* larvae were reduced in the absence of cells. By ultrastructural investigations, it has been demonstrated that synthesis of the fourth stage cuticle in some larvae was completed by 60 h of culture.[108] In the same culture system, *fourth stage* larvae recovered from host animals 74 h after implantation, increased by 11 to 14% in length during 4 d of cultivation. Growth during this relatively short period was comparable to that of control larvae recovered from host animals.[109]

Summary: *D. immitis* third stage larvae molted to the fourth stage in a variety of culture media, beginning after 2 to 4 d, which is comparable to the *in vivo* situation. Consistently, molting rates and survival have been enhanced when 10 to 20% of an appropriate serum have been added to the basic medium; certain cell lines also proved beneficial. In none of the culture systems, however, was distinct further development of the fourth stage larva observed, except with limited growth[100,104,106,109] and early alteration of the larval shape at the anterior and posterior end.[100,108] In this connection, it should be recalled that the development of *D. immitis* fourth stage larvae in the host is relatively slow (see Table 2).

B. *BRUGIA* SPP.

In vivo, infective larvae of *Brugia* spp. develop to adult worms within about 3 to 5 weeks. The molt to the fourth larval stage occurs 6 to 10 d after infection (see Table 2).

1. Larvae Preconditioned *In Vivo*

Cocultivated with dog sarcoma cells, *vector-derived B. pahangi* third stage larvae failed to molt, although they survived for about 20 d and increased in length by 24% (medium 199 + 10% newborn calf serum). Only such larvae molted which before culture had been *preconditioned* or "*triggered*" for at least 3 d in the mammalian host.[110] In a comparable study, triggered third stage larvae recovered from jirds 4 d after infection developed to the fourth stage when cultured with dog kidney cells in medium MEM plus 2 to 10% iFBS. Molting rates (up to about 90%) and growth (up to 44% increase in length) of the larvae were positively correlated with the cell number of the starter inoculum forming the cell monolayer and with the serum concentration. In these experiments, during the 6-d cultivation period, the medium was not changed.[111]

2. Vector Derived Larvae

With more beneficial cell medium-serum combinations, development of untriggered larvae, placed in culture immediate after recovery from the vector, has been achieved.[107,112-114] With only two of ten cell lines tested (bovine embryonic kidney or dog skeletal muscle; medium MEM + 10% FBS), *B. pahangi* larvae molted to the fourth stage on average after 11 or 12 d in culture. Average survival was 22 and 23 d, respectively.[107] *B. malayi* third stage larvae molted beginning on day 10, when cocultured with jird testis cells. Maximum survival was 38 d with an increase in larval length of 33%.[112]

Rhesus monkey kidney cell line LLCMK$_2$ — Mak et al. reported cultivation of vector-derived *B. pahangi* and *B. malayi* infective larvae over a layer of LLCMK$_2$ cells in medium RPMI 1640 supplemented with 10% inactivated human AB serum.[113,114] It was in this culture system that complete development of a filarial worm from the third to the adult stage was observed *in vitro*, i.e., two consecutive molts were recorded. With *B. malayi*, molting to the fourth larval stage began in the second week of culture and 100% of larvae had molted by 28 d. Subsequently, the genital apparatus started to develop in fourth stage larvae of both sexes. In male larvae, spicules and gubernaculum testis were formed; in female larvae, development of the vulva and uteri were observed. From the sixth week (44 d) onward, a total of 20% of the larvae molted to adult worms. Unfortunately, the sex of these adults was not reported. After 44 d in culture, the average length of worms was 3448 μm, which corresponds to a 2.1-fold increase from the infective stage, i.e., growth and development in culture were distinctly delayed, but morphologically comparable to the *in vivo* situation. With *B. pahangi*, 100% of the third stage larvae molted to the fourth stage as well; however, mortality of larvae was high, leaving only 19% alive by 28 d of culture; no final molt to the adult occurred.[114] In a subsequent study with *B. pahangi*, after 22 d a molting rate of 78% and a 50% survival were obtained, using the same cell line and medium, but 10% iFBS instead of human serum.[115] In the meantime, the culture system described by Mak et al.[113,114] has been used by several other workers, and the LLCMK$_2$ cell line has been comparatively successfully applied to other filarial species.[115,122,128,146]

C. *ACANTHOCHEILONEMA VITEAE* (SYN. *DIPETALONEMA VITEAE*) (FIGURE 3)

With *A. viteae* third stage larvae, besides culture media, sera, and other supplements,[116-120] the effect of various gas phases and the resulting oxygen and carbon dioxide tensions in the medium has been investigated in detail.[118]

Under a gas phase of 90% N$_2$/5% CO$_2$/5% O$_2$, six culture media have been tested with sera of hamsters and jirds or FBS (5 to 20%).[116] In no case did *vector-derived* larvae develop to the fourth stage. When cultures were initiated with larvae *triggered* for 2 to 6 d in jirds, only the 5- or 6-d-old larvae, which most probably were already close before their "natural" molt in the host animal (see Table 2), molted in culture (medium BHK 21 + 10% tryptose-phosphate broth + 10 or 15% jird serum).[116,117] After 15 d in culture, up to 50% were fourth stage larvae, some had grown distinctly. Larvae removed from such cultures after 6 to 8 d (third and fourth stage) were still capable of normal development. In hamsters, they developed further to fertile adult worms producing a microfilaremia in the host animal. Interestingly, the *in vivo* triggering of larvae could be partly replaced by cocultivation of irradiated hamster kidney cells, and 10% of vector-derived larvae molted completely to the fourth stage when these cells and jird serum (10%) were present in the culture system.[116]

Low oxygen tension — Franke and Weinstein demonstrated, that low oxygen tensions (optimum, 30 to 50 mmHG), obtained when cultures were gassed with 5% CO$_2$ in nitrogen, were a prerequisite for the development and molting of vector derived *A. viteae* third stage larvae in *cell free* culture systems.[118] Under this gas phase, in various culture media supplemented with 10 or 20% FBS, 63 to 68% of the larvae molted to the fourth stage and development of the reproductive system as well as considerable growth to an average length of 4 to 6 mm occurred.[119] This is about three to four times the length of a third stage larva. The most beneficial media were 1:1 mixtures of either NCTC 135 + RPMI 1640 or NCTC 135 + IMDM (10% FBS). With the latter, up to 50% of the larvae even developed further than the fourth stage, i.e., they molted to adult worms after 42 d of culture.[120] This was the first report that infective larvae of a filarial worm developed to adults in a *cell-free* culture system. Morphogenesis of the reproductive system of male and female fourth stage larvae were described in detail.[120] All but one of the adult worms from culture were males which had grown to a mean length of 12 mm (approximately eightfold increase compared to third stage larvae). The worms were "in excellent morphological condition" when the cultures were terminated after 65 to 71 d.[120] Although growth and development *in vitro* lagged behind that of worms developing in host animals, morphological changes were comparable. As the authors emphasized, considerable differences between sera existed in the ability to support larval development, and a suitable sample of FBS used was essential for the success of their cultures.[119,120]

D. *LITOMOSOIDES CARINII*

In vivo triggered third stage larvae, recovered from multimammate rats 7 d after infection, were cultured to the fourth stage (55 to 84%) in a cell free system (medium L-15, plus glucose and L-glutamin, 10% FBS or gamma globulin-free horse serum).[121] First molting was observed after 3 to 4 d in culture, but no growth occurred and after 10 d all larvae were dead. Taken into account that after the long trigger period the larvae were already close to a molt (see Table 2), the development *in vitro* was poor. This is consistent with the finding that larvae taken from culture later than 1 d had already lost their capability to further develop in a host animal.[121]

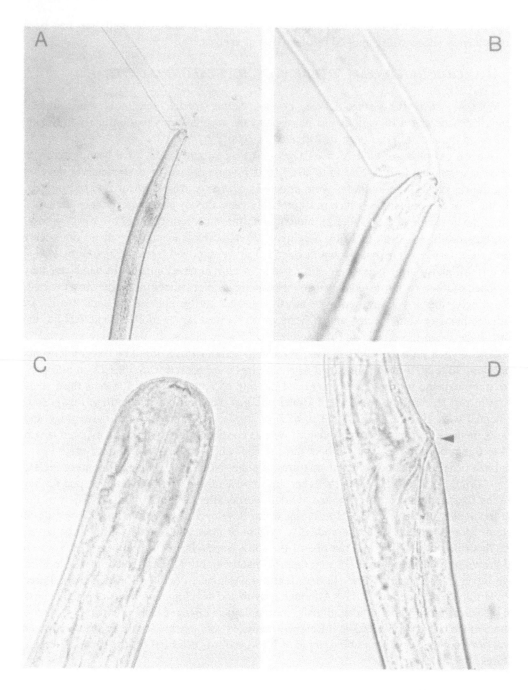

FIGURE 3.   Fourth stage larvae of *Acanthocheilonema viteae* grown in culture. (A) Molt, the fourth stage larva is emerging from the third stage cuticle; (B) posterior end of the fourth stage larva; the typical tail projections and the ribbed structure of the old cuticle are visible; (C) anterior end of a fourth stage larva with typical stoma and esophageal region; (D) posterior end of a male fourth stage larva with the cloaca (black arrow) and the developing spicules (white arrow).

E. *WUCHERERIA BANCROFTI*

The cultivation of *W. bancrofti* third stage larvae is a good example for the adaptation of successful culture systems which have been previously developed with other filarial species. Over a layer of rhesus monkey kidney cells (LLCMK$_2$) in medium RPMI 1640 plus 10% human AB serum, approximately 60% of vector derived larvae molted to the fourth stage after 10 to 16 d. However, growth was limited and survival was poor, with less than 50% of the larvae being alive by 19 d of culture.[122]

In a cell-free culture system (medium NCTC 135 + IMDM, 1:1), with 10% human serum, up to 100% of the larvae developed to the fourth stage and were "still motile and in excellent morphological condition" when the cultures were terminated after 35 d. The larvae had grown to an average length of 3.5 to 4.5 mm (about three times the length of a third stage larva). The maximum length of 5.5 (male larva) and 5.6 mm (female larva) obtained was comparable to that of larvae from monkeys, which served as surrogate hosts of *W. bancrofti* (see Table 2). Human serum yielded better results than FBS or goat serum; molting rates and growth were increased with 10% serum compared to 5%. Interestingly, in cultures with human serum from infected individuals, molting rates were not altered, but the mean length of larvae was significantly less than with serum from uninfected persons. No significant differences were found between cultures gassed with 5% CO_2 in air or 8% CO_2 in air.[123]

F. *ONCHOCERCA* SPP.

The course of development of most *Onchocerca* spp. infective larvae to adult worms in their definite hosts is not known in detail, but the long prepatency periods might indicate that this development is rather slow (see Table 2). On the other hand, the molt to the fourth larval stage occurs already after a relatively short period, namely, after 2 to 5 d with *O. lienalis, in vivo*[31] and *in vitro*,[124-126] and after 3 to 10 d with *O. volvulus in vitro*.[125,129-131] With the latter, after 72 h *in vitro*, the new cuticle had been partially formed and the process of exsheathment had commenced.[130]

In various *cell-free* culture systems, vector derived *O. lienalis* third stage larvae molted to the fourth stage. In medium MEM plus 10% iFBS, the mean molting rate was 13%;[124] in medium F12K, L-15, or NCTC 135 + IMDM (1:1), all supplemented with 20% FBS, 45 to 56% of the larvae developed to the fourth stage.[125] A gas phase of 5% CO_2 in nitrogen instead of 5% CO_2 in air did not generally improve the frequency of molting, but in some cases survival of the fourth stage larvae was enhanced (maximum, over 43 d).[125] In a recent study,[127] a molting rate of 72% was achieved using the culture system described by Franke and Weinstein (5% CO_2 in N_2, 20% FBS). *O. volvulus* third stage larvae molted in the same culture system (14 or 30%, depending on the serum used[125]); longest survival was over 28 d.[129] In another series of experiments with *O. volvulus*, molting rates of 73 and 79% were obtained, using "a CO_2-incubator" (NCTC 135 + IMDM, 20% iFBS).[131] Obviously, with *Onchocerca* spp. third stage larvae, low oxygen tensions are not a prerequisite for molting *in vitro*, as it proved to be with *A. viteae* larvae in cell-free cultures.[118]

Cocultivated with *mammalian cells*, especially the survival time of the larvae was considerably prolonged compared to cell-free systems. With two jird cell lines (spleen or kidney cells), on average 32 and 33% of *O. lienalis* larvae molted to the fourth stage.[126] Although the survival time varied considerably, in one experiment 5% of the larvae were still alive and healthy after 88 d of culture; however, they had not increased in size.[126] Cultivated with the monkey kidney cell line LLCMK$_2$, vector derived third stage larvae (length, 445 μm) developed to late fourth stages, measuring 600 to 700 μm, and survived longer than 160 d in culture.[128] With medium NCTC 135 + 10% iFBS, the molting rate was highest (79%) and one larva attempted to molt to the adult stage, but died during the molt.[128]

VIII. CULTIVATION OF ADULT WORMS

Criteria for the evaluation of adult worm culture have been the survival and the reproduction of worms. Survival mostly has been assessed based on motility only. Certain physiological parameters such as glucose uptake or lactate production, for example, have been used as an indicator for viability in a few studies.[136,147] The most easily assessible measure for reproduction is the release of microfilariae into the culture medium by female worms. However, this parameter may vary considerably between different worms and filarial species. Therefore, investigations on oogenesis and the intrauterine development of microfilariae may provide valuable further information.[19]

Cell-free culture systems have been used for adult worms, except in a few studies.[91,146] If not stated otherwise, the incubation temperature was 35 to 37°C and the gas phase was 5% CO_2 in air or pure air.

A. *DIROFILARIA* SPP.

1. *Dirofilaria immitis*

In Modified Eagle's Medium plus 10% inactivated horse serum, adult *D. immitis* survived up to 65 d. However, microfilariae were released for only 4 to 7 d, and besides microfilariae, numerous early embryonic stages were extruded by the females.[69] Similarly, for only 1 or 2 d, microfilariae were extruded by females cultures in Ringer's or Tyrode's solution supplemented with 25% dog serum plus chick embryo extract and glucose; male and female worms survived for only 7 d.[70]

In terms of microfilarial release, better results were achieved in a study with medium RPMI 1640 plus 10% FBS.[133] During the first and second day in culture, on average 2200 and 4500 microfilariae were released per female, respectively. From day 3 to day 9 in culture, the average output was 7,000 to 10,000 microfilariae per female per day. Subsequently the numbers decreased, and after 11 to 12 d, the release of viable microfilariae ceased completely (compare *A. viteae* and *L. carinii*). The average survival time of females was 25 d; the longest was 36 d in this culture system.[133]

In another study, it was shown that serum from dogs *immunized* against microfilariae, distinctly reduced microfilarial output *in vitro*, compared to serum from uninfected dogs (10% serum, "modified Taylor's culture medium",[70] 32°C).[132]

2. *Dirofilaria uniformis*

D. uniformis adults isolated from rabbits survived for 10 d in medium NCTC 109 alone, but survived for 3 to 4 weeks when 5, 10, or 20% rabbit serum was added. Microfilariae were released for 13 to 16 d with a peak output during days 2 and 3 of culture. On average, 17,000 to 21,000 microfilariae per female were released during the whole period of maintenance.[134]

B. *BRUGIA PAHANGI*

In Click's medium plus 10% horse serum, adult female worms survived up to 35 d and released microfilariae for 14 to 18 d; on average 600 to 1500 microfilariae were produced per female per day during the first week in culture.[135] With medium MEM + 10% iFBS, similar amounts of microfilariae were extruded, namely, on average 4900 microfilariae per female during 5 d of maintenance.[136]

Investigations on glucose consumption revealed that by day 3 of incubation without medium change (MEM + 10% iFBS), 50% of this available glucose had been taken up by an individual female; by day 5, all available glucose had been utilized (2 mg) and coincidently glycogen contents of worms had decreased distinctly. Although glucose in the medium was depleted, motility of worms had not decreased and was maintained longer than 5 d. Neither glucose utilization nor motility of the worms were affected by reduced oxygen tensions obtained by gassing cultures with 5% CO_2 in N_2. If an excess of glucose was added to the culture medium, a single female worm on average utilized 3.6 mg; a male utilized 0.8 mg of glucose during a 5-d incubation period. Great differences in glucose consumption *in vitro*

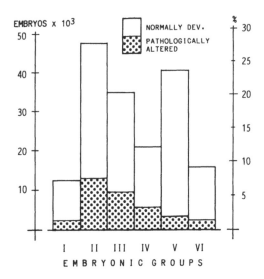

FIGURE 4. Embryogram of adult female *Acanthocheilonema viteae* immediately after recovery of worms from host animals 14 to 20 weeks after infection (n = 46 ♀♀; mean length, 55.0 ± 8.7 mm). The intrauterine developing stages are classified in six groups: (I) eggs, two- and four-cell stages; (II) small morulae; (III) big morulae; (IV) advanced embryos; (V) ring and pretzel stages; (VI) stretched microfilariae. (Based on data from Mössinger, J. and Barthold, E., *Parasitol Res.*, 74, 84, 1988.)

was observed between *B. pahangi* and *A. viteae*. On a wet weight basis, *B. pahangi* females utilized over three times more glucose,[136] which has been shown to be taken up via the transcuticular route.[49] Most interestingly, no evidence for an oral ingestion of materials has been obtained from worms *in vitro*.[49]

With the culture system described by Franke and Weinstein,[120] considerable survival of *B. pahangi* adults was achieved. Under low oxygen tensions in medium NCTC 135 + IMDM (1:1) plus 10% iFBS, worms were maintained for 3.5 up to almost 7 months and microfilariae collected from culture after 3 3/4 months were still capable to develop in mosquitoes.[137] However, with the same culture system, in another study, survival of worms was poor. Motility was often significantly reduced already after 5 d of culture.[136] Besides the parasite specimen themselves, the other component being obviously different in these two cultivation experiments was the sample of FBS used.

C. ACANTHOCHEILONEMA VITEAE (SYN. DIPETALONEMA VITEAE)

A. viteae belongs to those filarial species of which the intrauterine development of microfilariae *in vivo* has been studied in detail, both qualitatively and quantitatively.[18,20] After mating of the sexes, inseminated eggs developed to microfilariae within approximately 20 d and the first microfilariae were released about 2 months after infection (see Table 2).[18,20] Worms recovered from jirds 4 to 5 months after infection contained an average number of 173,000 intrauterine stages of all developmental forms per female (Figure 4).[20] The average theoretical microfilarial production of such females was calculated as approximately 7000 microfilariae per female per day, based on the number of *normally developed* intrauterine stages (137,000 per female) divided by the duration of embryogenesis.[20]

In vitro, adult worms survived up to 3 weeks in a culture medium consisting of 50% Hank's salt solution, 20% medium 199, and 30% FBS.[138] Occasionally, up to 18 d, microfilariae were released. The highest daily output was 15,000 microfilariae per female; average numbers during day 2 to day 4 of culture ranged from 4000 to 6500 microfilariae per female. Microfilarial release was distinctly reduced when 10% serum from *immunized* host animals

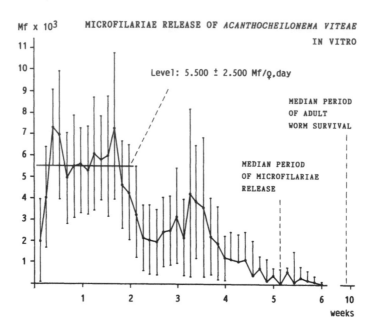

FIGURE 5. Course of microfilariae release of female *Acanthocheilonema vi-teae* maintained *in vitro* (n = 21 ♀♀; mean length, 58.2 ± 5.8 mm). Worms were transferred to fresh medium daily. Values represent arithmetic mean microfilariae counts ± standard deviation (±SD).

was used.[138] In medium RPMI 1640 supplemented with 20% iFBS, adult worms survived up to 100 d, and up to 75 d microfilariae were released.[84] Unfortunately, in this study, microfilarial release was not followed in detail; only a daily release of "some dozens to several thousands microfilariae" has been reported.[84]

In one of our recent studies,[139] female worms survived for 67 d (range, 50 to 83 d); males survived for 64 d (range, 55 to 80 d). The medium used consisted of 70% RPMI 1640, 20% MEM, and 10% newborn calf serum plus 400 μg/ml bovine serum albumin and MEM vitamins. Microfilariae were released for an average period of 36 d (range, 30 to 64 d), but only during the first two weeks, the number of microfilariae extruded was in the theoretically expected order of magnitude (5500 microfilariae per female per day, Figure 5). Investigations on the uterine contents of females at the end of maintenance revealed that worms were almost completely depleted of developmental stages, especially almost exclusively pathologically altered early embryonic forms (morulae) were found. Obviously, the production of viable eggs and possibly insemination of eggs had ceased rather soon *in vitro*. In total, during the whole period of maintenance, 1 female released an average number of 120,000 microfilariae, which corresponds to about 70% of the *total* number of intrauterine stages in control worms. Based on the information obtained from the embryogram (see Figure 4), it may be concluded that only such stages completed their development to microfilariae, which were already "big morulae" when the culture was started, i.e., *in vitro* no real production of new microfilariae occurred, but only preformed embryos completed in the intrauterine development to microfilariae. Under a gas phase of 5% CO_2 in N_2 instead of 5% CO_2 in air, survival of worms was extended to 90 to 100 d, but embryogenesis of microfilariae was not improved.[160]

D. *LITOMOSOIDES CARINII*

The reproduction of *L. carinii in vivo* has been sufficiently investigated to have available a good baseline for the evaluation of *in vitro* experiments.[19,140,141] The duration of embryogenesis of microfilariae is about 18 d; the expected *theoretical average fecundity* has been estimated as approximately 20,000 microfilariae per female per day.[19]

Maintained *in vitro* in Ringer's plus 25% horse serum, during the first 19 to 22 h the number of microfilariae released varied from 4,000 to 43,000 (average 18,000) per female worm; maximum survival of adults was 14 d.[140] With medium 199 + 33% rat serum, maximum survival of male and female worms was 23 d; for 18 d microfilariae were extruded. The number of microfilariae released fluctuated considerably with the highest output occurring during day 1 to day 3 of maintenance (approximately 40,000 microfilariae per 2 d).[70] In total, over the whole period of maintenance, on average 8300 microfilariae per female per day were released. The addition of glucose, chick embryo extract, or nutrient agar did not improve results and neither did the replacement of rat serum by rat plasma or horse or cotton rat serum.[70]

In addition to survival and microfilarial output, the rate of pathologically altered intrauterine stages after *in vitro* maintenance was used as a parameter for the evaluation of culture conditions.[19] With a mixture of medium RPMI 1640 (70%) plus MEM (20%), distinctly less intrauterine stages degenerated compared to medium 199 (both supplemented with 10% FBS).[19]

In a recent study with medium RPMI 1640 (70%), MEM (20%), plus 10% inactivated cotton rat serum, female worms survived up to 28 d; males survived up to 30 d. Within the first 2 weeks, a total number of 300,000 microfilariae (approximately 21,000 per day) were released per female worm.[139] However, at the end of maintenance, worms contained only a few remnant intrauterine stages which were almost exclusively pathologically altered. As with *A. viteae* females, *in vitro* approximately 60 to 70% of the uterine stages, which were already present at the start of maintenance, developed to microfilariae, but no new embryonic stages were produced.[160] Oogenesis or further fertilization of eggs did not occur in other studies with medium 199 supplemented with various sera (horse, jird, multimammate rat, cow, man).[142,143] Young adult worms did not mate *in vitro*, although on average they survived for 18 d, and females grew in length by 28% (medium 199 + 33% cotton rat serum).[143]

E. *ONCHOCERCA* SPP.

Compared with the filariae of laboratory animals, it is relatively difficult to obtain homogeneous samples of *Onchocerca* spp. adults for the initiation of cultures. First of all, the human or bovine hosts are naturally infected in the field and therefore one individual usually may harbor parasites of considerably different age. Furthermore, high rates of female worms may be reproductively inactive due to the occurrence of reproduction cycles[144] or possible seasonal changes in reproductivity.[146] Finally, the enzymatical isolation of worms with collagenase, which is particularly necessary to obtain undamaged females, may cause further variability. Enzyme concentration, incubation temperature, and duration may considerably influence the viability of worms.[91,145-147]

1. *Onchocerca volvulus*

O. volvulus females maintained in a biphasic culture system (blood agar overlaid with Hank's balanced salt solution) survived for only 1 to 7 d after enzymatical isolation. In contrast, two male worms which had spontaneously emerged from excised nodules remained alive for 22 and 81 d.[91] In another study, enzymatically isolated male and female worms survived on average for 11 (maximum 28 d) and 14.5 d (maximum 42 d), respectively, when incubated in medium 199 plus 10% human serum.[145] The use of human serum yielded better results than FBS or cow serum. The output of microfilariae was highly variable ranging from 0 to 4000 microfilariae per female per day. The longest period of microfilarial release was 25 d. Usually only less than 10% of the microfilariae contained in a female were released. Interestingly, it was observed that *O. volvulus* microfilariae emerged actively from the vulva one after another.[145] This might be different from other filarial species, e.g., *L. carinii*, where groups of microfilariae are expelled *in vitro* by contractions of the vulva.[148]

In recent studies with other basic media, longer survival of worms has been achieved.[147] In medium NCTC 135 + IMDM (1:1) *without* serum, female worms survived up to 4 weeks;[149] with 10% FBS + 5% human serum added, survival was extended to up to 6 weeks (gas phase, 5% O_2, 5% CO_2, 90% N_2; 34°C). Besides motility, lactate excretion was determined as a parameter for viability.[147,149]

2. Onchocerca gutturosa

O. gutturosa adults have been maintained in various cell-free or cell culture systems.[146] Male worms were isolated by dissection; females were isolated by collagenase digestion of the nuchal ligament connective tissue of cattle. With cell-free culture, good results were obtained using medium NCTC 135 + 10% iFBS. Males on average survived for 39 d (range, 25 to 41 d); female worms survived for 4 to 60 d. Under an atmosphere of 5% CO_2 in N_2, survival was marginally better than with 5% CO_2 in air.

Incorporation of bovine kidney or bovine trachea cells as a feeder layer slightly enhanced survival of worms. Exceptionally long survival was achieved when rhesus monkey kidney cells (LLCMK$_2$) were cocultivated (medium MEM + 10% iFBS, 5% CO_2 in air). Males survived for 6 to 7 months; the survival of females varied considerably from 18 d to more than 7 months, which was possibly due to their differing exposures to collagenase during isolation. Only 5 out of a total of 18 females released microfilariae, for a maximum period of 32 d. The output was low, ranging from 1 to 863 microfilariae per female per day. It was thought that the low reproductive activity of the worms in culture possibly was due to the fact that most females had been obtained during the winter months when reproduction might be on a naturally low level as no transmission of the infection occurs.[146]

IX. GENERAL EVALUATION

A. SUMMARY

In vitro culture of filarial worms has made significant progress during the last years, particularly with the cultivation of the infective third larval stage. However, there are certain developmental steps in the filarial life cycle which have not yet been overcome *in vitro* (Figure 6).

Microfilariae of most filarial species cultivated developed to the sausage form, but not beyond. Although in some culture systems formation of the larval gut, synthesis of a new cuticle, and separation of the old one occurred, the larvae did not grow and develop further to the second or even the third stage. An exception to this was one study where an insect cell culture system had been used, resulting in development of microfilariae distinctly beyond the sausage form.[92] However, this degree of development has never been achieved with other filarial species, although similar culture systems have been applied.

It has been impossible so far to reproduce *in vitro* the complex *intracellular* environment in which these parasites develop in their vectors. Even when whole organs recovered from naturally infected vectors, i.e., with the microfilariae *in situ*, were cultivated, the larvae did only develop to the late second stage, except in a single report.[98]

Generally, whole blood cultures, insect cell cultures, and cell-free culture systems yielded qualitatively comparable results. At present, only a limited number of factors which were essential or beneficial for early microfilarial development to the sausage form *in vitro* can be described. Incubation temperatures below 30°C proved to be essential as did a serum supplement to a basic culture medium. Among numerous cell culture media tested, Ham's F12 medium has been repeatedly beneficial. Reducing agents such as glutathione enhanced development as did organic acids and sugars or Grace's insect medium. Cocultivation of

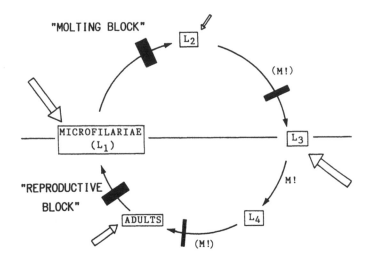

FIGURE 6.   Scheme of *in vitro* experimentation with filarial worms and the corresponding results. White arrows point to stages the cultures have been initiated with. Black bars represent developmental blocks which have not been overcome *in vitro*. The most distinct developmental barriers are, firstly, between the microfilariae and the second stage larvae (L_2, molting block) and, secondly, in the sexual reproduction of the parasite ("reproduction block"). Abbreviations: M!, molt occurred repeatedly with different species in various studies; (M!), molt occurred only in a few exceptional studies; L_3, third stage larva; L_4, fourth stage larva.

insect cells and arthropod hemolymph with microfilariae yielded higher rates of sausage forms and extended their survival in several studies, but not in general. With those microfilariae possessing a sheath, exsheathment, either induced artificially or occurring spontaneously, is a prerequisite for development.

Infective, third stage larvae of most filarial species cultivated, developed to the fourth larval stage *in vitro*, i.e., a complete molt occurred with the third stage cuticle being cast off. It is noteworthy that with *D. immitis* and *Onchocerca* spp. larvae, this molt occurred distinctly earlier than with the other species cultivated which corresponds with the *in vivo* situation.

The fourth larval stages usually survived up to several weeks or months and from various studies limited growth or partial development of the reproductive tract had been reported. Most encouraging, complete development of vector derived third stage larvae to young adult worms *in vitro* has been achieved with *B. malayi* and *A. viteae* larvae. However, this high degree of development currently has still to be regarded as an exception.

Cell-free culture systems and those with mammalian cell lines incorporated have been applied with comparable success. Consistently, origin and sample of the serum supplement added to the basic medium proved to be a critical factor in all culture systems. Cocultivation of certain cell lines improved culture conditions considerably in various studies, particularly when the rhesus monkey kidney cell line $LLCMK_2$ had been used. Low oxygen tensions of 30 to 50 mmHG in the culture medium have been shown to be essential for the development of *A. viteae* third stage larvae in cell-free culture systems.

Adult worms have been maintained *in vitro* up to several weeks or months. However, continuous production of microfilariae by female worms has not been observed. With *A. viteae* and *L. carinii* adults, approximately 60 to 70% of preformed embryonic stages completed their intrauterine development to microfilariae, which had been released, but no new oocytes had been produced *in vitro*. Furthermore, mating has not been observed. As with most other parasitic helminths, reproduction *in vitro* also seems to be a critical developmental step with filarial worms.

B. BASIC PROBLEMS AND RESEARCH NEEDS

It is a regular observation that parasite stages grown *in vitro* show signs of degeneration or malformations, and even if they develop to be morphologically normal, they are usually stunted, compared to specimens grown in natural host animals. Abnormal or retarded development is most probably due to a lack of specific nutritives or insufficient physicochemical culture conditions, or a combination of those factors.

Media — Several commercially available cell culture media proved to be suitable basic media for the cultivation of filarial worms, but none of them alone supported distinct growth or development. Results obtained with a certain medium may largely depend on the supplements added.

Sera — A serum supplement to the basic medium has been an indispensable component of any successful culture system. The quality of different lots of serum may vary considerably which probably is the most serious problem not only with filarial culture, but with *in vitro* cultivation of parasites or cells in general. Thus, obtaining consistent, reproducible results is not only impeded between different working groups, but even within one laboratory if a certain batch of serum has run out of stock. Replacement of serum by less complex and more standardized supplements as commercial serum substitutes and serum fractions or, which would be the optimum case, by defined substances is highly desirable. This does also apply to other supplements as arthropod hemolymph or organ extracts. Complex components which have been shown to enable or enhance development *in vitro* should be analyzed for their constituents to get possible clues which are the essential factors required for normal development. Whether or not heat inactivation of serum is beneficial should be tested with any sample used. Some of the exceptionally good results have been achieved with non-inactivated sera.

Cells — Especially if only less suitable sera have been available, cocultivation of cells and filarial stages improved culture conditions considerably. However, with this approach, further variables are introduced to the culture system and it is still to be investigated which are the beneficial factors deriving from the cells.

Antibiotics — Filarial worms have been found to harbor intracellular microorganisms.[150,151] These symbionts have not yet been definitely described and the kind of relationship between them and the filarial hosts is not yet understood. Although there is no strong evidence at present, the use of certain antibiotics, to which these microorganisms might be sensitive, possibly could influence the results of cultivation experiments.

Physicochemical conditions — Temperature, gas phase, pH, and osmolarity used should represent the natural conditions as closely as possible, if corresponding data are available. In regard of the gas phase, this concept has been particularly successful with the cultivation of third stage larvae. The spatial relations and substrates encountered by the parasite stages in the currently used culture systems, which are almost exclusively stationary cultures with a thinly liquid medium, certainly do not represent natural conditions. Solid material as glass wool has been shown to aid shedding of cuticles.[119] Possibly, the use of extracellular matrix substances such as collagen, for example, or of more viscous culture media could be advantageous. Eventually, unnaturally vigorous movements of adult worms[19] and their failure to ingest solid material *in vitro*[49] might be due to inadequate physicochemical conditions.

Defined supplements — The possible effects of defined supplements such as growth factors, hormones, or certain nutritives may interfere with, or be masked by, the serum supplement used. On the other hand, if the serum is omitted, certain substances may be ineffective as they may act only in a synergistic manner. This is a serious obstacle in research towards a chemically defined culture medium.

Nevertheless, to develop completely defined culture systems should be a research objective. In consideration of the complexity of the filarial life cycles and their habitats, this certainly is not an easy task to undertake. To improve present culture systems on a logical basis, more analytical studies on nutrition and physiology of filarial worms, *in vivo* and *in vitro* as well, are urgently needed.

C. APPLICATIONS

Potential research applications of culture systems are numerous, especially if the parasites' development and physiology *in vitro* are normal. This is particularly important if regulatory or metabolic studies are to be performed. Current studies with filarial worms are predominantly concerned with drug action and screening or collection of excretory/secretory antigens from *in vitro* cultures.[147,152,153] Many of these studies have been short-term incubation experiments lasting only several hours to a few days. For screening purposes, it is essential to define reliable, objective measures for the evaluation of drug efficacy. Besides worm motility,[154,155] inhibition of molting,[111,127] and suppression of microfilariae release,[136,156,157] various biochemical parameters such as glucose consumption,[136] lactate output,[147,149] and leakage of incorporated radiolabeled adenine[158,159] have been investigated.

X. ACKNOWLEDGMENTS

I am grateful to Professor P. Wenk and Drs. H. Schulz-Key and P. T. Soboslay for invaluable discussions on the subject. Furthermore, I wish to thank Professor P. P. Weinstein for making available his extensive reprint collection during a research scholarship at the Department of Biological Sciences, University of Notre Dame, Notre Dame, Indiana. Thanks are also due to Mrs. Petra Eulzer for the sometimes laborious processing of the text and to Mrs. Christa Eck and Dr. Sabine Kläger for taking photographs (Figure 3) of cultured larvae. Throughout the period of manuscript preparation, I received support from the German Ministry for Research and Technology (Grant FKZ 0318865 A). I am particularly grateful to Professor J. D. Smyth for giving me the opportunity to contribute to this volume.

REFERENCES

1. **Taylor, A. E. R. and Baker, J. R.,** *The Cultivation of Parasites In Vitro*, Blackwell Scientific, Oxford, 1968, 301.
2. **Weinstein, P. P.,** A review of the culture of filarial worms, in *H. D. Srivastava Commemoration Volume*, Izatnagar, Uttar Pradesh, 1970, 493.
3. **Pudney, M. and Varma, M. G. R.,** Present state of knowledge of *in vitro* cultivation of filariae, in *The In Vitro Cultivation of the Pathogens of Tropical Diseases*, Schwabe, Basel, 1980, 367.
4. **Douvres, F. W. and Urban, J. F.,** Nematoda except parasites of insects, in *In Vitro Methods for Parasite Cultivation*, Taylor, A. E. R. and Baker, J. R., Eds., Academic Press, Cambridge, England, 1987, 326.
5. **Smyth, J. D.,** *Introduction to Animal Parasitology*, 2nd ed., Hodder and Stoughton, London, 1976, 349.
6. **Schacher, J. F.,** Laboratory models in filariasis: a review of filarial life-cycle patterns, *Southeast Asian J. Trop. Med. Public Health*, 4, 336, 1973.
7. **Taylor, A. E. R.,** Studies on the microfilariae of *Loa loa, Wuchereria bancrofti, Brugia pahangi, Dirofilaria immitis, D. repens* and *D. aethiopis, J. Helminthol.*, 34, 13, 1960.
8. **Schacher, J. F.,** Morphology of the microfilariae of *Brugia pahangi* and of the larval stages in the mosquitoes, *J. Parasitol.*, 48, 679, 1962.
9. **Laurence, B. R. and Simpson, M. G.,** The microfilaria of *Brugia*: a first stage nematode larva, *J. Helminthol.*, 45, 23, 1971.
10. **Nelson, G. S.,** Observations on the development of *Setaria labiatopapillosa* using new techniques for infecting *Aedes aegypti* with this nematode, *J. Helminthol.*, 36, 281, 1962.
11. **Ewert, A.,** Exsheathment of the microfilariae of *Brugia pahangi* in susceptible and refractory mosquitoes, *Am. J. Trop. Med. Hyg.*, 14, 260, 1965.
12. **Scott, J. A.,** A description of certain details of growth and development of the filarial worms of cotton rats and the value of these observations in the study of immunity. Libro Homenaje Dr. Ed. Caballero y C., Mexico, 1960, 501.

13. **Taylor, A. E. R.,** The development of *Dirofilaria immitis* in the mosquito *Aedes aegypti, J. Helminthol.,* 34, 27, 1960.
14. **Renz, A. and Wenk, P.,** Intracellular development of the cotton rat filaria *Litomosoides carinii* in the vector mite *Ornithonyssus bacoti, Trans. R. Soc. Trop. Med. Hyg.,* 75, 166, 1981.
15. **Orhiel, T. C.,** Morphology of the larval stages of *Dirofilaria immitis* in the dog, *J. Parasitol.,* 47, 251, 1961.
16. **Schacher, J. F.,** Developmental stages of *Brugia pahangi* in the final host, *J. Parasitol.,* 48, 693, 1962.
17. **Ash, L. R. and Riley, M.,** Development of *Brugia pahangi* in the jird, *Meriones unguiculatus,* with notes on infections in other rodents, *J. Parasitol.,* 56, 962, 1970.
18. **Johnson, M. H., Orhiel, T. C., and Beaver, P. C.,** *Dipetalonema viteae* in the experimentally infected jird, *Meriones unguiculatus.* I. Insemination, development from egg to microfilariae, reinsemination, and longevity of mated and unmated worms, *J. Parasitol.,* 60, 302, 1974.
19. **Mössinger, J. and Wenk, P.,** Fecundity of *Litomosoides carinii* (Nematoda, Filarioidea) *in vivo* and *in vitro, Z. Parasitenkd.,* 72, 121, 1986.
20. **Mössinger, J. and Barthold, E.,** Fecundity and localization of *Dipetalonema viteae* (Nematoda, Filarioidea) in the jird *Meriones unguiculatus, Parasitol. Res.,* 74, 84, 1988.
21. **Worms, M. J. and Terry, R. J.,** *Dipetalonema witei,* filarial parasite of the jird, *Meriones libycus.* I. Maintenance in the laboratory, *J. Parasitol.,* 47, 963, 1961.
22. **Bain, O.,** Morphologie des stades larvaires d'*Onchocerca volvulus* chez *Simulium damnosum* et redescription de la microfilaire, *Annales de Parasitologie (Paris),* 44, 69, 1969.
23. **Schulz-Key, H. and Wenk, P.,** The transmission of *Onchocerca tarsicola* (Filarioidea: Onchocercidae) by *Odagmia ornata* and *Prosimulium nigripes* (Diptera: Simuliidae), *J. Helminthol.,* 55, 161, 1981.
24. **Ham, P. J., Smail, A. J., and Groeger, B. K.,** Surface carbohydrate changes on *Onchocerca lienalis* larvae as they develop from microfilariae to the infective third-stage in *Simulium ornatum, J. Helminthol.,* 62, 195, 1988.
25. **Lichtenfels, J. R., Pilitt, A., Kotani, T., and Powers, K. G.,** Morphogenesis of developmental stages of *Dirofilaria immitis* (Nematoda) in the dog, *Proc. Helminthol. Soc. Wash.,* 52, 98, 1985.
26. **Edeson, J. F. B. and Buckley, J. J. C.,** Studies on filariasis in Malaya: On the migration and rate of growth of *Wuchereria malayi* in experimentally infected cats, *Ann. Trop. Med. Parasitol.,* 53, 113, 1959.
27. **Ash, L. R. and Riley, J. M.,** Development of subperiodic *Brugia malayi* in the jird, *Meriones unguiculatus,* with notes on infections in other rodents, *J. Parasitol.,* 56, 969, 1970.
28. **Scott, J. A., MacDonald, E. M., and Terman, B.,** A description of the stages in the life cycle of the filarial worm *Litomosoides carinii, J. Parasitol.,* 37, 425, 1951.
29. **Schulz-Key, H.,** personal communication, 1988.
30. Epidemiology of onchocerciasis, Report of a WHO expert committee, *W. H. O. Tech. Rep. Ser.,* 597, 25, 1976.
31. **Bianco, A. E. and Muller, R. L.,** Experimental transmission of *Onchocerca lienalis* to calves, in *Parasites — Their World and Ours,* Abstr. 5th Int. Congress of Parasitology, Toronto, August 1982; *Mol. Biochem. Parasitol.,* Suppl. 349, 1982.
32. **Ash, L. R. and Schacher, J. F.,** Early life cycle and larval morphogenesis of *Wuchereria bancrofti* in the jird, *Meriones unguiculatus, J. Parasitol.,* 57, 1043, 1971.
33. **Cross, J. H., Partono, F., Hsu, M.-Y. K., Ash, L. R., and Oemijati, S.,** Experimental transmission of *Wuchereria bancrofti* to monkeys, *Am. J. Trop. Med. Hyg.,* 28, 56, 1979.
34. **Howells, R. E.,** Filariae: dynamics of the surface, in *The Host Invader Interplay,* Van den Bossche, H., Ed., Elsevier/North-Holland, Amsterdam, 1980, 69.
35. **Jaffe, J. J. and Doremus, H. M.,** Metabolic patterns of *Dirofilaria immitis* microfilariae *in vitro, J. Parasitol.,* 56, 254, 1970.
36. **Ando, K., Mitsuhashi, J., and Kitamura, S.,** Uptake of amino acids and glucose by microfilariae of *Dirofilaria immitis in vitro, Am. J. Trop. Med. Hyg.,* 29, 213, 1980.
37. **Kharat, I. and Harinath, B. C.,** Uptake of ^{14}C-labelled sugars and amino acids by *Wuchereria bancrofti* microfilariae *in vitro, Ind. J. Exp. Biol.,* 23, 118, 1985.
38. **Lehane, M. J.,** The first stage larva of *Brugia pahangi* in *Aedes togoi*: an ultrastructural study, *Int. J. Parasitol.,* 8, 202, 1978.
39. **Kan, S.-P. and Ho, B.-C.,** Development of *Brugia pahangi* in the flight muscles of *Aedes togoi.* Ultrastructural changes in the infected muscles fibres and the infecting filarial larvae, *Am. J. Trop. Med. Hyg.,* 22, 179, 1973.
40. **Lehane, M. J. and Laurence, B. R.,** Flight muscle ultrastructure of susceptible and refractory mosquitoes parasitized by larval *Brugia pahangi, Parasitology,* 74, 87, 1977.
41. **Beckett, E. B. and Boothroyd, B.,** Mode of nutrition of the larvae of the filarial nematode *Brugia pahangi, Parasitology,* 60, 21, 1970.
42. **Collin, W. K.,** Ultrastructural morphology of the esophageal region of the infective larva of *Brugia pahangi* (Nematoda: Filarioidea), *J. Parasitol.,* 57, 449, 1971.

43. **Chen, S. N. and Howells, R. E.**, The uptake *in vitro* of dyes, monosaccharides and amino acids by the filarial worm *Brugia pahangi, Parasitology*, 78, 343, 1979.
44. **Rao, U. R., Chandrashekar, R., Parab, P. B., and Subrahmanyam, D.**, The effect of *p*-aminobenzoic acid and folic acid on the development of infective larvae of *Brugia malayi* in *Aedes aegypti, Acta Trop.*, 41, 61, 1984.
45. **Travi, B. L. and Orihel, T. C.**, Development of *Brugia malayi* and *Dirofilaria immitis* in *Aeda aegypti*: effect of the host's nutrition, *Trop. Med. Parasitol.*, 38, 19, 1987.
46. **Sneller, V.-P. and Dadd, R. H.**, *Brugia pahangi*: development in *Aedes aegypti* reared axenically on a defined synthetic diet, *Exp. Parasitol.*, 51, 169, 1981.
47. **Yoeli, M., Upmanis, R. S., and Most, H.**, Studies on filariasis. II. The relation between hormonal activities of the adult mosquito and the growth of *Dirofilaria immitis, Exp. Parasitol.*, 12, 125, 1962.
48. **Gwadz, R. W. and Spielman, A.**, Development of the filarial nematode, *Brugia pahangi*, in *Aedes aegypti* mosquitoes: nondependence upon host hormones, *J. Parasitol.*, 60, 134, 1974.
49. **Howells, R. E. and Chen, S. N.**, *Brugia pahangi*: feeding and nutrient uptake *in vitro* and *in vivo, Exp. Parasitol.*, 51, 42, 1981.
50. **Howells, R. E.**, Dynamics of the filarial surface, in *Filariasis*, Evered, D. and Clark, S., Eds., Ciba Foundation Seminar 1987, John Wiley & Sons, Chichester, 1987, 94.
51. **Chen, S. N. and Howells, R. E.**, *Brugia pahangi*: uptake and incorporation of adenosine and thymidine, *Exp. Parasitol.*, 47, 209, 1979.
52. **Chen, S. N. and Howells, R. E.**, The uptake *in vitro* of monosaccharids, disaccharide and nucleid acid precursors by adult *Dirofilaria immitis, Ann. Trop. Med. Parasitol.*, 75, 329, 1981.
53. **Howells, R. E., Mendis, A. M., and Bray, P. G.**, The mechanisms of amino acid uptake by *Brugia pahangi in vitro, Z. Parasitenkd.*, 69, 247, 1983.
54. **Howells, R. E., Bray, P. G., and Allan, D.**, An analysis of glucose uptake by *Brugia pahangi, Trans. R. Soc. Trop. Med. Hyg.*, 78, 273, 1984.
55. **Bueding, E.**, Studies on the metabolism of the filarial worm, *Litomosoides carinii, J. Exp. Medicine*, 89, 107, 1949.
56. **Comley, J. C. W. and Mendis, A. H. W.**, Advances in the biochemistry of filariae, *Parasitol. Today*, 2, 34, 1986.
57. **Stürchler, D., Wyss, F., and Hanck, A.**, Retinol, onchocerciasis and *Onchocerca volvulus, Trans. R. Soc. Trop. Med. Hyg.*, 75, 617, 1981.
58. **Storey, D. M.**, Vitamin A deficiency and the development of *Litomosoides carinii* (Nematoda, Filarioidea) in cotton rats, *Z. Parasitenkd.*, 67, 309, 1982.
59. **Barrett, J.**, *Biochemistry of Parasitic Helminths*, McMillan, London, 1981, 233.
60. **Franz, M. and Büttner, D. W.**, The fine structure of adult *Onchocerca volvulus*. V. The digestive tract and the reproductive system of the female worm, *Tropenmed. Parasitol.*, 34, 155, 1983.
61. **Sawyer, T. K. and Weinstein, P. P.**, Studies on the microfilariae of the dog heartworm *Dirofilaria immitis*: separation of parasites from whole blood, *J. Parasitol.*, 49, 39, 1963.
62. **Obeck, D. K.**, Blood microfilariae: new and existing techniques for isolation, *J. Parasitol.*, 59, 220, 1973.
63. **Feldmeier, H., Bienzle, U., Schuh, D., Geister, R., and Guggenmoos-Holzmann, I.**, Detection of *Dirofilaria immitis* microfilariae in peripheral blood. A quantitative comparison of the efficiency and sensitivity of four techniques, *Acta Trop.*, 43, 131, 1986.
64. **Devaney, E. and Howells, R. E.**, The exsheathment of *Brugia pahangi* microfilariae under controlled conditions *in vitro, Ann. Trop. Med. Parasitol.*, 73, 227, 1979.
65. **Aoki, Y.**, Exsheathing phenomenon of microfilaria *in vitro* (I), *Trop. Med. (Nagasaki)*, 13, 134, 1971.
66. **Aoki, Y.**, Exsheathing phenomenon of microfilaria *in vitro* (II), *Trop. Med. (Nagasaki)*, 13, 170, 1971.
67. **Irungu, L. W.**, Studies on the *in vitro* exsheathment of *B. pahangi*. II. The *in vitro* exsheathment of *B. pahangi* microfilariae incubated with mosquito tissues and cells, *Insect Sci. Its Appl.*, 8, 49, 1987.
68. **Schulz-Key, H., Albiez, E. J., and Büttner, D. W.**, Isolation of living adult *Onchocerca volvulus* from nodules, *Tropenmed. Parasitol.*, 28, 428, 1977.
69. **Earl, P. R.**, Filariae from the dog *in vitro, Ann. N.Y. Acad. Sci.*, 77, 163, 1959.
70. **Taylor, A. E. R.**, Maintenance of filarial worms *in vitro, Exp. Parasitol.*, 9, 113, 1960.
71. **Sawyer, T. K. and Weinstein, P. P.**, Survival of *Dirofilaria immitis* microfilariae in modified physiological saline solutions, *J. Parasitol.*, 47(4, Sect. 2), 24, 1961.
72. **Sawyer, T. K. and Weinstein, P. P.**, The *in vitro* development of microfilariae of the dog heartworm *Dirofilaria immitis* to the "sausage-form", *J. Parasitol.*, 49, 218, 1963.
73. **Sawyer, T. K. and Weinstein, P. P.**, Morphologic changes occurring in canine microfilariae maintained in whole blood cultures, *Am. J. Vet. Res.*, 24, 402, 1963.
74. **Klein, J. B. and Bradley, R. E.**, Induction of morphological changes in microfilariae from *Dirofilaria immitis* by *in vitro* culture techniques, *J. Parasitol.*, 60, 649, 1974.

75. **Cupp, E. W.,** Development of Filariae in Mosquito Cell Cultures, unpublished document, WHO/FIL/ 73.111, World Health Organization, Geneva, 1973.
76. **Devaney, E. and Howells, R. E.,** Culture systems for the maintenance and development of microfilariae, *Ann. Trop. Med. Parasitol.,* 73, 139, 1979.
77. **Ando, K., Chinzei, Y., and Kitamura, S.,** Condition of *in vitro* culture for development of microfilariae of *Dirofilaria immitis, Jpn. J. Parasitol.,* 6, 483, 1980.
78. **Sneller, V.-P. and Weinstein, P. P.,** *In vitro* development of *Dirofilaria immitis* microfilariae: selection of culture media and serum levels, *Int. J. Parasitol.,* 12, 233, 1982.
79. **Abraham, D., Lauria, S., Mika-Grieve, M., Lok, J. B., and Grieve, R. B.,** Survival and viability of *Dirofilaria immitis* microfilariae in defined and undefined culture media, *J. Parasitol.,* 72, 776, 1986.
80. **Lok, J. B., Mika-Grieve, M., and Grieve, R. B.,** Cryopreservation of *Dirofilaria immitis* microfilariae and third-stage larvae, *J. Helminthol.,* 57, 319, 1983.
81. **Ando, K. and Kitamura, S.,** Osmotic pressure-dependent development of microfilariae of *Dirofilaria immitis in vitro, Jpn. J. Parasitol.,* 31, 219, 1982.
82. **Devaney, E. and Howells, R. E.,** The development of exsheathed microfilariae of *Brugia pahangi* and *Brugia malayi* in mosquito cell lines, *Ann. Trop. Med. Parasitol.,* 73, 387, 1979.
83. **Garrigues, R. M., Hockmeyer, W. T., and Balinas, J. C.,** Development of Larval Forms of *Brugia pahangi In Vitro,* Abstr. 96, American Society of Parasitology, Annual Meeting, New Orleans, LA, 1975.
84. **Miegeville, M., Bouillard, C., Marjolet, M., Vermeil, C., and Avranche, P.,** Nouvelle contribution a l'étude de *Dipetalonema viteae* maintien en survie des adultes *in vitro, Bull. Soc. Pathol. Exot.,* 74, 207, 1981.
85. **Weinstein, P. P.,** Development *in vitro* of the microfilariae of *Wuchereria bancrofti* and of *Litomosoides carinii* as far as the sausage form, *Trans. R. Soc. Trop. Med. Hyg.,* 57, 236, 1963.
86. **Kharat, I., Satyanarayana, U., Ghirnikar, S. N., and Harinath, B. C.,** *In vitro* cultivation of *Wuchereria bancrofti* microfilariae, *Ind. J. Exp. Biol.,* 18, 1245, 1980.
87. **Kharat, J. and Harinath, B. C.,** *In vivo* and *in vitro* development of *Wuchereria bancrofti* microfilariae, *Ind. J. Med. Res.,* 82, 127, 1985.
88. **Pudney, M.,** The use of arthropod cell cultures for the *in vitro* study of filariae (*Onchocerca* sp.), in *Invertebrate Systems In Vitro,* Kurstak, E., Maramorosch, K., and Dübendorfer, A., Eds., Elsevier/North-Holland, Amsterdam, 1980, 317.
89. **Devaney, E. and Howells, R. E.,** The differentiation of microfilariae of *Onchocerca lienalis in vitro, Ann. Trop. Med. Parasitol.,* 77, 103, 1983.
90. **Pollack, R. J., Lok, J. B., and Donnelly, J. J.,** Analysis of glutathione-enhanced differentiation by microfilariae of *Onchocerca lienalis* (Filarioidea: Onchocercidae) *in vitro, J. Parasitol.,* 74, 353, 1988.
91. **Schiller, E. L., Turner, V. M., Figueroa Marroquin, H., and D'Antonio, R.,** The cryopreservation and *in vitro* cultivation of larval *Onchocerca volvulus, Am. J. Trop. Med. Hyg.,* 28, 997, 1979.
92. **Wood, D. E. and Suitor, E. C.,** *In vitro* development of microfilariae of *Macacanema formosana* in mosquito cell cultures, *Nature,* 211, 868, 1966.
93. **Dhar, D. N., Basu, P. C., and Pattanayak, S.,** *In vitro* cultivation of *Dirofilaria repens* microfilariae up to the sausage stage, *Ind. J. Med. Res.,* 55, 915, 1967.
94. **Devaney, E.,** The development of *Dirofilaria immitis* in cultured malpighian tubules, *Acta Trop.,* 38, 251, 1981.
95. **Ogura, N., Kobayashi, M., and Yamamoto, H.,** Studies on filariasis. IV. Critical period of molting from the second to the third stage in filarial larvae of *Brugia pahangi, Dokkyo J. Med. Sc.,* 8, 74, 1981.
96. **Simpson, M. G. and Laurence, B. R.,** Histochemical studies on microfilariae, *Parasitology,* 64, 61, 1972.
97. **Weber, P.,** Electron microscope study on the developmental stages of *Wuchereria bancrofti* in the intermediate host: structure of the digestive tract, *Trop. Med. Parasitol.,* 36, 109, 1985.
98. **Lehane, M. J.,** The first stage larva of *Brugia pahangi* in *Aedes togoi*: an ultrastructural study, *Int. J. Parasitol.,* 8, 23, 1978.
99. **Weinstein, P. P.,** personal communication (1976), in *Methods of Cultivating Parasites In Vitro,* Taylor, A. E. R. and Baker, J. R., Eds., Academic Press, London, 1978, 246.
100. **Yoeli, M., Upmanis, R. S., and Most, H.,** Studies on filariasis. III. Partial growth of the mammalian stages of *Dirofilaria immitis in vitro, Exp. Parasitol.,* 15, 325, 1964.
101. **Sawyer, T. K.,** *In vitro* culture of third-stage larvae of *Dirofilaria immitis, J. Parasitol.,* 49(5, Sect. 2), 59, 1963.
102. **Sawyer, T. K.,** Molting and exsheathment *in vitro* of third-stage *Dirofilaria immitis, J. Parasitol.,* 51, 1016, 1965.
103. **Sawyer, T. K. and Weinstein, P. P.,** Third molt of *Dirofilaria immitis in vitro* and *in vivo, J. Parasitol.,* 51(2, Sect. 2), 48, 1965.
104. **Abraham, D., Mok, M., Mika-Grieve, M., and Grieve, R. B.,** *In vitro* culture of *Dirofilaria immitis* third- and fourth-stage larvae under defined conditions, *J. Parasitol.,* 73, 377, 1987.

105. **Kumar, H., Sahai, R., and Rao, C. K.**, *In vitro* cultivation of infective larvae of *Dirofilaria immitis*, *J. Commun. Dis.*, 16, 77, 1984.

106. **Lok, J. B., Mika-Grieve, M., Grieve, R. B., and Chin, T. K.**, *In vitro* development of third- and fourth-stage larvae of *Dirofilaria immitis*: comparison of basal culture media, serum levels and possible serum substitutes, *Acta Trop.*, 41, 145, 1984.

107. **Wong, M. M., Knighton, R., Fidel, J., and Wada, M.**, *In vitro* cultures of infective-stage larvae of *Dirofilaria immitis* and *Brugia pahangi*, *Ann. Trop. Med. Parasitol.*, 76, 239, 1982.

108. **Devaney, E.**, *Dirofilaria immitis*: the moulting of the infective larva *in vitro*, *J. Helminthol.*, 59, 47, 1985.

109. **Delves, C. J. and Howells, R. E.**, Development of *Dirofilaria immitis* third stage larvae (Nematoda: Filarioidea) in micropore chambers implanted into surrogate hosts, *Trop. Med. Parasitol.*, 36, 29, 1985.

110. **Chen, S. N. and Howells, R. E.**, The *in vitro* cultivation of the infective larvae and the early mammalian stages of the filarial worm *Brugia pahangi*, *Ann. Trop. Med. Parasitol.*, 73, 473, 1979.

111. **Court, J. P. and Lees, G. M.**, Improvement of *in vitro* culture conditions of *Brugia pahangi* four day old developing larvae for use in an antifilarial drug assay, *Tropenmed. Parasitol.*, 34, 162, 1983.

112. **Huijun, Z., Zhenghou, T., Wengfan, C., and Xiaohui, Z.**, *In vitro* culture of infective larvae of periodic *Brugia malayi*, *J. Parasitol. Parasitic Dis.*, 2, 107, 1984.

113. **Mak, J. W., Lim, P. K. C., Sim, B. K. L., and Liew, L. M.**, *In vitro* cultivation of *Brugia malayi* and *Brugia pahangi* infective larvae to fourth and fifth stage, *Trans. R. Soc. Trop. Med. Hyg.*, 76, 702, 1982.

114. **Mak, J. W., Lim, P. K. C., Sim, B. K. L., and Liew, L. M.**, *Brugia malayi* and *Brugia pahangi*: cultivation *in vitro* of infective larvae to the fourth and fifth stages, *Exp. Parasitol.*, 55, 243, 1983.

115. **Fujimaki, Y., Shimada, M., Kimura, E., and Aoki, Y.**, DEC-inhibited development of third-stage *Brugia pahangi in vitro*, *Parasitol. Res.*, 74, 299, 1988.

116. **Tanner, M.**, *Dipetalonema viteae* (Filarioidea): development of the infective larvae *in vitro*, *Acta Trop.*, 38, 241, 1981.

117. **Tanner, M. and Weiss, N.**, *Dipetalonema viteae* (Filarioidea): evidence for a serum-dependent cytotoxicity against developing third and fourth stage larvae *in vitro*, *Acta Trop.*, 38, 325, 1981.

118. **Franke, E. D. and Weinstein, P. P.**, *In vitro* cultivation of *Dipetalonema viteae* third-stage larvae: effect of the gas phase, *J. Parasitol.*, 70, 493, 1984.

119. **Franke, E. D. and Weinstein, P. P.**, *In vitro* cultivation of *Dipetalonema viteae* third-stage larvae: evaluation of culture media, serum, and other supplements, *J. Parasitol.*, 70, 618, 1984.

120. **Franke, E. D. and Weinstein, P. P.**, *Dipetalonema viteae* (Nematoda: Filarioidea): culture of third-stage larvae to young adults *in vitro*, *Science*, 221, 16, 1983.

121. **Nelson, P. D., Weiner, D. J., Stromberg, B. E., and Abraham, D.**, *In vitro* cultivation of third-stage larvae of *Litomosoides carinii* to the fourth stage, *J. Parasitol.*, 68, 971, 1982.

122. **Zaraspe, G. and Cross, J. H.**, Attempt to culture *Wuchereria bancrofti in vitro*, *Southeast Asian J. Trop. Med. Public Health*, 17, 579, 1986.

123. **Franke, E. D., Riberu, W., and Wiady, I.**, *In vitro* cultivation of third stage larvae of *Wuchereria bancrofti* to the fourth stage, *Am. J. Trop. Med. Hyg.*, 37, 370, 1987.

124. **Court, J. P., Bianco, A. E., Townson, S., Ham, P. J., and Friedheim, E.**, Study on the activity of antiparasitic agents against *Onchocerca lienalis* third stage larvae *in vitro*, *Trop. Med. Parasitol.*, 36, 117, 1985.

125. **Lok, J. B., Pollack, R. J., Cupp, E. W., Bernardo, M. J., Donnelly, J. J., and Albiez, E. J.**, Development of *Onchocerca lienalis* and *O. volvulus* from the third to fourth larval stage *in vitro*, *Tropenmed. Parasitol.*, 35, 209, 1984.

126. **Pudney, M., Lichtfield, T., Bianco, A. E., and Mackenzie, C. D.**, Moulting and exsheathment of the infective larvae of *Onchocerca lienalis* (Filarioidea) *in vitro*, *Acta Trop.*, 45, 67, 1988.

127. **Lok, J. B., Pollack, R. J., and Donnelly, J. J.**, Studies of the growth-regulating effects of ivermectin on larval *Onchocerca lienalis in vitro*, *J. Parasitol.*, 73, 80, 1987.

128. **Townson, S., Connelly, C., and Ham, P. J.**, Cultivation of *Onchocerca lienalis* developing larvae *in vitro*, *Trop. Med. Parasitol.*, 37, 94, 1986.

129. **Strote, G.**, Successful use of a simple anaerobic system for the *in vitro* cultivation of infective larvae of *Onchocerca volvulus*, *Trans. R. Soc. Trop. Med. Hyg.*, 81, 174, 1987.

130. **Strote, G.**, Morphology of third and fourth larvae of *Onchocerca volvulus*, *Trop. Med. Parasitol.*, 38, 73, 1987.

131. **Strote, G.**, Studies on the activity of the Ciba Geigy compounds CGP 6140, 20376, 20309 and 21833 against third and fourth stage larvae of *Onchocerca volvulus*, *Trop. Med. Parasitol.*, 40, 51, 1989.

132. **Wong, M. M.**, Studies on microfilaremia in dogs. II. Levels of microfilaremia in relation to immunologic responses of the host, *Am. J. Trop. Med. Hyg.*, 13, 66, 1964.

133. **Tamashiro, W. K. and Palumbo, N. E.**, Diethylcarbamazine: *in vitro* inhibitory effects on microfilarial production by *Dirofilaria immitis*, *J. Parasitol.*, 71, 381, 1985.

134. **Weinstein, P. P. and Sawyer, T. K.**, Survival of adults of *Dirofilaria uniformis in vitro* and their production of microfilariae, *J. Parasitol.*, 47(4, Sect. 2), 23, 1961.

135. **Furman, A. and Ash, L. R.**, Characterization of the exposed cabohydrates on the sheath surface of *in vitro*-derived *Brugia pahangi* microfilariae by analysis of lectin binding, *J. Parasitol.*, 69, 1043, 1983.

136. **Court, J. P., Martin-Short, M., and Lees, G. M.**, A comparison of the response of *Dipetalonema viteae* and *Brugia pahangi* adult worms to antifilarial agents *in vitro*, *Trop. Med. Parasitol.*, 37, 375, 1986.

137. **Weinstein, P. P.**, Filariasis: problems and challenges, *Am. J. Trop. Med. Hyg.*, 35, 221, 1986.

138. **Weiss, N.**, Parasitologische und immunbiologische Untersuchungen über die durch *Dipetalonema viteae* erzeugte Nagetierfilariose, *Acta Trop.*, 27, 217, 1970.

139. **Mössinger, J., Wenk, P., and Schulz-Key, H.**, *In vitro* maintenance of adult *Dipetalonema viteae* and *Litomosoides carinii* (Nematoda, Filarioidea): fecundity and survival in cell-free culture systems, *Zentralbl. Bakteriol. Parasitenkd. Infektionskr. Hyg. Abt. Orig. Reihe A*, 267, 303, 1987.

140. **Hawking, F.**, The reproductive system of *Litomosoides carinii*, a filarial parasite of the cotton rat. III. The number of microfilariae produced, *Ann. Trop. Med. Parasitol.*, 48, 382, 1954.

141. **McFadzean, J. A. and Smiles, J.**, Studies of *Litomosoides carinii* by phase-contrast microscopy: the development of larvae, *J. Helminthol.*, 30, 25, 1956.

142. **Wenk, P., Illgen, B., and Seitz, H. M.**, *In vitro*-Versuche mit *Litomosoides carinii* (Nematoda: Filarioidea). I. Haltung von adulten Weibchen und Mikrofilarien sowie Ausstoß von Mikrofilarien in verschiedenen Kulturmedien, *Z. Parasitenkd.*, 55, 63, 1978.

143. **Illgen, B.**, *In vitro*-Versuche mit *Litomosoides carinii* (Nematoda: Filarioidea). II. Einfluß verschiedener Seren auf die Freisetzung von Mikrofilarien sowie Versuche zur Kopulationsbereitschaft der Würmer, *Z. Parasitenkd.*, 67, 227, 1982.

144. **Schulz-Key, H. and Karam, M.**, Periodic reproduction of *Onchocerca volvulus*, *Parasitol. Today*, 2, 284, 1986.

145. **Engelbrecht, F. and Schulz-Key, H.**, Observations on adult *Onchocerca volvulus* maintained *in vitro*, *Trans. R. Soc. Trop. Med. Hyg.*, 78, 212, 1984.

146. **Townson, S., Connely, C., and Muller, R.**, Optimization of culture conditions for the maintenance of *Onchocerca gutturosa* adult worms *in vitro*, *J. Helminthol.*, 60, 323, 1986.

147. **Walter, R. D.**, *In vitro* maintenance of *Onchocerca volvulus* for harvest of excretory and secretory products, *Trop. Med. Parasit.*, 39, 448, 1988.

148. **Mössinger, J.**, personal observation, 1986.

149. **Walter, R. D., Arias, A. E., Rathaur, S., and Diekmann, E. F.**, Secretory enzymes from *Onchocerca volvulus* during *in vitro*-maintenance, *Trop. Med. Parasitol.*, 38, 339, 1987.

150. **McLaren, D. J., Worms, M. J., Laurence, B. R., and Simpson, M. G.**, Micro-organisms in filarial larvae (Nematoda), *Trans. R. Soc. Trop. Med. Hyg.*, 69, 509, 1975.

151. **Kozek, W. J. and Marroquin, H. F.**, Intracytoplasmatic bacteria in *Onchocerca volvulus*, *Am. J. Trop. Med. Hyg.*, 26, 663, 1977.

152. **Maizels, R. M., Denham, D. A., and Sutano, I.**, Secreted and circulating antigens of the filarial parasite *Brugia pahangi*: analysis of in vitro released components and detection of parasite products in vivo, *Mol. Biochem. Parasitol.*, 17, 277, 1985.

153. **Devaney, E.**, The biochemical and immunochemical characterisation of a 30 kilodalton surface antigen of *Brugia pahangi*, *Mol. Biochem. Parasitol.*, 27, 83, 1988.

154. **Nowak, M., Hutchinson, G. W., and Copeman, D. B.**, *In vitro* drug screening in isolated male *Onchocerca gibsoni* using motility suppression, *Trop. Med. Parasitol.*, 38, 128, 1987.

155. **Townson, S., Connelly, C., Dobinson, A., and Muller, R.**, Drug activity against *Onchocerca gutturosa* males *in vitro*: a model for chemotherapeutic research on onchocerciasis, *J. Helminthol.*, 61, 271, 1987.

156. **Mössinger, J. and Wenk, P.**, Efficacy of ivermectin on adults and microfilarie of *Dipetalonema viteae* and *Litomosoides carinii* in vitro, *Trop. Med. Parasitol.*, 39, 78, 1988.

157. **Tekwani, B. L., Tripathi, L. M., Shukla, O. P., and Ghatak, S. P.**, Suppression of release of microfilariae in vitro from *Setaria cervi* (Nematoda: Filarioidea) by chlorpromazine, *J. Parasitol.*, 74(5), 893, 1988.

158. **Comley, J. C. W., Rees, M. J., and O'Dowd, A. B.**, Leakage of incorporated radiolabelled adenine — a marker for drug-induced damage of macrofilariae in vitro, *Trop. Med. Parasitol.*, 39, 221, 1988.

159. **Comley, J. W. C. and Rees, M. J.**, The assessment of in vitro worm death: a comparison of 7 non-subjective parameters, *Trop. Med. Parasitol.*, 40, 82, 1989.

160. **Mössinger, J.**, Fecundity of the Filarial Worms *Litomosoides carinii* and *Acanthocheilonema viteae In Vivo* and During Maintenance in Culture Systems, Ph.D. thesis, University of Tübingen, Tübingen, FRG, 1989.

Chapter 7

NEMATODES: OTHER THAN FILARIOIDEA

J. D. Smyth

TABLE OF CONTENTS

I. BASIC PROBLEMS OF INTESTINAL NEMATODE CULTURE

A. GENERAL COMMENTS

1. Maintenance of Adult Worms

In general, the establishment of cultures of nematodes — intestinal species in particular — have proved to be somewhat simpler than with trematodes and cestodes, probably because nematodes, with their thick cuticle, are generally more "robust" than the platyhelminths with their "naked" cytoplasmic tegument. This is particularly true of adult worms and there are some excellent *in vitro* techniques available which enable some species to be "maintained" in a reasonable physiological condition, for long periods, as adjudged by the normality of their egg production (see Figure 2). Inevitably, however, such cultures degenerate and eventually die. "Maintenance" then is not a major problem and nematodes — especially large species, such as *Ascaris*, have been widely used for biochemical and physiological studies.

2. Problems of Growth and Differentiation

Reproducing the appropriate conditions in which nematodes can be grown from the egg to the adult worm *in vitro*, however, has proved to be much more demanding and just as difficult as with trematodes and cestodes with which it, more or less, shares the same basic problems. With some species, it has been necessary to develop two- or even three-step culture systems to deal with the varying requirements of the different, free-living and parasitic, stages in the life cycle. Although it has proved to be relatively easy to grow the free-living larval stages from the egg, the parasitic stages have proved to be much more demanding, particular problems being encountered with (1) transformation (molting) of L_3 larvae to L_4 and of L_4 to adult worms and (2) the induction of copulation and fertilization with the production of fertile eggs. Even when fertile eggs have been formed, they have only been in relatively small numbers and, with most species, producing conditions which encourage copulation *in vitro* remains a major problem.

3. Aims of This Chapter

In reviewing the *in vitro* culture of this group, no attempt is made to review the many early (largely unsuccessful) attempts to culture many species. Most of these have been comprehensively reviewed by other authors (see below). Early work of special historical interest is quoted, however, where appropriate. The aim of this chapter is to give an account of the most successful *in vitro* techniques for major representatives of each superfamily with some evaluation of the success and/or limitations of the system used. Except in the case of very well-known species, details of the natural life cycle are also provided so that comparison between *in vitro* and *in vivo* development can be readily made. Particular attention is paid to those species which have been cultured to ovigerous adults.

B. LITERATURE REVIEW

Early work has been reviewed by Leyland,[1] Silverman,[2] Smyth,[3] and Hansen and Hansen,[4] and Taylor and Baker.[5] More recent work has been comprehensively reviewed by Douvres and Urban.[6]

II. ANCYLOSTOMOIDEA

A. *NECATOR AMERICANUS*

1. Natural Host, Basic Biology

Man is the natural host of this species, the adult worms occurring largely in the second and third portions of the jejunum and occasionally in the duodenum. Its distribution is limited to the tropics and subtropics. The general biology, morphology, and epidemiology is too well known to be described here.

2. Experimental Hosts, Life Cycle

From the experimental point of view, it has been difficult to find a suitable animal host to maintain this species in the laboratory and this has made it one of the least studied of the common nematodes of man. Dogs, rabbits, and adult hamsters are reasonably good hosts, but not for regular and long-term passage.[7]

Neonatal hamsters (*Mesocricetus auratus*), 3 to 5 d old, however, have proved to be excellent hosts[8] and readily produce adequate supplies of adult worms. The time scale of development in this host (Figure 1) as given by Behnke et al.[7] is as follows:

Day 0:	Filariform (L3) infective larvae (grown from fecal egg cultures by the standard techniques) are placed on the skin.
Day 1 to 2:	Most worms stay in skin for 48 h.
Day 3 to 4:	Migration to lungs commences.
Day 3 to 8:	Some larvae may develop to L4(?).
Day 7+:	Migration to intestine via trachea and esophagus occurs.
Day 11 to 18:	Most larvae molt to L4 in intestine.
Day 17 to 21:	Final molt to adults occurs.
Day 34+:	Eggs may appear as early as day 34, but continuous egg production does not commence until day 42 and reaches a peak about day 70.

3. Technique: *In Vitro* Culture

Egg to Filariform Larva — Filariform larvae (L3) were grown from eggs under sterile conditions in an early study by Weinstein[9] using the technique which he found successful for *Ancylostoma* spp. based on chick embryo extract (CEE$_{50}$) or liver extract.

Filariform larva to adult — This has not been achieved with this species. Using methods successful for *Nippostrongylus brasiliensis* (= *N. muris*), Weinstein and Jones[10] obtained partial growth and differentiation from filariform larvae. By day 3, some 70% of larvae had exsheathed and growth and differentiation became evident on day 5. On day 34, when cultures were terminated, the mouth parts were clearly differentiated and the genital tract was in the early stages of formation, but adult worms were not obtained.

4. Technique: *In Vitro* Maintenance of Adult Worms

Although it has not yet been found possible to grow this species from infective larva to adult, it has been possible to "maintain" male and females for short periods (for the collection of antigens) or for long periods, in excess of a month.

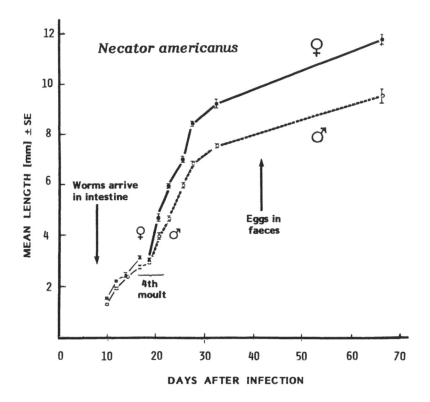

FIGURE 1. *Necator americanus*: growth of male and female worms in neonatal hamsters. (After Behnke, J. M., Paul, V., and Rajasekariah, G. R., *Trans. R. Soc. Trop. Med. Hyg.*, 80, 146, 1986. With permission.)

Short-term maintenance (for antigen collection)[11] — This technique was successfully used for short-term (48-h) maintenance of adult *Necator americanus*, obtained from neonatal hamsters by this method of Sen and Seth,[8] and used for the collection of excretory/secretory (E/S) antigens.

Procedure:

1. Wash worms for 4 h in 0.9% (w/v) NaCl + 100 IU ml^{-1} of penicillin and 100 μg ml^{-1} of streptomycin.
2. Transfer to maintenance medium, consisting of Eagles' MEM supplemented with:
   - 3% (v/v) normal hamster serum
   - 2 mM of L-glutamine
   - Antibiotics, as above
   - 100 μCi ml^{-1} [³H] leucine (specific activity >120 Ci mmol^{-1})
   - 250 μCi ml^{-1} [³⁵S] methionine (specific activity > 1000 Ci mmol^{-1})
3. Incubate at 37°C in a gas phase of 95% air/5% CO_2 for 48 h.

This incubation resulted in a wide range of protein species being excreted into the medium, antigens of 15, 30, 33, 34, 44, 46, and 49 kDa being identified.

Long-term maintenance[12] — Xianxiang and Jiajun[12] maintained female and male worms for 52 and 43 d, respectively, in a medium of heat-inactivated puppy serum and 5 to 50% Parker 199; pH 7.43; 37°C in air. Cultures consisted of 2 to 10 worms in 1.5 ml of medium or 30 worms in 3 ml. The medium was changed daily; 64.2% of worms survived for 2 weeks and 21% survived for >1 month. That this medium was reasonably successful in maintaining worms in a good physiological condition is reflected in the fact that copulation was observed and fertilized as well as unfertilized eggs were found in the medium. Full details of the culture technique are only available in Chinese, for which the original paper should be consulted.

5. Evaluation

Although satisfactory techniques for short-term maintenance of adult worms are available, attempts to culture the adults from filariform larvae has only had limited success and needs further experimentation. No recent work on this species appears to have been attempted.

B. *ANCYLOSTOMA* SPP.

Although a number of attempts have been made to culture various species of *Ancylostoma* from egg to adult, this aim has not yet being achieved. Nevertheless, some useful techniques have been established and infective larvae of several species have been cultured from eggs, and egg-laying adult *Ancylostoma duodenale* and *A. caninum* have been maintained *in vitro* for long periods.

1. Technique: *In Vitro* Culture: Eggs to Filariform Larvae

Preparation of eggs in sterile condition presents little difficulty, as their resistant nature allows the use of strong antiseptic reagents such as 95% ethanol, 10% formol, or 0.1% mercuric chloride or mixtures such as White's solution, followed by washing in antibiotics.

In early experiments, Weinstein[13] cultured the L3 filariform larvae of *A. duodenale* and *A. caninum* from eggs using complex media involving 50% CEE_{50} (chick embryo extract) and/or rat liver extract. CEE_{50} is prepared by homogenizing 1 g of tissue with 1 ml of BSS. The term is open to misinterpretation. To avoid misunderstanding, it should be made clear that the supernatant prepared by centrifuging the above 1:1 homogenate is considered to be 100% homogenate.[10]

In the best results, more than 80% of larvae developed to the filariform stage. Details of his technique have been summarized by Taylor and Baker.[5] The larvae of *A. tubaeforme*[4] have also been similarly cultured. For practical details the original papers should be consulted.

2. Technique: *In Vitro* Culture: Filariform Larvae to Adults

A number of workers have attempted to grow the adult worm from free-living larvae, but with little success. For a review of this early work, see Hansen and Hansen.[4] The best result was probably that of Banerjee[14] using a modification of Leland's Ae medium, which is a relatively simple medium consisting of 50 ml of CEE_{50}, 19 ml of serum (from helminth-free lambs or calves), 15 ml of cystine-fortified 2% sodium caseinate, 5 ml of 2% Sigma liver concentrate (0.1 g/100 ml of medium), 6 ml of vitamin mixture (Eagle's 100X), 1 ml of antiobiotic mixture, and 9 ml of Earle's BSS. The culture temperature was 25°C at which the limit of larval survival was 25 d, but when the culture temperature was raised to body temperature (37°C), all worms died within 48 h.

3. Technique: Maintenance of Adult Worms[15]

Source of material — Specimens of *Ancylostoma duodenale* were collected from artificially infected puppies, and those of *A. caninum* were collected from naturally infected dogs. Most experiments were carried out with *A. duodenale*.

Culture media and conditions — The following media were used: Tyrode's saline, human serum, young and adult dog serum, CEE_{50}, placenta extract, and placenta extract plus young dog serum. Cultivation was carried out in 3.5-cm culture dishes at 37°C, presumably in a gas phase of air. Organisms were transferred daily to a new culture dish with fresh medium. The size of the worms was randomly measured from time to time.

Results — "Young" *A. duodenale* survived somewhat better than adult worms, best results being obtained in young dog serum. The smaller the size of the worm, the better it appeared to adapt to the culture environment. Most (86%) adult worms did not survive for a month and only 8% survived for 1 to 4 months, the longest survival time being less than 200 d; 41% of young worms survived for less than 1 month and 55% survived for 1 to 4

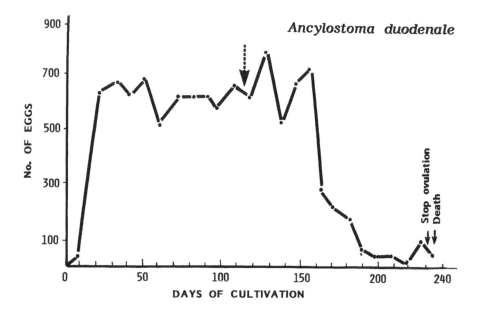

FIGURE 2. *Ancylostoma duodenale*: output of eggs from adult worms maintained *in vitro*; before arrow: average output from two to five worms; after arrow: output from one worm. (After Fenglin, W., Rongjun, H., Xianxiang, H., and Xihui, Z., *Chin. Med. J.*, 92, 839, 1979. With permission.)

months, the longest survival time being more than 200 d. Adult worms laid eggs (Figure 2) from the outset, with most eggs being fertilized during the early days of culture. As culture continued, the numbers of fertilized eggs gradually diminished and were replaced by unfertilized eggs, the worms eventually dying. Four pairs of worms were observed to copulate. These authors (apparently?) obtained comparable results with *A. caninum*, but copulation did not take place, although earlier work had found that copulation took place in worms cultured in human serum.[5]

4. Evaluation

Shortage of experimental material has probably restricted work on this species. A survival time of 200 d for adult worms means that many aspects of its physiology *in vitro* could be studied as well as providing a useful drug-screening protocol. A suitable technique for growing L3 larvae to adults still awaits development.

III. ASCARIDOIDEA

A. *ASCARIS SUUM*
1. General Comments

The ready availability of this species from slaughter houses has made it a favorite model for nematode *in vitro* cultivation. Early attempts to culture this species, which met with varying degrees of success have been reviewed by Taylor and Baker[5] and Hansen and Hansen[4] and will not be discussed further here. More recent — and more successful — experiments have been succinctly reviewed by Douvres and Urban,[6] who themselves have been responsible for much progress in the whole field of nematode culture. No attempt is made here to review all the variety of techniques used for *A. suum*, but an account is given of the most successful techniques now available for each phase of development of this species.

2. Life Cycle

In natural pig host (Figure 3) — Infection is brought about by ingestion of viable

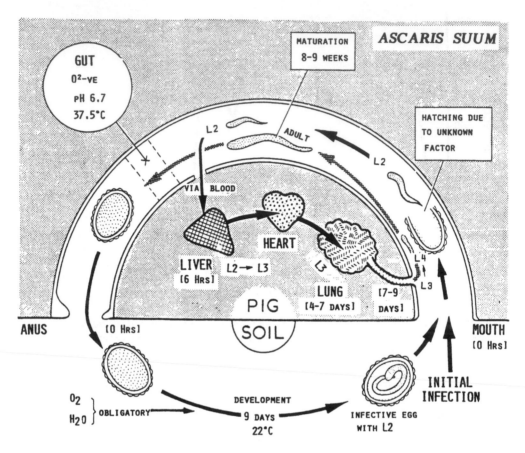

FIGURE 3. *Ascaris suum*: life cycle. Times of development are based on Douvres et al.[18] (From Smyth, J. D., *An Introduction to Animal Parasitology*, 2nd ed., Edward Arnold, London, 1976, 328. With permission.)

(embryonated) eggs which hatch largely in the duodenum, but some hatching takes place in the stomach. The released larva is either a second-stage larva (L2) (i.e., it has molted once within the egg) or a larva still within the sheath of its first molt. On hatching, larvae burrow into the mucosa, penetrate blood vessels, and appear as second-stage larvae in the liver within 6 h post-infection (PI). They remain in the liver for a few days and develop to the early third-stage larva (L3). Larvae then continue to the heart and are carried via the pulmonary arteries to the lungs, appearing there within 4 to 7 d; 7 or 8 d PI, they break out of capillaries into the alveoli and finally work their way up the trachea and reach the small intestine of day 8 to 10 PI. Within the intestine, larvae begin the next molt on the ninth day and are in the fourth stage (L4) by day 10 PI. The prepatent period of *A. suum* in pigs is 40 to 53 d, which is less than that of *A. lumbricoides* (54 to 61 d) in pigs.[3]

In experimental hosts — In rabbits, development is practically identical with that in pigs in that early L4 appear in the intestine on 11 d PI.[16] In mice, infection terminates 4 d PI with L3 in the liver. In guinea pigs, L3 are found in the lungs on day 7, with a few migrating to the intestine.

3. Morphology of Larval Stages

In order to assess the success — or otherwise — of development *in vitro*, it is important to have information of the morphology of the various larval stages. The most useful characters for differentiating the late fourth and early fifth stages are the labial, cervical, and cuticular structures.[17] These have been worked out in detail by Douvres et al.[18] some of which are shown in Figure 4. The original paper also contains some superb micrographs of these to which reference can be made.

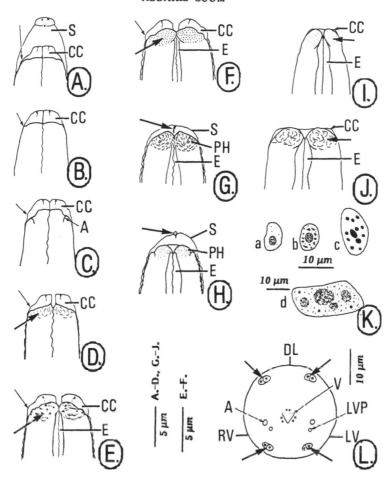

FIGURE 4. *Ascaris suum*: morphology of larvae from first molt to L3. (A) Larva in first molt pressed from infective egg; (B, C) L2 from intestine, lateral and ventral views, respectively; (D) late L2 from liver, 10 h PI (postinfection); (E, F) lateral view of late L2 from liver, 24 and 18 h PI, respectively. (In A to F, note that small arrows point to bulge in labial cuticle marking location of one of the large double papillae; large arrows point to cephalic tissues.) (G, H) Larvae in early and late second molt, respectively, from liver 28 h PI, lateral views; note separation of labial cuticle, herewith identified as sheath (S), from underlying provisional head. Large arrows point to distinctive features associated with second molt, namely, the knot-like thickening (in G) formed from the cuticular lining of the vestibule of L2 and the oil droplet (in H) appearing to emanate from the vestibule opening. (I, J) L3 from liver, 36 and 48 h PI, respectively, lateral views. Arrows point to cephalic tissues. (K) Development of excretory sinus nucleus (a, b, c, d); (L) *en face* view of late L3 from lungs, 10 days PI. Large arrows point to papillae. Abbreviations: A, amphid; CC, labial (cephalic) cuticle; DL, dorsal lip; E, esophagus; LV, left subventral lip; LVP, lateroventral papilla; LV, left subventral lip; PH, provisional head; RV, right subventral lip; S, sheath; V, vestibule opening. (From Douvres, F. W., Tromba, F. G., and Malakatis, G. M., *J. Parasitol.*, 55, 689, 1969. With permission.)

4. Eggs

Extraction from worms — Eggs (after embryonation, see below) can be safely stored in 0.1 to 2% formalin for up to 8 months at 5°C. Eggs (unembryonated; Figure 5) can be obtained from the uteri of females and separated and hatched by the methods of Costello et al.[19] and Urban et al.[20] which provides clean, decoated eggs in sterile condition, as follows:

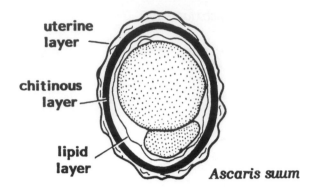

FIGURE 5. *Ascaris suum*: egg (unembryonated). (From Smyth, J. D., *An Introduction to Animal Parasitology,* 2nd ed., Edward Arnold, London, 1976, 305. With permission.)

1. Resect uteri from fresh worms and place in 0.5% *N* NaOH.
2. Transfer to a glass Dounce (or equivalent) tissue grinder, with a clearance of 0.12 to 0.17 mm between tube and pestle.
3. Homogenize by hand.
4. Leaving the pestle inserted in the mortar, pour off the homogenate containing the eggs; the undisrupted tissue remains at the bottom of the tube.
5. Decoat the eggs by centrifuging in 0.5 *N* NaOH at 200 × g; three times is usually sufficient.
6. Wash five times with distilled water.

Embryonation and hatching eggs *in vitro* — Early methods for hatching *Ascaris* eggs are somewhat lengthy and not generally appropriate for providing larvae suitable for cultivation *in vitro*. The following method[20] has been specifically designed for *in vitro* culture work (Figure 6).

1. Suspend eggs in 0.1% formalin in a separating funnel with continuous aeration from a small pump at 25°C for 28 to 30 d. Check for development by pressing larvae free on a slide under a coverslip; first molt or second stage larva should be seen.
2. Infective eggs may be stored, if desired, in 0.1% formalin at 4°C until required for use.
3. For hatching, wash five times by centrifugation in 40-ml suspension in 0.85% NaCl at 200 × g for 1 min with no braking of the motor.
4. Suspend pellet of eggs in 20 ml of a solution of 5.25% sodium hypochlorite (undiluted commercial Chlorox®) at 37°C followed by ten washes in 0.85% saline. This removes most of the chitinous layer of the egg (see Figure 6).
5. Suspend eggs in 4 to 8 ml of Earle's BSS (without phenol red) in a flat-bottomed 125-ml flask with a layer of glass beads (4- to 6-mm diameter) and a spinner bar. Place flask in incubator at 37°C over a magnetic stirrer, at a speed to slowly move the beads, for 30 to 40 min. Examine microscopically; hatching generally occurs in 15 to 30 min. Generally 80 to 90% of eggs hatch and few larvae are damaged.
6. Transfer larval suspension to a centrifuge tube and wash by centrifugation three times in Earle's BSS at 37°C to remove eggs debris.
7. Layer larval suspension over a pad of cotton gauze (dry weight of 2.5 g) immersed in Earle's BSS in a 10-cm funnel and maintained at 37°C for 1.5 to 2 h.
8. Collect (aseptically), in a 5-ml aliquot, larvae which settle in the funnel neck and wash ten times in warm, sterile Dulbecco's phosphate buffered saline. About 80 to 90% of larval suspension will consist of L2 larvae; the remainder will be first-molt larvae.
9. A total of 95% of larvae should be undamaged and mobile and suitable for *in vitro* and *in vivo* experiments. A packed egg volume of 0.5 to 1.0 ml can be expected to yield approximately 5×10^5 to 1×10^6 viable larvae.

5. Technique: *In Vitro* Culture: L2 (From Eggs) to Adults

Douvres and Urban[21] have successfully grown sexually mature adult worms from L2, which were hatched from eggs by using the embryonated/hatching technique described above.

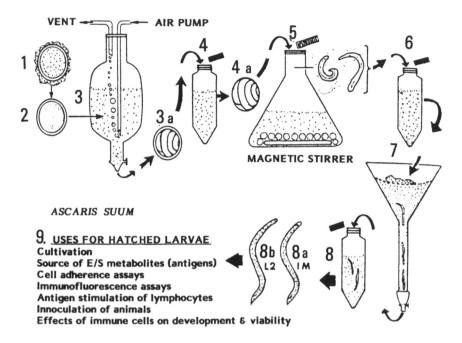

FIGURE 6. *Ascaris suum*: technique for embryonation of infective eggs and procedure for hatching: (1) fertilized egg; (2) proteinaceous (uterine) coat removed; (3) embryonation of eggs in aerated vessel; (3a) infective egg; (4) hypochlorite treatment and washing method; (4a) chitinous layer removed; (5) hatching method using glass beads and magnetic stirrer; (6) removal of egg debris; (7) filtration through cotton; (8) aseptic washing procedure; (8a) first-molt larva (IM); (8b) second-stage larvae (L2); (9) list of uses for larvae. (After Urban, J. F., Douvres, F. W., and Tromba, F. G., *Proc. Helminthol. Soc. Wash.*, 48, 241, 1981. With permission.)

This represented a considerable achievement and involved the development of complex media, KW-2, API-1 (see Appendix) and AP-18, in a two- or three-step culture system.

Starting material — The initial inocula consisted of 95 to 98% L2 and the remainder of the first molt.

Media — Details of the composition and preparation of these complex media, have been given by Douvres and Urban.[6,21] The basic media were supplemented with bovine hemin and/or L-cysteine.

Culture conditions — Culture vessels were either 1.0- or 1.8-liter glass bottles containing 80 ml of medium rotated at 1 revolution every 1.5 min at 39°C. Two-step and three-step systems were used involving the use of different media and two different gas phases as follows:

Step 1. Medium KW-2 + 10 mM L-cysteine; gas phase 95% N_2/5% CO_2 for 4 d, then 85% N_2/10% CO_2/5% O_2 onward

Step 2. Medium API-18 from day 11 to 18

Step 3. Medium API-1 + 24 μg ml^{-1} hemin from day 18 until termination

Results — Advanced development from hatched L2 larvae was obtained in all four culture systems used (Table 1). In step 1 cultures, L2 larvae advanced to L3 and third molt. In step 2 cultures, larvae advanced further to L4. In step 3 cultures, a few larvae developed to sexual maturity, one mature male and two mature females developing after 67 to 80 d. In the male, spermatozoa were produced and paired, weakly sclerotinized spicules appeared. In the females, oviposition of unfertilized eggs took place. One female produced some 1,356,000 eggs from days 67 to 125.

TABLE 1

Ascaris suum: Development of Second-Stage Larvae (L2) in a Three-Step Culture System[a]

Days in culture	Number of cultures	Total inoculum alive (%)	Live worms (%) in stage[b,c]						
			Third			Fourth			Young adult
			Early to mid	Late	Third molt	Early	Late	Fourth molt	
			Step 1 (Medium KW-2 + cysteine)						
4	8	81—98 (92)	59—93 (78)[d]	0	0	0	0	0	0
11	10	59—87 (77)	83—98 (90)[e]	3—10 (7)	+	0	0	0	0
			Step 2 (Medium API-18)						
18	7	30—60 (42)	55—80 (66)[e]	15—39 (24)	2—15 (8)	1—3 (2)	0	0	0
			Step 3 (Medium API-1 + hemin)						
25	5	13—25 (16)	11—33 (24)[f]	28—50 (36)	14—50 (36)	4—14 (9)	+ +	0	0
40	5	1—16 (3)	0	3—16 (8)	16—50 (36)	17—50 (40)	7—21 (15)	0.4—2 (1)	±0.4 (0.08)

[a] Data based on ranges (and averages) of estimated percentages or actual counts.
[b] Not shown for inocula were (average) 8% L2 and 14% 2M, on day 4, and 3% 2M, on day 11.
[c] + = five or fewer worms/culture; + + = six to ten worms/culture.
[d] 100% were early third stage.
[e] Majority were early third stage.
[f] Majority were mid third stage.

From Douvres, F. W. and Urban, J. F., *J. Parasitol.*, 69, 549, 1983. With permission.

6. Evaluation

The result obviously represents a major step forward in nematode *in vitro* culture and is the culmination of many years of persistent work by Douvres and co-workers. The next stage will clearly be to induce copulation *in vitro* and the production of fertile eggs. It is interesting to note that when cultures are initiated with L3 larvae, copulation and fertile eggs result, as described below.

7. Technique: *In Vitro* Culture of L3 Larvae (From Rabbits) to Adults[22]

It is clearly likely that cultures starting with the L3 stage derived from an animal host are likely to grow more successfully than those starting from the egg-derived L2 stage, and so it has turned out in practice. Using L3 derived from rabbits, Douvres and Urban[22] have grown adult worms which copulated *in vitro* and produced fertile eggs.

Material — New Zealand White rabbits were infected orally with 75,000 decoated, embryonated eggs and the lungs were removed every 7 d after infection. Larvae were recovered from the host tissue by the method of Urban and Douvres,[23] 95% being mid-to-late L3 and the rest being early L3.

Media — The following media were utilized: API-1, API-18, KW-2, and RMPI 1640 plus various supplements — serum, glutathione (reduced), and hemin — in a two-step culture system (see Table 2 and Appendix).

Culture conditions — Ten roller tube cultures were set up at 39°C with a gas phase of 85% N_2/5% O_2/10% CO_2 and inocula were transferred to fresh media and a new culture bottle on days 7, 14, 21, and 28.

Results — The most successful culture systems (J, Table 2) consisted of medium AP-18 for 7 d, followed thereafter by AP-18 supplemented with hemin. In this system, L3 developed to fourth molt at 14 d, to young adults at 20 d, and to mature adults at 53 d; the largest male was 110 mm long and the female was 140 mm long. Copulation was observed in one culture (system E). In cultures containing mature males and females, both fertilized and unfertilized eggs were found, 73,000 eggs being found in culture J on day 70. In general, morphogenesis was comparable to that obtained *in vivo*, but — as is common in other *in vitro* systems — the rate of development was slower.

8. Evaluation

This clearly is a most successful culture system as adjudged by the fact that copulation took place and fertilized eggs were produced *in vitro*. The only other cultured nematode in which copulation has been observed *in vitro* appears to be *Nematospiroides dubius*.[24]

9. Technique: *In Vitro* Antigen Collection

Antigen collection from cultured larvae is an obvious application from the above system and this has been used in the collection of E/S antigens. A difficulty is that the media which permit growth to mature worms always contain an undefined component, e.g., serum.

However, many nematode larvae — which include *Ascaris suum* — can survive for long periods in defined media, although only undergrowing limited development. Urban and Douvres[23] have developed a simple, stationary, multicell system utilizing RMPI 1640, supplemented with pyruvate and glutathione, in which L3 larvae will transform to L4. This system has been utilized for the collection of E/S antigens in an attempt to develop a vaccine against *A. suum* in pigs. Somewhat similar systems have been also used for screening anthelminthics and for nutritional and metabolic studies. Douvres and Urban[6] have briefly reviewed recent developments in this area.

B. OTHER ASCARIDS: ANISAKINE NEMATODES

For early work on the culture of anisakine nematodes, see Hansen and Hansen[4] and Taylor and Baker.[5] The biology of the "herring worm" *Anisakis simplex*, and related species,

TABLE 2

Ascaris suum: Protocols Used for the *In Vitro* Culture of Third-Stage Larvae (L3)[a] in Roller Culture Systems, Utilizing One or Two Media With or Without Supplements of Serum (S), Glutathionine (Reduced) (G) or Hemin (H) at 39°C[b]

Culture system[c]	Number of trials (and cultures)[d]	Nonsupplemented and supplemented media used during the following periods of incubation		Age (d) when cultures terminated
		Days 0—7	Day 7 on	
A	2 (4)	RPMI 1640 + S	RPMI 1640 + S + H	21,35
B	2 (4)	KW-2	KW-2 + H	30
C	3 (6)	API-18	API-18 + H	21,40,48
D	2 (4)	API-1 + G	API-1	32,35
E	7 (40)	API-1 + G	API-1 + H	26,32,45,54,65,80,112
F	2 (4)	API-1 + G	API-22 + H	30,40
G	4 (9)	API-1 + G	API-23 + H	30,77,80,97
H	2 (5)	RPMI 1640 + S	API-1 + H	49,54,123
I	2 (4)	KW-2	API-1 + H	30,123
J	6 (16)	API-18	API-1 + H	32,42,49,70,108, 116,123,140

[a] Recovered from the lungs of rabbits 7 d after oral inoculation with eggs.

[b] Cultures were prepared with 40 ml of medium for the first 7 d and 80 ml of medium from day 7 on, with a gas phase of 85% nitrogen/5% oxygen/10% carbon dioxide, in roller bottles that were rotated at 1 revolution per 1.5 min.

[c] Systems A through J were tested with inocula of 5000 L3; systems E and J were also tested with inocula of 2500 and 10,000 L3.

[d] A trial consisted of two or more cultures per trial.

From Douvres, F. W. and Urban, J. F., *Proc. Helminthol. Soc. Wash.*, 53, 256, 1986. With permission.

has been comprehensively reviewed by Reimer.[25] In an interesting and important application of *in vitro* culture, Carvajal et al.[26] cultured unknown species of *Anisakis* and *Phocanema* larvae from the Chilean hake, *Merluccius gayi*, using an undefined medium, and were successful in obtaining adult worms. These were identified as *A. simplex* and *Phocanema decipiens*; 42% of the former and 25% of the latter reached sexual maturity. For details of the media used, the original paper should be consulted. Almost identical results were obtained by Hurst[27] with larvae from New Zealand marine teleosts; *A. simplex* and *Pseudoterranovo decipiens* were identified from the adults grown from larvae.

IV. TRICHINELLOIDEA

A. *TRICHINELLA SPIRALIS*

This well-known nematode is the causative agent of trichinosis (trichinellosis) in domestic animals and man throughout the world. A vast literature exists on all aspects of its biology; a valuable recent synthesis and analysis of this is given in the monograph by Campbell.[28]

1. Life Cycle

The best known stage is the encapsulated third-stage (L3) larva which occurs in nature in the muscles of mammals, especially rats, but many hundreds of species can act as intermediate hosts. On ingestion by another mammal (including man), larvae are released in the duodenum and (being somewhat progenetic) rapidly become sexually mature. Copulation probably takes place within 30 to 36 h postinfection (PI).[3] The ovoviviparous females release large numbers of L2 larvae which penetrate the mucosa and migrate to striated muscle via connective tissue and the blood, encapsulation in muscles commencing about day 19

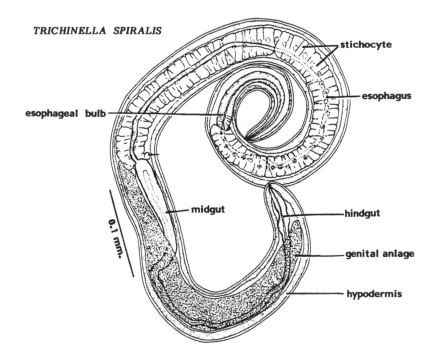

TRICHINELLA SPIRALIS

stichocyte

esophagus

esophageal bulb

0.1 mm.

midgut

hindgut

genital anlage

hypodermis

FIGURE 7.   *Trichinella spiralis*: muscle-stage larva (L2) extracted from cyst. (Based on Villela.[35])

PI.[29] The first molt occurs in the interuterine stage and two further molts take place during the encapsulation phase in the muscle.

2. Morphology

The morphology of the encysted larva in muscle and the adult male and female worms are shown in Figures 7 and 8.

3. Extraction of Muscle Larvae

Valuable practical details of techniques for obtaining and harvesting the various stages of the life cycle — adult worms from the intestine, newborn larvae and larvae from muscles — have been given by Blair.[30] To obtain muscle larvae suitable for *in vitro* culture, Berntzen[29] utilized the following "rapid" technique, the essential stages of which are as follows; for full details the original paper should be consulted.

- Add 2 g (homogenized) of infected rat muscle to 100 ml of 2% pepsin (1:10,000) in 1% HCl and digest at 37°C for 6 h.
- Pass the digest through No. 24-, 32-, and 42-mesh sieves into a Baerman apparatus and allow to settle for 1 h. Suspend in BSS in 50-ml conical centrifuge tubes and centrifuge at 1000 rpm for 6 min.
- Wash ten times in BSS plus antibiotics; transfer to culture system, with or without antibiotics.

NOTE: Some workers follow the pepsin treatment by a further digestion in a trypsin/pancreatin mixture and there are many variations of the above treatment in use.

4. Technique: *In Vitro* Culture: Encysted Larvae to (Near) Adults[31]

General comments — Early work on the cultivation of *T. spiralis* has been reviewed by Denham,[32] Sakamoto,[31] Taylor and Baker,[5] and Hansen and Hansen.[4] The most successful reported *in vitro* technique is that of Berntzen,[29] who used a complex continuous flow method, a number of different media, and a gas phase of 85% N_2/5% CO_2/10% O_2. He reported obtaining sexually differentiated adults which copulated, the females producing

TRICHINELLA SPIRALIS

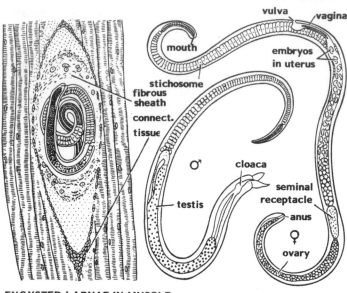

ENCYSTED LARVAE IN MUSCLE

FIGURE 8. *Trichinella spiralis*: morphology of male and female and an encysted larva in muscle fibres. (From Smyth, J. D., *An Introduction to Animal Parasitology*, 2nd ed., Edward Arnold, London, 1976, 317. With permission.)

"normal-appearing" L2 larvae. However, other workers have been unable to produce his results,[32] and as there is some doubt as to the merits of this system, it is not further discussed here. For details see the original paper of Taylor and Baker.[5] Although no other worker has succeeded in obtaining sexually mature adults *in vitro* from excysted larvae, promising results have been obtained by Sakamoto[31] using the following system:

Source of material — Muscle (L4) larvae (originally obtained from a polar bear, *Thalarctos maritimus*), were aseptically collected from guinea pigs, homogenized and excysted by a digestion procedure similar to that described above.

Medium and conditions — The medium consisted of NCTC 135 + 0.5% lactalbumen hydrolysate, 0.1% yeast extract, and 20% inactivated calf serum + antibiotics, in a gas phase of 90% N_2/5% CO_2/5% O_2. The culture system utilized 30-ml flasks or 30-ml roller tubes with 5 ml of medium at 37°C, pH 7.4, the medium being changed daily. On the third day of cultivation, worms were agitated in Rinaldini's solution with 0.2% trypsin + 1.0% pancreatin at 37°C using a magnetic stirrer for agitation. After treatment, worms were passed through Berntzen's desheathing apparatus (Figure 9).

Results — Within 24 h of culture, worms formed a thin sheath and some larvae exsheathed naturally. Considerable differentiation of genitalia occurred within 48 h with appearance of the caudal appendage in males and the anlagen of the ovary, uterus, vagina, and vaginal plate in females. No further differentiation took place after 3 to 4 weeks, but when further treated in trypsin/pancreatin and passage through the desheathing apparatus (see Figure 9) sexual differentiation was resumed. Sheathed males developed copulatory papillae, seminal vesicles and testes (without sperm) and females developed a vulva, uterus, and ovary (without embryos). Spermatogenesis and embryogenesis were, however, found in a few males and females and copulation was occasionally seen. Degeneration began at 3 weeks and death occurred at 4 weeks.

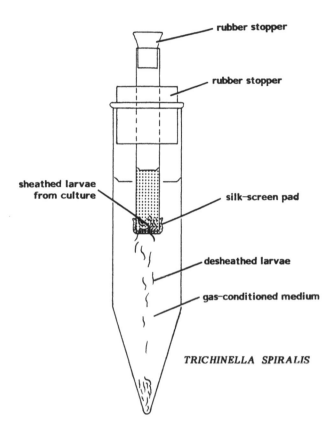

FIGURE 9.   Desheathing apparatus of Berntzen. Sheathed larvae from culture are placed in the inner tube and become desheathed after passing through the silk-screen pad. (Modified from Berntzen, A. K., *Exp. Parasitol.*, 16, 74, 1965. With permission.)

5. Technique: *In Vitro* Culture of Larvae in Muscle-Cell Tissue Cultures

Sakamoto[31] also successfully obtained the development of larvae in cell cultures of mouse muscle. Using flask with a collagen-covered surface, muscle cells at a concentration of 10^6 cells per milliliter were introduced into the flask. When a complete sheet of muscle cells were established, newborn larvae (isolated from adults by filtration using a 200-mesh sieve) were introduced. Larvae penetrated the muscle cells and began coiling within 5 d of introduction; stichocytes of the larvae became visible. Muscle cells which had been penetrated by larvae fell off the culture plate surface and eventually died.

6. Technique: *In Vitro* Culture of 24-h (*In Vivo*) Worms to Adults

As might be expected, worms allowed to develop in the intestine for 24 h before culture readily matured to adults *in vitro* in the same culture system as above. On day 2 of cultivation, males developed a seminal vesicle with abundant sperm, and eggs were present in the uterus of females. Fully developed larvae appeared on day 4 and large numbers were released into the medium. Adults died after 14 to 25 d of culture.

7. Evaluation

All the above systems developed by Sakamoto[31] look promising as tools for further work on the biology of this species. The muscle-cell system, in particular, could provide valuable data on the mechanism of cell penetration. The basic larva/adult system, too, looks promising and suggests that perhaps a minor improvement in the medium could lead to the development

TABLE 3

Medium Used for the *In Vitro* Maintenance of *Nippostrogylus brasiliensis* and *Trichinella spiralis* for Screening of Anthelmintics[33,34]

Tryptic digest of casein (Difco)	2.0 g		
Yeast extract (Difco or B.B.L.)	1.0 g		
D-Glucose	0.5 g		
Di-potassium hydrogen orthophosphate	0.08 g	Basal	
Potassium dihydrogen orthophosphate	0.08 g	medium	
Distilled water, add to	100.0 ml		Complete medium
Adjust pH to 7.2 with 10 *N* KOH			
Autoclave at 20 lb for 15 min			
Sodium benzylpenicillin	60 mg		
Streptomycin sulfate	100 mg		
5-Fluorocytosine	1 mg		
Calf serum (heat inactivated)	18 ml		
Bovine erythrocyte lysate (blood cells lysed in an equal volume of sterile water)	2.4 ml		

Note: Final pH = 7.2; osmolality = 268 milliosmoles.

From Jenkins, D. C., Armitage, R., and Carrington, T. S., *Z. Parasitenkd.*, 63, 261, 1980. With permission.

of fully mature males and females. The fact that copulation took place is encouraging and suggests that the physical conditions of cultivation approach those pertaining *in vivo*.

8. Technique: *In Vitro* Drug Screening

A simple *in vitro* screening system for drugs showing activity against the tissue stages of *T. spiralis* has been developed by Jenkins and Carrington.[33] The system makes use of a medium designed for *Nippostrongylus brasiliensis* (Table 3) which is capable of supporting the partial development of the worms. Cultivation was carried out in plastic multichambered dishes ("Repli" Sterlin) using a final volume of 2 ml per 50 larvae at 37°C in air. Larvae "survived" well under these conditions and the system is claimed to be a reliable drug screening system, giving very few irrelevant or misleading results.

V. STRONGYLOIDEA

A. *OESOPHAGOSTOMUM RADIATUM*
1. Morphology and Life Cycle

The morphology of strongyles is well documented in standard textbooks and will not be given in detail here. This species is known as the "nodular" worm of cattle. The adults are stout worms, the species *radiatum* being characterized by a rounded mouth-collar, a large cephalic vesicle constricted behind its middle, and lack of an extra leaf-crown. The life cycle has been described in detail by Andrews and Maldonado.[36] Adults occur in the colon of cattle and partly embryonated eggs are passed in the feces. These hatch into L1 rhabditiform larvae which pass through the usual molts to become sheathed third-stage (L3) larvae. Cattle become infected by ingesting larvae with the forage and the larvae exsheath in the upper part of the small intestine and invade the mucosal and submucosal wall of the ileum and cecum where, within cysts (= nodules) formed by the host, they molt to L4 larvae. After some days, larvae leave the intestine, undergo another molt, and migrate to the cecum and colon and develop into adults. Larvae which fail to escape from nodules become encased in caseous or calcified masses which remain as permanent nodules.

2. Technique: *In Vitro* Culture: L3 to Young Adults[37]

Early attempts to culture this species, which only resulted in the development to the fourth stage larva, have been reviewed by Hansen and Hansen.[4] Successful culture to young adults was finally achieved by Douvres[37] by the technique given below.

Source of material — Eggs were embryonated in sphagnum moss cultures, freed from debris with a Baermann apparatus, and cleaned with 0.85% NaCl followed by Earle's BSS containing antibiotics. Larvae were stored in tap water at 5°C and, when required, were exsheathed in 1.25% Chlorox® and washed five times in BSS. Suspensions of 80,000 larvae in 1.0 ml were used per culture.

Culture media and conditions — The culture media consisted of API-1 (see Appendix, this chapter) used alone or supplemented with either glutathione or heme. A special preparation of bovine heme was prepared; for details, see Douvres[37] or Douvres and Urban.[6] Larvae developed best in a one-step roller culture system (1 revolution per 1.5 min), utilizing 1.8-liter screw-top glass bottles with 40 ml of medium up to day 14 and 80 ml thereafter; the gas phase was 5% O_2/10% CO_2/85% N_2; pH was 6.8; media was changed weekly.

Results — Seven different combinations of media components were used, and young adults were obtained in six of these (Tables 4 and 5). Generally, morphogenesis of larval and young adults closely followed the *in vivo* pattern, although the onset of oogenesis and spermatogenesis was retarded *in vitro*. Comparison of growth in size *in vivo* and *in vitro* is shown in Table 5. The best culture system produced yields of up to 25% young adults, after 28 to 42 d of culture, from an initial inocula of 80,000 larvae.

3. Evaluation

This is clearly a promising system, although the fact that eggs were not obtained indicates that *in vivo* nutritional and/or cultural requirements were not yet fully satisfied. In this respect, the results are not as satisfactory as some brief early experiments with the related species *O. quadrispinulatum* in which mature adults laying infertile eggs were produced.[4,38]

B. *STRONGYLUS* SPP.

To date, the culture of larval *Strongylus* to adults has not been completed for any species. The most successful experiments to date have been those of Farrar and Klei,[39] who cultured the horse strongyle, *S. edentatus*, from L3 to L4. They used various combinations of media, sera, buffers, and organ explant cultures at 37°C in 95% air/5% CO_2. Similar promising results were obtained with *S. vulgaris* and *S. equinus*.

VI. TRICHOSTRONGYLOIDEA

A. GENERAL COMMENTS

As well as containing a number of genera of veterinary importance, such as *Cooperia, Dictyocaulus, Haemonchus, Ostertagia*, and *Trichostrongylus*, this superfamily contains the rat "hookworm", *Nippostrongylus brasiliensis*, which was the first parasitic nematode to be cultured to sexual maturity *in vitro*. This was first achieved by Weinstein and Jones[40-42] and their work proved to be a great stimulus to other workers. As this species is readily maintained in the laboratory, it is a most valuable experimental model and for this reason, its *in vivo* as well as its *in vitro* culture are described below. Although this work was carried out more than 30 years ago, the original techniques are still useful and *N. brasiliensis* remains a valuable model for someone entering the field of nematode culture for the first time.

B. *NIPPOSTRONGYLUS BRASILIENSIS* (= *MURIS*)
1. Life Cycle: Natural and Experimental

The morphology and life cycle were first described by Yokagama[43,44] under the synonym

TABLE 4
Oesophagostomum radiatum: Times of Development of L3 Larvae *In Vitro* in Different Culture Systems[a] Compared with Development *In Vivo* (in Cattle)

Culture system	Time (d) of earliest development to stage					
	Parasitic third	Third molt[c]	Early fourth[c]	Fourth molt[c]	Young adult[b]	
					Males	Females
C	1	7	9	22	—[d]	—[d]
S-24	1	7	9	22	33	33
S-48	1	7	9	20	31	31
S-96	1	7	9	21	33	35
F-24	1	7	9	22	34	34
F-48	1	7	9	19	25	25
F-96	1	7	9	19	25	26
In cattle	2	3	5	17	19	19

[a] On basis of data obtained from two to four cultures per system.
[b] Includes adults that were partly or completely free of the sheath of fourth ecdysis.
[c] Data apply to males and females.
[d] Denotes stage not attained.

From Douvres, F. W., *J. Parasitol.*, 69, 570, 1983. With permission.

TABLE 5
Oesophagostomum radiatum: Comparison of Growth of L3 Larvae to Young Adults *In Vitro* and *In Vivo*[a]

Stage of development	Body length (mm) of worms grown[b]	
	In vitro	In cattle
Parasitic third	0.9	1.1
Third molt		
Males	1.5	1.4
Females	1.5	1.4
Early fourth		
Males	1.8	1.7
Females	1.9	1.8
Fourth molt		
Males	3.6	3.5
Females	3.8	4.0
Early, young adults		
Males	3.1	3.3
Females	4.5	4.4
Late, young adults[c]		
Males	7.0	8.1
Females	7.6	10.0

[a] Data for *in vivo*-grown worms (original).
[b] Data for each measurement based on averages for 10 to 20 specimens grown *in vitro* and comparable data for *in vivo*-grown stages.
[c] Classification based on adults with reproductive systems showing evidence of spermatogenesis and oogenesis.

From Douvres, F. W., *J. Parasitol.*, 69, 570, 1983. With permission.

of *N. muris*. Its morphology and life cycle are summarized in Smyth.[3] The morphology is clear from Figure 10. The adult worm occurs in the intestine of rats where it feeds on blood and tissue cells; the blood-gorged intestines of the worms themselves make them readily recognizable as blood-red worms at autopsy.

Eggs are passed in the feces and require abundant oxygen and moisture for their embryonation and further development. These conditions can be readily provided for laboratory maintenance by mixing eggs with charcoal or alumina and spreading them on a moist filter paper (Figure 11). Hatching of the L1 (rhabditiform) larva takes place at room temperature (18 to 22°C) in about 18 to 24 h. The first molt occurs in about 48 h and the second molt occurs in 4—5 d to give the L3 (filariform) larva. This normally exsheaths while still on the surface of the alumina (or charcoal) culture — unlike many other strongyloid larvae which only exsheath when taken into the host gut.

The L3 larvae, which show strong negative geotropism, migrate to the edge of the filter paper and, on reaching the highest peak, extend themselves in the air and weave slowly back and forth. This apes the behavior of the natural state where they likewise climb to the top of soil or herbage and await a suitable host. Larvae show strong thigmotactic and thermotactic reactions which greatly improve their chances of infecting the rat host.

Experimental infection in the laboratory is accomplished by placing larvae on the (shaved) abdomen of a rat and allowing them to penetrate, about 5000 larvae giving a heavy, but not fatal, infection. Hypodermic injection of larvae through the skin is even more effective, but feeding larvae via the mouth is rather ineffective. After entering the bloodstream, larvae are carried to the lungs, via the heart, in about 11 to 20 h (usually about 15 h).

Within the lungs, larvae feed on whole blood and molt to the fourth stage larvae (L4). The latter are carried by ciliary action up the bronchi and trachea and down the pharynx to the intestine, about 50% arriving there within 45 to 50 h. Here, the fourth and final molt occurs and adult male and females develop. There is evidence that the female attracts the male by releasing a pheromone.[45] Eggs appear in the feces on the 6th day and reach a peak of production in about 10 d. From the point of view of continuous laboratory maintenance, it is important to note that rats infected with *N. brasiliensis* rapidly develop immunity to it and toward the second week of infection, a "self-cure" reaction occurs and as a result, a high percentage of worms are expelled.

2. Technique: *In Vitro* Culture: Egg to Third-Stage Larva[42]
Preparation of egg inoculum — Eggs were concentrated from feces by centrifugal flotation in saturated salt solution (specific gravity, 1.2) and refloated by the same procedure. Eggs were then axenized by treatment with 1.25% (v/v) Chlorox® (a commercial product, 5.25% (w/v) sodium hypochlorite) for 15 min followed by several washes in sterile BSS. Sterility was tested for with thioglycollate broth.

Media and culture conditions — Two different media were used successfully to grow third-stage filariform larvae from eggs. Medium 1 (CEE_{50}-vitamin mixture) consisting of chick embryo extract (CEE_{50}) supplemented with vitamin mixture (Table 6) and Medium 2 (CEE_{50}, liver concentrate) made up of CEE_{50} with 2.6 mg/ml of liver concentrate. Mass cultures of eggs were prepared in 125-ml Erlenmeyer, rubber-stoppered flasks with 10 ml of medium at 26°C. Medium 1 cultures were terminated at 7 d and Medium 2 cultures were terminated at 14 d. In Medium 1, 68% of the hatched larvae developed to the filariform stage and slightly more (73%) developed in Medium 2. The average length of 40 larvae in Medium 1 was 503 + 31 μm and was 5219 + 31 μm in Medium 2. These axenically grown larvae were then used as initial culture material and grown to adults as described below.

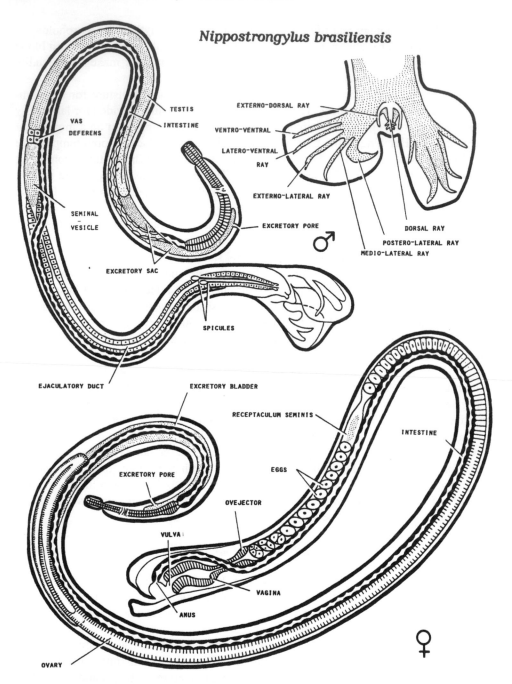

FIGURE 10. *Nippostrongylus brasiliensis*: morphology of male and female. (From Smyth, J. D., *An Intro-duction to Animal Parasitology*, 2nd ed., Edward Arnold, London, 1976, 335. With permission.)

3. Technique: *In Vitro* Culture: Third-Stage Larvae to Adults[42]

Growth of L3 larvae to adults was first achieved by Weinstein and Jones[40] and further improved later.[42] They used third-stage larvae either from axenic cultures (as above) or from filter paper cultures (see Figure 11). Although sexually mature male and female worms developed *in vitro*, copulation was not observed and only infertile eggs were obtained.

A variety of culture media and conditions were successfully used to grow adult worms and for comprehensive details of the these, the original paper[42] should be consulted. Briefly, the basic medium consisted essentially of 70% CEE_{50} and 20% rat or human serum in Earle's

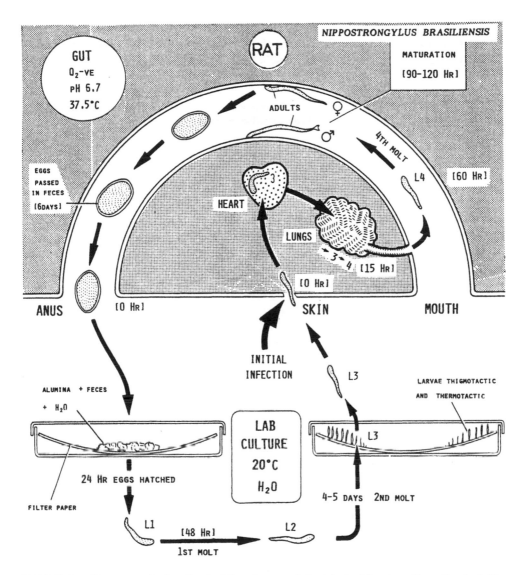

FIGURE 11. *Nippostrongylus brasiliensis*: life-cycle as carried out in the laboratory. (From Smyth, J. D., *An Introduction to Animal Parasitology*, 2nd ed., Edward Arnold, London, 1976, 336. With permission.)

saline supplemented with vitamin or liver concentrate. Cultivation was carried out in 16 × 150 mm culture tubes in 2 ml of medium in a roller tube system (12 revolutions per hour); medium was changed three times a week.

4. Results

It was early demonstrated that a high concentration of CEE_{50} was essential for good growth and differentiation and a level of 70% was used in the most successful cultures (Figure 12 and Table 7). As mentioned, although sexually mature males containing sperm and females containing ova developed, copulation was never observed and the small number of eggs which were produced in culture were infertile.

Various attempts to improve the culture media or conditions were unsuccessful. Replacing the basic media with defined media, such as Eagle's, Morgan 199, or NCTC 109, gave poorer results[42] and supplementing the media with dialyzed yeast extract did not improve results.

TABLE 6
Nippostrongylus brasiliensis: Developmental Response *In Vitro* to Basal Medium (70% CEE_{50}, 20% Rat Serum, in Earle's BSS) Plus Vitamin Mixture or Liver Concentrate

Medium	Supplement (final concentration in medium)	Percentage of fifth-stage worms recovered[a]
70% chick embryo homogenate, 20% rat serum	None	13
	Vitamin mix II[b] (onefold)	25
	Vitamin mix II[b] (twofold)	23
	Vitamin mix II[b] (fourfold)	24
	Liver conc[c] (0.1%)	23
	Liver conc[c] (0.3%)	12
	Liver conc[c] (0.5%)	3

[a] Based on total number of immature plus mature fifth-stage worms present at termination of the cultures.
[b] Composition given in text.
[c] Product of Sigma Chemical Co., St. Louis, MO.

From Weinstein, P. P. and Jones, M. F., *Ann. N.Y. Acad. Sci.*, 77, 137, 1959. With permission.

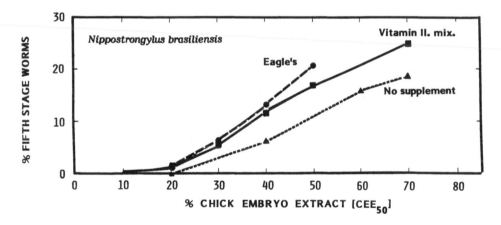

FIGURE 12. *Nippostrongylus brasiliensis*: effects of different concentrations of chick embryo extract (CEE_{50}), with different supplements, on *in vitro* development; all media contained 20% rat serum. (From Weinstein, P. P. and Jones, M. F., *Ann. N.Y. Acad. Sci.*, 77, 137, 1959. With permission.)

One of the difficulties in utilizing media containing CEE_{50} is that it shows variability in the hands of different workers, probably due to differences in preparation. It also shows a tendency to precipitate.[1] Nevertheless, it is an essential component for stimulating growth and differentiation of third-stage larvae to become sexually mature. This applies not only to *N. brasiliensis*, but also to other species and it has been an important component of the "Ae" medium which has been used successfully for the culture of other trichostrongyles, such as *Ostertagia*.[46]

5. Evaluation

Although it is 30 years since Weinstein and Jones[42] grew *Nippostrongylus brasiliensis* through its life cycle from egg to adult, no worker appears to have succeeded in markedly improving the original technique sufficiently to obtain copulation and fertile eggs *in vitro*. Phillipson[47] has provided experimental evidence that fertilization only occurs in this species when the adult can make contact with a suitable surface such as the mucosa. This situation

TABLE 7
Nippostrongylus brasiliensis: Development *In Vitro* of Fifth-Stage Worms from Filariform Larvae Derived from Axenic Egg Cultures

Medium		Antibiotics[a] (per ml of medium)	Fifth-stage worms (%)[b]	Mature fifth-stage worms (%)
From which filariform larvae derived	For parasitic cultures			
Embryo homogenate-vitamin mixture II	Basal[c] A	5 U P 5 µg S	8	3
Embryo homogenate-vitamin mixture II	Basal A + 2.5% laked rat rbc[d]	5 U P 5 µg S	12	7
Embryo homogenate-liver conc.	Basal B	None	4	1
Embryo homogenate-liver conc.	Basal B	5 U P 5 µg S	7	2
Embryo homogenate-liver conc.	Basal B + vitamin mixture II (onefold)	None	11	7
Embryo homogenate-liver conc.	Basal B + vitamin mixture II (onefold)	5 U P 5 µg S	15	8
Embryo homogenate-liver conc.	Basal B + vitamin mixture II (onefold)	300 U P 300 µg S	10	6
Embryo homogenate-liver conc.	Basal B + vitamin mixture II (twofold)	5 U P 5 µg S	12	9

[a] P, penicillin; S, streptomycin.
[b] Immature plus mature fifth-stage worms.
[c] 70% chick embryo homogenate; 20% rat serum. A and B were prepared from two different samples of homogenate-serum.
[d] rbc, red blood cell.

From Weinstein, P. P. and Jones, M. F., *Ann. N.Y. Acad. Sci.*, 77, 137, 1959. With permission.

is reminiscent of the situation in some cestodes, e.g., *Schistocephalus solidus*, where it has been shown that *in vitro* compression within a cellulose tube (see Chapter 5, Figure 6) is essential for insemination and fertilization to take place. A similar surface contact appears to be essential for sexual differentiation in *Echinococcus granulosus*; probably insemination and fertilization in this species and *E. multilocularis* may also require a surface contact of some kind. It is somewhat surprising that with so many new, defined media now commercially available that an improved technique for the culture of this important laboratory model has not been developed.

6. Technique: *In Vitro* Maintenance of Adult Worms
A number of workers have used the techniques developed above, or various modifications of them, to maintain the adult worms *in vitro* for various studies, such as screening anthelmintics;[34] synthesis of cuticular proteins;[48] genetic control of differentiation;[49] antigen production;[50] pheromone production;[51] and production and release of acetylcholinesterase.[52] Other applications have been reviewed by Hansen and Hansen[4]; and Douvres and Urban.[6]

C. *OSTERTAGIA OSTERTAGIA*
1. Life Cycle
O. ostertagia is an important parasite of cattle and the successful culture of L3 larvae to adults with fertile egg production by Douvres and Malakatis[53] represents a major achievement in nematode culture. Early, less successful work on this species has been reviewed by Hansen and Hansen.[4] The adult worms live in the abomasum of cattle where they cause bovine ostertagiosis. The early life cycle follows the typical trichostrongyle pattern: eggs

develop in moist situations and L1 larvae hatch, feed, and molt to L2 and feed and molt again to (sheathed) infective L3, filariform, larvae, Douvres[54] has given a valuable, illustrated, account of the morphology of these various larval stages. The larvae show negative geotropism and climb blades of grass from which they are readily ingested in the forage. The sheath is lost in the forestomachs and the larvae settle in the abomasum. An important feature of bovine osteragiosis is that the larvae (about 800 μm) spend a considerable time within the glands of the abomasal mucosa and do not emerge until they are young adults about 1 cm in length. This causes pressure necrosis on the (specialized) gland cells which are replaced by the (less specialized) mucus cells. In some countries (e.g., Australia), some strains of *O. ostertagia* become *hypobiotic*, i.e., they remain in the mucosa as early fourth stage larvae during a period when the external environmental conditions are unsuitable for the development of eggs to free-living stages.

2. Technique: *In Vitro* Culture: L3 to Adults[53]

This represents one of the few *in vitro* systems in which adult nematodes grown from larvae have produced fertile eggs.

Source of material — L3 larvae were obtained from 14-d sphagnum moss egg cultures and freed of debris using a Baermann apparatus with passage through a 25-μm nylon cloth. Inocula of 100,000 and 200,000 were used, either immediately or after storage in tap water at 5°C for up to 14 d. Larvae were cleaned by centrifuging (700 × g) in sterile 0.85% NaCl and Earle's BSS — five washes each. Larvae were allowed to remain in the last wash for 30 to 60 min.

Exsheathment of larvae — Larvae were exsheathed by a modification of the process used by Cypess et al.[55] for exsheathing *Nematospiroides dubius* larvae. This involved placing larvae in 10 ml of RFN medium in a 50-ml, screw-topped Erlenmeyer flask and incubating it in a water bath shaker (120 strokes per minute) at 37°C under gas phases of (1) 40% CO_2/60% N_2 (delivered at 200 ml/min) for 10 min; (2) a nongassing phase, with shaking for 30 min; (3) an aerating phase with compressed air with shaking for 5 min; and (4) a nongassing phase in a sealed flask with shaking for 25 min. These exsheathed larvae then formed the inocula for *in vitro* culture, the population of which were approximately 15 to 30% partially exsheathed, 25 to 50% totally exsheathed, and the rest still unsheathed, but with loose sheaths.

Culture media and conditions — A two-step culture system consisting of media API-1 and RFN was used. As the composition of these media is exceptionally complex, the original paper should be consulted for their preparation.[53] A condensed account of this is also given in Douvres and Urban.[6] Step 1 consisted of culturing 100,000 to 200,000 larvae in 100 ml of medium RFN with a gas phase of 95% air/5% CO_2 at pH 7.3. Step 2 consisted of medium API-1 + pepsin at pH 4.5 from day 3 to day 9 and thereafter at pH 6.0. Cultures were transferred to new media and new culture vessels every 6 to 7 d.

3. Results

Results are summarized in Tables 8 and 9 and Figures 13 and 14. In all experiments, larvae grew and developed, best results being obtained in the four cultures of Experiment 3 (Table 8) using 100,000 larvae per culture. Mature adult males and females developed in 28 to 29 d and egg-laying commenced in 34 to 41 days (see Table 9). Spermatogenesis and oogenesis could be detected as early as day 24. However, development *in vitro* was considerably slower than that *in vivo* (see Table 9; Figure 13).

When cultures were terminated on day 52, 3 had fertile eggs in the 2- to 16-cell stage and in the morula stage, free in the medium; 21,500 eggs were released altogether, of which 75% were infertile and 25% were fertile. Both fertile and infertile eggs were smaller, but otherwise identical, to those *in vivo*. Copulation was not actually observed, but this may

TABLE 8
Ostertagia ostertagia: *In Vitro* Development of L3 Larvae to Adult Worms in Two-Step Roller Culture Systems[a]

Experiment	Age (d) of larvae	Number of cultures	Larvae[b] per culture (× 1000)	RFN at pH 7.3	API-1 + pepsin[c] at pH 4.5	API-1 + pepsin[c] at pH 6.0	Age (d) when cultures terminated[d]
1	28[e]	2	200	2	6	21	29 (1)[f]
				2	6	73	81 (1)
2	14	4	200	2	6	12	20 (2)
				2	6	42	50 (2)
3	14	4	100	3	6	43	52 (4)

(Number of days incubated in media)

[a] Cultures prepared with 100 ml of medium in roller bottle (maximum capacity, 2.8 liter) which was rotated at 1 revolution/1.5 min, on a (cell production) roller apparatus.
[b] Larvae stimulated to initiate exsheathment by exposure to medium RFN and carbon dioxide, in 70-min process.
[c] Medium at pH 6.0 and culture vessel changed every 6 or 7 d.
[d] Figures in parentheses represent number of cultures carried to termination date.
[e] Includes storage period of 14 d.
[f] Ended because of fungus contaminant.

From Douvres, F. W. and Malakatis, G. H., *J. Parasitol.*, 63, 520, 1977. With permission.

TABLE 9
Ostertagia ostertagia: Time of Development of L3 Larvae to Adult Worm *In Vitro*[a] Compared with That Required *In Vivo*

Stages of development[b]	In vitro	In vivo[c]
Parasitic third	1	2
Third molt ♂, ♀	7	3
Fourth		
Early ♂, ♀	8	4
Mid-late ♂, ♀	9—13	6—8
Fourth molt ♂, ♀	15—19	10
Adults		
Young ♂, ♀	19—20	13
Mature ♂	28	16
Mature ♀	29	16
Egg-laying females	34—41	18

(Time in days of earliest development)

[a] On the basis of data obtained from four cultures.
[b] Classified according to descriptions of Ransom[56] and Douvres.[54]
[c] From Douvres[54] (unpublished observations).

From Douvres, F. W. and Malakatis, G. H., *J. Parasitol.*, 63, 520, 1977. With permission.

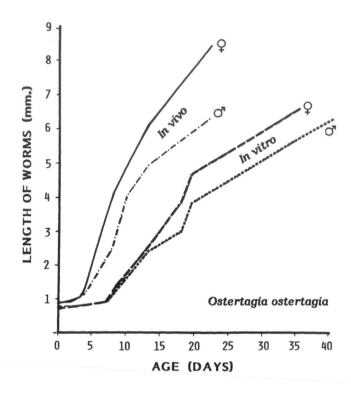

FIGURE 13.   *Ostertagia ostertagia*: comparison of growth in length of L3 larvae *in vitro* and *in vivo*. (From Douvres, F. W. and Malakatis, G. M., *J. Parasitol.*, 63, 520, 1977. With permission.)

have been related to the fact that, from the fourth molt onward, most populations of larvae anc adults were clumped and intertwined making close observations impossible. The overall yield of worms was low, only 10 to 20% mature worms finally developing.

4. Evaluation

This is a most elegant *in vitro* system and one of the few in which fertile eggs were obtained and it clearly has much potential for future work. However, the low yield of adults and low percentage of fertile eggs indicate that further improvements are necessary before yields approaching those *in vivo* are likely to be obtained.

D. *HAEMONCHUS CONTORTUS*

Earlier work on the *in vitro* culture of this species has been reviewed by Taylor and Baker[5] and Hansen and Hansen.[4] Much of this work was promising, but adult worms were not obtained. More recent techniques, discussed below, have resulted in the development of sexually mature, ovipositing worms.

1. Life Cycle

H. contortus is a well-known parasite of the abomasum of the sheep, but it is also common in goats and cattle. It is now well recognized that this species exists in many different strains and various morphological variants are well documented.[57] Its normal life cycle (Figure 15) is similar to that of *Ostertagia ostertagia* and will only be briefly summarized here. Eggs are laid in the morula stage in the abomasum lumen and are voided in the feces. In a suitable warm (20 to 26°C) and moist environment, the eggs embryonate and hatch to L1 larvae and, following two successive molts, reach the L3 stage in about 8 d. On ingestion in the forage, the L3 exsheaths and molts in the abomasum to the L4 stage

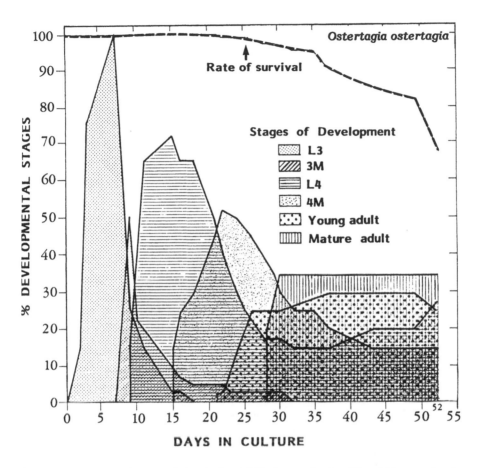

FIGURE 14. *Ostertagia ostertagia*: survival and yield of worms developed from L3 larvae cultured *in vitro*. Data based on averages of estimated percentages or actual counts from four cultures begun with 100,000 larvae per culture; L3 = third stage; 3M = third molt; L4 = fourth stage; 4M = fourth molt. (From Douvres, F. W. and Malakatis, G. M., *J. Parasitol.*, 63, 520, 1977. With permission.)

and again to give the fifth stage which grows to the adult worm. Sexual maturity is reached in about 18 d. The species has a high fecundity, a female producing 5,000 to 10,000 eggs per day — much higher than any of the other gastrointestinal (GI) nematodes of sheep.

There is a possibility that an alternative cycle may exist. This hypothesis has arisen from recent work on the *in vitro* culture of mature adults, discussed further below, by Ayalew and Murphy.[58] In their experiments they found that *in vitro* some eggs *in utero* exhibited viviparity and normal development to L4 larvae took place. These authors speculated that viviparity, similarly, might also occur *in vivo* with subsequent risk of autoinfection occurring. If this does in fact occur in nature — and is not just an incidental *in vitro* phenomenon — it could explain some of the more puzzling clinical and epidemiological features of hemonchosis.

2. Technique: *In Vitro* Culture: L3 to Adults[59,60]

Source of material — L3 larvae of the BPL isolate of *H. contortus* was recovered from fecal cultures, freed of debris, cleaned with 0.85% NaCl, exsheathed by 1.25% Chlorox®,[37] and washed with Earle's BSS + antibiotics.

Culture media and conditions — The basic culture techniques were described first by Stringfellow[59] and elaborated later.[60] The basic medium was API-1 (see Appendix) as developed by Douvres and Malakatis[53] for *O. ostertagia* — with various supplements, such

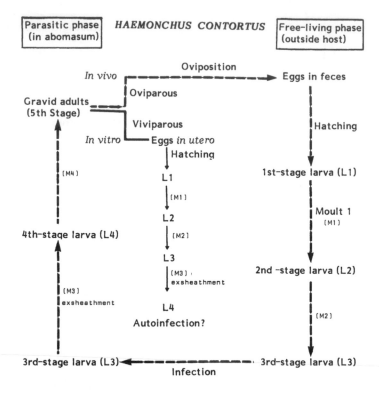

FIGURE 15.   *Haemonchus contortus*: the classical, natural life-cycle pattern (oviposition) is indicated by broken arrows. An alternative pattern, observed during *in vitro* culture of gravid worms, of eggs hatching *in utero* and subsequently developing to L4 larvae is indicated by small, solid arrows. It is speculated that if *in utero* hatching should occur *in vivo*, autoinfection could result. (From Ayalew, L. and Murphy, B. E. P., *Parasitology*, 93, 371, 1986. With permission.)

as Fildes reagent, abomasal extracts, and gastric contents from cattle and sheep, as shown in Table 10, and ten culture systems were tested. For a detailed account of these, the original papers should be consulted. Cultures were inoculated with 40,000 fresh, exsheathed larvae per 40- to 42-ml culture and transfers were made every 7 d. The culture vessels were 1.8-liter glass bottles rotating at 1 revolutions per 1.5 min at 39°C in a gas phase of 5% O_2/10% CO_2/85% N_2. Optimum results were obtained in a one-step system with the pH maintained at 6.4 for days 1 to 7 and 6.8 thereafter.

3. Results

Development of *H. contortus* ranged from early L4 larvae (in culture systems 3 to 6) to egg-laying females in system 10 (see Table 10). In the best results (in system 10), larvae developed to mature, adult males in 28 d and to mature, ovipositing females in 36 d. The largest mature adult male and female worms measured 10 and 16 mm compared with adults male and female worms grown *in vivo* which measured 20 and 25 mm, respectively.

Mature females laid a mean of 16 eggs per culture. When laid, these were unsegmented, but a few developed as far as the two- to eight-cell stage. *In vitro* eggs were smaller (72 × 39 µm) than those *in vivo* (93 × 32 µm). Copulation was not seen *in vitro*, but spermatozoa were abundant in the seminal vesicles of males, although they were not seen in the uteri of females.

TABLE 10

Haemonchus contortus: *In Vitro* Development of Exsheathed L3 Larvae in Medium API-1 Supplemented with Fildes' Reagent, Bovine (MBE), or Sheep (OME) Abomasal Extract and Bovine (BGC) or Sheep (OGC) Gastric Contents

	Time (d) development to stage							
					Young adult		Mature adult	
Culture system	Parasitic third	Third molt	Early fourth	Fourth molt	Male	Female	Male	Female
API-1								
1	0	3—5	3—5	21	—	—	—	—
API-1 + Fildes								
2	0	3—5	3—5	17	21	21	—	—
API-1 +								
3 (BME)	0	3—5	3—5	21	—	—	—	—
4 (OME)	0	3—5	3—5	—	—	—	—	—
5 (BGC)	0	3—5	3—5	—	—	—	—	—
6 (OGC)	0	3—5	3—5	—	—	—	—	—
API-1 + Fildes +								
7 (BME)	0	3—5	3—5	14	21	21	—	—
8 (OME)	0	3—5	3—5	17	17	—	—	—
9 (BGC)	0	3—5	3—5	10	17	17	28	—
10 (OGC)	0	3—5	3—5	17	17	17	28	36
In vivo								
Calf	3	3—5	3—5	10	16	16	18—20	18—20
Sheep	3	3—5	4—7	8—12	9—13	9—13	18—20	18—20

From Stringfellow, F., *J. Parasitol.*, 72, 339, 1986. With permission.

4. Evaluation

This is clearly a promising *in vitro* system, but as with many helminths of all groups, the (apparent) failure of fertilization to take place suggests a lack of some nutritional and/or physical factor restraining copulation. The authors speculated that the eggs which were segmenting were doing so parthenogenetically, although the high temperature of the culture (39°C) may have been an inhibiting factor to further development. As mentioned earlier, *H. contortus* exists in a number of strains, and it is important to recognize that a culture technique which is successful for one strain may not be suitable for another, as each strain may have different nutritional or metabolic requirements. An outstanding example of this is seen in cestodes (Chapter 5) in *Echinococcus granulosus*, in which larvae (protoscoleces) isolated from sheep grow readily to adults *in vitro*, whereas those from horse failed to develop.

5. Technique: *In Vitro* Maintenance of Adult Worms

Techniques for the maintenance of adult *H. contortus* have been developed by Douvres and Fetterer[61] and Ayalew and Murphy[58] and will only be briefly described here as only short periods of maintenance were carried out. Douvres and Fetterer[61] cultured adult worms from the sheep abomasum in Medium RBS, S-Def:API-27 or API-28 + 24 μg of bovine hemin per milliliter, at pH 6.8, 39°C, in a gas phase of 10% CO_2/5% O_2/85% N_2. Cultures consisted of 100 to 250 worms in 20 ml of medium, without changes, in screw-top roller bottles (205 × 15 mm). Worms survived for 3 d during which time they continued to oviposit fertile eggs and show high viability.

In another series of experiments, Ayalew and Murphy[58] cultured adult worms in 3 ml of Ham's F 10 Medium in Roux culture flasks at 39 to 40°C in a gas phase of 5% CO_2/95% air. Adults laid eggs for 2 d, but stopped after 2 to 5 d. By the eighth day, only intermittent movements were seen. However, this work produced a most interesting result in that, after 12 h of culture, large numbers of eggs within most females hatched into L1 larvae. Many of these larvae, still within the uterus, continued to develop normally and L2 were observed by 24 h followed by L3 within 2 to 3 d of incubation. The fact that these larvae hatched viviparously led the authors to speculate that the phenomenon may sometimes occur *in vivo* and, if so, this could result in autoinfection of the host sheep (see Figure 15). This unusual hypothesis obviously requires confirmation by further *in vivo* studies. However, it is perhaps more likely that the unnatural conditions *in vitro* "triggered" this unusual (parthenogenetic?) development and that similar conditions may never occur *in vivo*. The phenomenon clearly requires further examination.

E. *COOPERIA PUNCTATA*

1. General Comments

Much of the early work on *in vitro* culture of nematodes was carried out on species of *Cooperia*, especially *C. punctata*, by Leland and colleagues.[1,46,62] This promising work was based on the development of a medium "Ae" which also proved useful for the culture of several other nematode species. Although highly successful in that numerous eggs were produced, only a small number of these proved to be fertilized.[63] These early techniques, which have been also reviewed by Taylor and Baker[5] and Hansen and Hansen,[4] do not appear to have been improved on and are the basis of the techniques given here. More recently, an unpublished technique by Douvres for another species, *C. oncophora*, has also been briefly referred to, but it is not considered in detail here.[6]

2. Life Cycle

C. punctata occurs in cattle and occasionally in sheep, being generally confined to the intestine but may be found in the abomasum. It has a typical trichostrongyle life cycle.[64] Eggs passed in the feces require moisture and a warm temperature to embryonate. At 20 to 26°C, eggs hatch in about 20 h into typical rhabditiform L1 larvae, and L2 larvae appear at 45 h at that temperature range, but only 30 h at 28°C. L3 larvae appear at 96 h at the lower temperature, but at only 75 h at 28°C. When the larvae are ingested in the forage, they exsheath within 24 h in the abomasum, but are found in the intestine within 3 d still mostly at the third stage. L4 larvae are readily found on day 4. The fourth molt takes place during the seventh day. By day 10, adult worms are approaching maturity.

3. Technique: *In Vitro* Culture: Eggs or L3 to Adults

Historical review — In the early experiments,[46] worms were cultured from L3 larvae to sexually mature adults, but although large numbers of eggs were produced, none hatched. This technique was substantially improved later by Leland[62] and resulted in the organism being cultured from "egg to egg", although again only infertile eggs were produced. The latter were produced by both virgin females and females from male-containing cultures. Later Zimmerman and Leland[63] carried out further experiments, using essentially the same system, which resulted in the production of fertile eggs, some of which embryonated and gave rise to a few infective L3 larvae. The following methods are largely based on Leland.[62]

Preparation of eggs for culture[62] — Eggs were obtained from fresh rectal collections by the gradient centrifugation technique of Marquardt,[65] washed, and sterilized by being subjected to White's solution[13] for 24 min, including 5 min for centrifugation and rewashing. Eggs were further washed in BSS plus antibiotics (penicillin, streptomycin, and mycostatin). If eggs were to be hatched before culture, they were suspended in 1- to 2-ml of sterile glass-distilled water in a roller system (1 revolution per 5 min) at 24°C for 24 h.

Preparation of third-stage (L3) larvae for culture — Infective L3 larvae were obtained from 7- to 14-d-old sphagnum moss cultures[66] and isolated by the Baermann technique.

Larvae were washed by centrifugation (at $100 \times g$) in distilled water and ''sterilized'' by 10-min exposure to 2% Chlorox® (5.25% sodium hypochlorite by weight; Chlorox Chemical Co., Oakland, CA) followed by centrifugation. This was further followed by three centrifugation washes in distilled water and finally suspension in Earle's BSS.

Preparation of fourth-stage (L4) larvae for culture — L3 larvae, produced as described above, were cultured in Medium Ae for 12 d and some L4 larvae developed. Some L4 females were isolated and all-female cultures (20 to 25 females per tube) were prepared.

General culture conditions[46,62] — Cultures were incubated in screw-capped culture tubes (16×125 mm) on a roller system at 22 to 26 or 38.5°C. Three media were largely used: Ae; A-S (= Ae with serum replaced by Earle's BSS); and AIS (= Ae with inactivated calf serum).

4. Results

Ae proved to be unsatisfactory as a medium for development of L3 larvae (at 24°C) from eggs, but both A-S and AIS supported this phase of development. Later development (at 38.5°C) of L3 to adults was best supported in Ae or AIS. Overall, in terms of number of eggs and adults produced, the best results were obtained with a two-step system consisting of Medium A-S for 7 to 14 d at 25°C, followed by Medium Ae at 38.5°C until adults developed. In this system, in initial egg cultures, L1 larvae were detected in 24 h; L2 larvae were detected in 3 d; L3 larvae were detected in 5 to 8 d; L4 larvae were detected in 14 to 28 d; L5 larvae were detected in 24 to 40 d; and eggs in the ovejectors of females were detected in 39 to 56 d. Some females continued laying eggs for more than 100 d. All eggs produced were infertile and copulation was not observed.

5. Later Work

In later experiments, using essentially the same techniques, Zimmerman and Leland,[63] starting with L3 larvae, cultured them to adult worms which laid large numbers of eggs. Although many of these hatched to L1 and L2, only a few of the latter developed to L3. Population density considerations played an important part in the result, and embryonation and hatching of eggs and subsequent development was only observed in cultures with a population density fewer than 1742 worms per 2 ml of medium, suggesting that a ''crowding effect'' is operating (Figure 16). It has also been recorded[6] that Douvres,[67] working with *C. oncophora*, has also obtained mature adults producing segmenting eggs, using a medium developed for the culture of *Trichostrongylus colubriformis*.[68]

6. Evaluation

It is clear from the above results that the entire life cycle of *C. punctata* can be reproduced *in vitro*. However, the various systems have not yet been sufficiently integrated to give consequential development. Moreover, there are clearly many factors which need further examination in order to improve the overall yield; these include the population density of cultures and the factors supporting copulation.

F. *TRICHOSTRONGYLUS COLUBRIFORMIS*

Early work on the *in vitro* cultivation of *Trichostrongylus* spp., much of which was not very successful, has been reviewed by Hansen and Hansen[4] and will not be further discussed here. More recent work, reviewed by Douvres and Urban,[6] although more promising, has still not achieved development to egg-producing adults. Most work has been carried out on *T. colubriformis* (small intestine) with some work on *T. axei* (abomasum).

1. Life Cycle

T. colubriformis is a common, widely distributed, hairlike nematode of the small intestine of ruminants. Infection takes place by ingestion of the L3 larvae in forage. Larvae pass

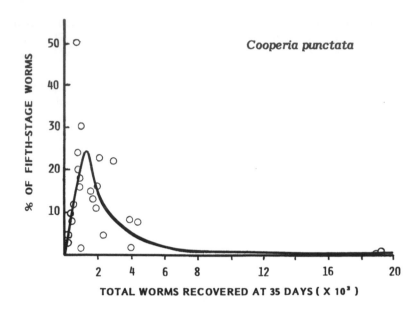

FIGURE 16.   *Cooperia punctata*: the effect of population density on *in vitro* development. The inhibited development is comparable with the well-known "crowding effect" seen *in vivo* helminth populations. (From Leland, S. E., *Adv. Vet. Sci. Comp. Med.*, 14, 29, 1970. With permission.)

through the abomasum without any development and, on reaching the intestine, burrow into the mucosa. The third molt takes place in the mucosa and the L4 larvae leave the mucosa to molt again to grow to adults. Sexual maturity is reached in about 17 d PI. Guinea pigs serve as useful experimental hosts.

2. Technique: *In Vitro* Culture: Eggs to L3 Larvae

This was first carried out by Khan and Dorman[69] using a medium containing tissue extract and serum and was later modified by Dorsman and Bijl[70] using a medium containing many supplements but free of tissue (embryo) extract and serum, thus making it more reproducible. The latter result represented the first time this has been achieved for those nematodes in vertebrates, whose free-living rhabditiform larvae are food dependent.

Preparation of eggs for culture[69,70] — Eggs were sterilized with White's sterilizing agent[71] and washed four times in tap water + antibiotics (kanamycin, 1000 µg/ml; penicillin, 1000 IU/ml) followed by a fifth washing with reduced antibiotics (kanamycin, 50 µg/ml; penicillin, 300 IU/ml) and finally a washing in sterile tap water plus media (see below). Eggs were hatched in 3 ml of the last washing fluid at 24°C[69] or 31 to 32°C[70] and washed again; 300 to 500 larvae were used per culture.

Media and conditions — Cultures were carried out in 16 × 120 mm screw-capped tubes and gassed with 5% CO_2 in air[70] and media were changed on the third day and thereafter every other day. Khan and Dorsman[69] used a variety of media based on NCTC 135, chick embryo extract (CEE), fetal calf serum, and either lactalbumin hydrolysate or yeast extract (Table 11). Dorsman and Bijl[70] utilized media free of serum and embryo extract, but containing hydrolyzed casein type I or II (amino nitrogen: total nitrogen ratios, 0.53 or 0.39), yeast extract, phosphatidycholine from soya bean, and a number of chemical defined substances such as saline solution, a sterol and an iron porphyrin (Table 12).

3. Results

Both of the above media produced L3 larvae. Best results by Khan and Dorsman[69] were obtained in Medium Tc5 (see Table 11), a yield of 10 to 20% L3 larvae being obtained.

TABLE 11

Trichostrongylus colubriformis: Media Used for the *In Vitro* Culture of Free-Living Larvae from Eggs

Medium code	CEE (%)	NCTC 135 (%)	Lactalbumin hydrolysate (%)	Casein hydrolysate (%)	Serum (%)	Yeast extract per 100 ml	Liver extract (%)	Distilled water (%)	Hemin (µg/ml)	Myoglobin (µg/ml)	Cholesterol (µg/ml)	Vitamin B6 (µg/ml)	Culture time (d)	Remarks
Ta		100											4	L_1, no development
Tb	25	50	25			0.5 g							10	Early L_2 only
Tcl	25	25	25		25 FC								73	Few L_3 after 20 d
Tc2	25	25	25		25 L	0.5 g							20	
Tc3	25	20		25	25 FC			5	100	500			20	L_2 only
Tca	25[a]	25	15		25 L	1.0 g		10	50		50		20	
Tga4	50	15	5		15 L	1.0 g	5	10	50		50		20	
Tc4	25	13.4			25 FC	25% B		11.6	100	500		30	7	10—20% L_3
Tc5	25[a]	25			25 FC	25% B							7	10—20% L_3

[a] Unfiltered chick embryo extract.

From Khan, Z. I. and Dorsman, W., *J. Parasitol.*, 64, 1024, 1978. With permission.

TABLE 12

Trichostrongylus colubriformis: Serum-Free Media Used for the *In Vitro* Culture of Free-Living Stages from Eggs

Medium[a]		SP-I	SP-II	CH1	CH2	CH3	CH4	CH5	CH6	CH7	CH8
Salt solution A	(%)	60									
Salt solution C	(%)		60	54.5	54.5	54.5	54.5	46	46	46	46
KHCO$_3$	(µg/ml)		1160	1160	1160	1160	1160	1160	1160	1160	1160
NaHCO$_3$	(µg/ml)		70								
CoCl$_2$ · 6H$_2$O	(µg/ml)	0.06									
SP Hy-Soy	(mg/ml)	40									
SP S/ELB	(mg/ml)		40								
PP	(mg/ml)			20	20						
CH AN:TN 0.53	(mg/ml)			20	20	40					
CH AN:TN 0.39	(mg/ml)						40	40	40	60	80
FYE type I	(%)	10	10								
FYE type II	(%)			20	20	20	20	20	20	20	20
Cytochrome C	(µg/ml)		340	340	340	340	340	340	340		
Myoglobin	(µg/ml)	500								500	500
β-Sitosterol	(µg/ml)	10									
Cholesterol	(µg/ml)		75	75	75	75	75	75	75	75	75
Tween 80	(mg/ml)	1	3	3	3	3	3	3	3	3	3
Glucose	(µg/ml)		1500								
Supplements (Table II)		–	+	+	+	+	+	+	+	+	+
Penicillin	(IU/ml)	300									
Econazole	(µg/ml)	3							10		
5-Fluorocytosine	(µg/ml)										
pH		7.0	7.5	7.2	7.3	7.3	7.5	7.3	7.3	7.5	7.4

[a] Abbreviations: SP, soy peptone; PP, proteose peptone; CH, casein hydrolysate; FYE, fresh yeast extract.

From Dorsman, W. and Bijl, A. C., *J. Parasitol.*, 71, 200, 1985. With permission.

Moreover, these cultured larvae were able to produce patent infections in guinea pigs. In the serum-free media used by Dorsman and Bijl,[70] the yield was lower (17%), but the infectivity to animals was not tested.

4. Evaluation
Even with the low yields obtained, the fact that the cultured larvae proved to be infective to laboratory animals makes the technique a useful one for experimental purposes. Further experimental work may make it possible to improve this yield.

5. Technique: *In Vitro* Culture: L3 to Adults[68]
As only young adults have so far been grown from L3 larvae, the technique for this is only briefly described here.

Source of material — Infective L3 larvae were recovered from 14-d-old sheep or calf fecal sphagnum moss cultures using a Baermann apparatus and stored in tap water. They were exsheathed by the method of Douvres and Malakatis[53] as used for *Ostertagia ostertagia* (see Section VI.C.2).

Culture media and conditions — Four different culture systems were tested, using roller-tube bottles. Best results were obtained in culture system IV — a two-step system at 39°C. Step 1 consisted of RFN (pH 6.8) in a gas phase of 5% CO_2/air for 2 d; Step 2 consisted of Medium API-16 (see Appendix), with pepsin and reduced glutathione at pH 6.8 in the same gas phase from day 2 to day 9, followed by API-16 supplemented with pepsin with a change of gas phase to 5%/10% CO_2/85% N_2.

6. Results
Although advanced development was obtained in all cultures, best results were obtained in system IV as described above. Development *in vitro* was substantially retarded compared with that *in vivo* (Table 13); only early sexual differentiation was observed and cultured worms were smaller than those grown *in vivo* (Table 14).

7. Evaluation
Failure of cultivated worms to reach maturity suggests that probably a substantial improvement in media and/or conditions may be necessary before this can be achieved.

8. Anthelminthic Testing
Rapson et al.[72] used fourth-stage larvae from experimental infections in *Meriones unguiculatus* to test the actions of various drugs; 50% of larvae transformed into young adults in a complex media over 7 d and the effect of drugs on development was examined. Molting was affected by as little as 0.01 μM of ivermectin.

G. *TRICHOSTRONGYLUS AXEI*
T. axei is a common, widely distributed parasite of abomasum of ruminants or the stomach of horses. Like *T. colubriformis*, it has been cultured to early sexual development.[73] However, with a maximum yield of <1% of young adults in populations of 100,000 *T. axei* by 21 to 28 d, the system is clearly not very satisfactory. For details, the original paper should be consulted.

TABLE 13
Trichostrongylus colubriformis: *In Vitro* Development of L3 Larvae in Different Two-Step, Roller Culture Systems[a] Compared with Development *In Vivo*

	Time (d) of earliest development to stage					
					Young adult[b]	
Culture system	Parasitic third	Third molt[c]	Early fourth[c]	Fourth molt[c]	Males	Females
I	1	8	9	18	21	21
II	1	6	9	15	21	27
III	1	6	11	18	21	22
IV	1	5	7	13	16	16
In cattle	1	3	5	6	10	10

[a]   On basis of data obtained from two to four cultures per system.
[b]   Includes adults that were partly or completely free of the sheath of fourth ecdysis.
[c]   Data apply to males and females.

From Douvres, F. W., *J. Parasitol.*, 66, 466, 1980. With permission.

TABLE 14
Trichostrongylus colubriformis: **Comparative Growth *In Vitro* in Two-Step, Roller Culture Systems and *In Vivo* in Cattle[a]**

	Body length (mm) of worms grown[b]	
Stage of development	*In vitro*	In cattle
Parasitic third	0.6	0.8
Third molt		
Males	0.8	0.9
Females	0.8	0.8
Early fourth		
Males	1.0	1.5
Females	1.1	1.5
Fourth molt		
Males	1.3	2.1
Females	1.5	2.3
Young adult		
Males	1.6	5.2
Females	2.0	6.6

[a]   Data for *in vivo*-grown worms from Douvres.
[b]   Data for each measurement based on averages for ten specimens grown *in vitro* and comparable data for *in vivo*-grown stages.

From Douvres, F. W., *J. Parasitol.*, 66, 466, 1980. With permission.

VII. APPENDIX

Composition of the API[a] Series of Media[b]

Media[c] stock components	Stock concentration	API media						
	(Volumes, ml, added to give 1.0 liter of medium)	1	6	16	22	23	27	28
1. Bovine calf serum	—	250	200	250	250	250	250	250
2. EMEM essential AA	50×	20	20	20	20	—	20	20
3. EMEM nonessential AA	100×	10	10	10	10	—	10	10
4. BME vitamins	100×	20	20	20	20	—	20	20
5. Vitamin B_{12}	1%	1	1	1	1	—	1	1
6. Vitamin A	0.025%	0.4	0.4	0.4	0.4	—	0.4	0.4
7. Vitamin K	0.025%	0.04	0.04	0.04	0.04	—	0.04	0.04
8. Vitamin E	0.025%	0.04	0.04	0.04	0.04	—	0.04	0.04
9. Vitamin D_3	1%	0.1	0.1	0.1	0.1	—	0.1	0.1
10. PABA	0.5%	0.75	0.75	0.75	0.75	—	0.75	0.75
11. NAG	5.0%	1.5	1.5	1.5	1.5	—	1.5	1.5
12. L-glutamine	2.9%	10	10	10	10	—	10	10
13. TCA mixture	5×	20	20	20	20	20	—	20
14. Nucleic acid mixture	1×	50	50	50	50	—	50	50
15. Fatty acid mixture	10×	10	10	10	10	10	—	—
16. Ascorbic acid	1%	5	—	5	5	—	5	5
17. Cysteine·HCl	2.5%	5	—	5	5	—	5	5
18. Maltose	8.5%	10	—	10	10	10	10	10
19. Dextrose	10%	7	20	2	7	7	7	7
20. Trehalose	7%	5	—	5	5	5	5	5
21. Fructose	3.6%	5	—	5	5	5	5	5
22. EBSS	10×	55.5	0.5	0.5	55.5	55.5	55.5	55.5
23. $NaH_2PO_4 \cdot H_2O$	8.1%	0.1	—	4.1	0.1	0.1	0.1	0.1

VII. APPENDIX (continued)

Composition of the API[a] Series of Media[b]

Media[c] stock components	Stock concentration (Volumes, ml, added to give 1.0 liter of medium)	API media						
		1	6	16	22	23	27	28
24. KCl	8.95%	—	—	10	—	—	—	—
25. NaHCO$_3$	8.4%	20	3.6	20	20	20	20	20
26. REE	20%	75	250	75	—	75	—	—
27. CLE	20%	70	100	70	—	70	—	—
28. Bacto-peptone	25.8%	20	—	20	—	20	—	—
29. Trypticase	20%	25	50	25	—	25	—	—
30. Yeast extract	30%	10	16.7	10	—	10	—	—
31. Casein	15%	10	—	10	—	10	—	—
32. Lipid mixture	—	—	—	12	—	—	—	—
33. tH$_2$O	—	283.6	215.4	317.6	493.6	407.4	523.6	503.6

[a] Animal Parasitology Institute, Agricultural Research Service, U.S. Department of Agriculture, Beltsville, MD.

[b] Developed by Dr. F. W. Douvres and colleagues.

[c] Antibiotics are added to all media to give a concentration of 1000 IU/ml penicillin; 1.0 mg/ml streptomycin; and 10 µg/ml of amphotericin B. The pH is also adjusted.

REFERENCES

1. **Leland, S. E.,** *In vitro* cultivation of nematode parasites important to veterinary medicine, *Adv. Med. Sci. Comp. Med.,* 14, 29, 1970.
2. **Silverman, P. H.,** *In vitro* cultivation procedures for parasitic helminths, *Adv. Parasitol.,* 3, 159, 1965.
3. **Smyth, J. D.,** *Introduction to Animal Parasitology,* 2nd ed., Edward Arnold, London, 1976, chap. 35.
4. **Hansen, E. L. and Hansen, J. W.,** Nematoda parasitic in plants and animals, in *Methods of Cultivating Parasites in Vitro,* Taylor, A. E. R. and Baker, J. R., Eds., Academic Press, London, 1978, 227.
5. **Taylor, A. E. R. and Baker, J. R.,** *The Cultivation of Parasites in Vitro,* Blackwell Scientific, Oxford, 1968.
6. **Douvres, F. W. and Urban, J. F.,** Nematoda except parasites of insects, in *In Vitro Methods for Parasite Cultivation* Taylor, A. E. R. and Baker, J. R., Eds., Academic Press, London, 1987, 318.
7. **Behnke, J. M., Paul, V., and Rajasekariah, G. R.,** The growth and migration of *Necator americanus* following infection of neonatal hamsters, *Trans. R. Soc. Trop. Med. Hyg.,* 80, 146, 1986.
8. **Sen, H. G. and Seth, D.,** Development of *Necator americanus* in golden hamsters *Mesocricetus auratus,* *Indian J. Med. Res.,* 58, 1356, 1970.
9. **Weinstein, P. P.,** The cultivation of free-living stages of *Nippostrongylus muris* and *Necator americanus* in the absence of living bacteria, *J. Parasitol.,* 40(Sect. 5, Suppl. 2), 14, 1954.
10. **Weinstein, P. P. and Jones, M. F.,** Development *in vitro* of some parasitic nematodes of vertebrates, *Ann. N.Y. Acad. Sci.,* 77, 137, 1959.
11. **Carr, A. and Pritchard, D. D.,** Identification of hookworm *Necator americanus* antigens and their translation *in vitro,* *Mol. Biochem. Parasitol.,* 19, 251, 1986.
12. **Xianxiang, H. and Jiajun, H.,** Studies on cultivation of adult *Necator americanus in vitro,* *J. Parasitol. Parasitic Dis.,* 3, 35, 1985 (in Chinese, English summary).
13. **Weinstein, P. P.,** The cultivation of the free-living stages of hookworms in the absence of living bacteria, *Am. J. Hyg.,* 58, 352, 1953.
14. **Banerjee, D.,** *In vitro* cultivation of third stage larvae of *Ancylostoma caninum* (Ercolani, 1859), *J. Commun. Dis.,* 4, 175, 1972.
15. **Fenglin, W., Rongjun, H., Xianxiang, H., and Xihui, Z.,** *In vitro* cultivation of *Ancylostoma caninum,* *Chin. Med. J. (Engl. Ed.),* 92, 839, 1979.
16. **Douvres, F. W. and Tromba, F. G.,** Comparative development of *Ascaris suum* in rabbits, guinea pigs, mice and swine in 11 days, *Proc. Helminthol. Soc. Wash.,* 38, 246, 1971.
17. **Pilitt, P. A., Lichtenfels, J. R., Tromba, F. G., and Madden, P. A.,** Differentiation of late and early fifth stages of *Ascaris suum* Goeze, 1782 (Nematoda: Ascaridoidea) in swine, *Proc. Helminthol. Soc. Wash.,* 48, 1, 1981.
18. **Douvres, F. W., Tromba, F. G., and Malakatis, G. M.,** Morphogenesis and migration of *Ascaris suum* larvae developing to fourth stage in swine, *J. Parasitol.,* 55, 689, 1969.
19. **Costello, L. C.,** A simplified method of isolating *Ascaris* eggs, *J. Parasitol.,* 47, 24, 1961.
20. **Urban, J. R., Douvres, F. W., and Tromba, F. G.,** A rapid method for hatching *Ascaris suum* eggs *in vitro,* *Proc. Helminthol. Soc. Wash.,* 48, 241, 1981.
21. **Douvres, F. W. and Urban, J. F.,** Factors contributing to the *in vitro* development of *Ascaris suum* from second-stage larvae to mature adults, *J. Parasitol.,* 69, 549, 1983.
22. **Douvres, F. W. and Urban, J. F.,** Development of *Ascaris suum* from *in vivo* derived third-stage larvae to egg-laying adults *in vitro,* *Proc. Helminthol. Soc. Wash.,* 53, 256, 1986.
23. **Urban, J. F. and Douvres, F. W.,** *In vitro* development of *Ascaris suum* from third- to fourth-stage larvae and detection of metabolic antigens in multi-well culture systems, *J. Parasitol.,* 67, 800, 1981.
24. **Sommerville, R. I. and Weinstein, P. P.,** Reproductive behaviour of *Nematosporoides dubius in vivo* and *in vitro,* *J. Parasitol.,* 50, 401, 1964.
25. **Reimer, L. W.,** Die Heringswürmer (*Anisakis simplex* und verwandte Arten), *Angew. Parasitol.,* 24(Suppl. 4), 1, 1983.
26. **Carvajal, J., Barros, C., Santander, G., and Alacalde, C.,** *In vitro* culture of larval anisakid parasites of the Chilean hake *Merluccuius gayi,* *J. Parasitol.,* 67, 958, 1981.
27. **Hurst, R. J.,** Identification and description of larval *Anisakis simplex* and *Pseudoterranova decipiens* (Anisakidae: Nematoda) from New Zealand waters, *N.Z. J. Mar. Freshwater Res.,* 18, 177, 1984.
28. **Campbell, W. C., Ed.,** *Trichinella and Trichinosis,* Plenum Press, New York, 1983.
29. **Berntzen, A. K.,** Comparative growth and development of *Trichinella spiralis in vitro* and *in vivo,* with redescription of the life cycle, *Exp. Parasitol.,* 16, 74, 1965.
30. **Blair, L. S.,** Laboratory techniques, in *Trichinella and Trichinosis,* Campbell, W. C., Ed., Plenum Press, New York, 1983, 563.
31. **Sakamoto, T.,** Development and behaviour of adult and larval *Trichinella spiralis,* *Mem. Fac. Agric. Kagoshima Univ.,* 15, 107, 1979.

32. **Denham, D. A.,** Applications of the *in vitro* culture of nematodes, especially *Trichinella spiralis, Symp. Br. Soc. Parasitol.,* 5, 49, 1967.
33. **Jenkins, D. C. and Carrington, T. S.,** An *in vitro* screening test for compounds active against the parenteral stages of *Trichinella spiralis, Tropenmed. Parasitol.,* 32, 31, 1981.
34. **Jenkins, D. C., Armitage, R., and Carrington, T. S.,** A new primary screening test for anthelminthics utilizing the parasitic stages of *Nippostrongylus brasiliensis, in vitro, Z. Parasitenkd.,* 63, 261, 1980.
35. **Villella, J. B.,** Life cycle and morphology, in *Trichinosis in Man and Animals,* Gould, S. E., Ed., Charles C Thomas, Springfield, IL, 1970, 19.
36. **Andrews, J. S. and Maldonado, J. F.,** The life history of *Oesophagostomum radiatum,* the common nodular worm of cattle, *P.R. Agric. Exp. Stn. Rio Piedras Res. Bull.,* 2, 1, 1941.
37. **Douvres, F. W.,** The *in vitro* cultivation of *Oesophagostomum radiatum,* the nodular worm of cattle. III. Effects of bovine heme on development in adults, *J. Parasitol.,* 69, 570, 1983.
38. **Schulz, H. P. and Dalchow, W.,** Kultivierung der parasitischen Larvenstadien von *Oesophagostomum quadrispinulatum* (Marcone, 1901) *in vitro, Berl. Munch. Tieraerztl. Wochenschr.,* 82, 143, 1969.
39. **Farrar, R. G. and Klei, T. R.,** *In vitro* development of *Strongylus edentatus* to the fourth larval stage with notes on *Strongylus vulgaris* and *Strongylus equinus, J. Parasitol.,* 71, 489, 1983.
40. **Weinstein, P. P. and Jones, M. F.,** The *in vitro* cultivation of *Nippostrongylus muris* to the adult stage, *J. Parasitol.,* 42, 215, 1956.
41. **Weinstein, P. P. and Jones, M. F.,** The development of a study on the axenic growth *in vitro* of *Nippostrongylus muris* to the adult stage, *Am. J. Trop. Med. Hyg.,* 6, 480, 1957.
42. **Weinstein, P. P. and Jones, M. F.,** Development *in vitro* of some parasitic nematodes of vertebrates, *Ann. N.Y. Acad. Sci.,* 77, 137, 1959.
43. **Yokagawa, S.,** A new nematode from the rat, *J. Parasitol.,* 7, 29, 1920.
44. **Yokagawa, S.,** Development of *Heligmosomum* (= *Nippostrongylus*) *muris* Yokagawa, a nematode from the intestine of the wild rat, *Parasitology,* 14, 127, 1922.
45. **Bone, L. W.,** *Nippostrongylus brasiliensis:* female incubation, release of fractionation of incubates, *Exp. Parasitol.,* 54, 12, 1982.
46. **Leland, S. E.,** Studies on the *in vitro* growth of parasitic nematodes. I. Complete or partial parasitic development of some gastrointestinal nematodes of sheep and cattle, *J. Parasitol.,* 49, 600, 1963.
47. **Phillipson, R. F.,** Extrinsic factors affecting the reproduction of *Nippostrongylus brasiliensis, Parasitology,* 66, 405, 1973.
48. **Bonner, T. P., Weinstein, P. P., and Saz, H. J.,** Synthesis of cuticular protein during the third molt in the nematode *Nippostrongylus brasiliensis, Comp. Biochem. Physiol.,* 40B, 121, 1971.
49. **Bolla, R. I., Bonner, T., and Weinstein, P. P.,** Genetic control of the postembryonic development of *Nippostrongylus brasiliensis, Comp. Biochem. Physiol.,* 41B, 801, 1972.
50. **Wilson, R. J. M.,** Homocytotropic antibody response to the nematode *Nippostrongylus brasiliensis* in the rat, *J. Parasitol.,* 53, 752, 1967.
51. **Ward, J. B. and Bone, L. W.,** *Nippostrongylus brasiliensis:* environmental influences on pheromone production by females, *Int. J. Parasitol.,* 13, 499, 1983.
52. **Burt, J. S. and Ogilvie, B. M.,** *In vitro* maintenance of nematode parasites assayed by acetylcholinesterase and allergen secretion, *Exp. Parasitol.,* 38, 75, 1975.
53. **Douvres, F. W. and Malakatis, G. M.,** *In vitro* cultivation of *Ostertagia ostertagia,* the medium stomach worm of cattle. I. Development from infective larvae to egg-laying adults, *J. Parasitol.,* 63, 520, 1977.
54. **Douvres, F. W.,** Morphogenesis of the parasitic stages of *Ostertagia ostertagia,* a nematode parasite in cattle, *J. Parasitol.,* 42, 626, 1956.
55. **Cypess, R. H., Pratt, E. A., and Van Zandt, P.,** Rapid exsheathment of *Nematospiroides dubius* infective larvae, *J. Parasitol.,* 59, 247, 1973.
56. **Ransom, B. H.,** The nematodes parasitic in the alimentary canal of cattle, sheep and other ruminants, *U.S. Dep. Agric. Bur. Anim. Ind. Bull.,* 127, 132, 1911.
57. **Le Jambre, L. F.,** Genetics of vulvar morph types in *Haemonchus contortus: Haemonchus contortus cayugensis* from the Finger Lakes region of New York, *Int. J. Parasitol.,* 7, 9, 1977.
58. **Ayalew, L. and Murphy, B. E. P.,** *In vitro* demonstration of *in utero* larval development in an oviparous parasitic nematode: *Haemonchus contortus, Parasitology,* 93, 371, 1986.
59. **Stringfellow, F.,** Effects of bovine heme on development of *Haemonchus contortus in vitro, J. Parasitol.,* 70, 989, 1984.
60. **Stringfellow, F.,** Cultivation of *Haemonchus contortus* (Nematoda: Trichostrongylidae) from infective larvae to adult male and egg-laying female, *J. Parasitol.,* 72, 339, 1986.
61. **Douvres, F. W. and Fetterer, R. H.,** Survival of *Haemonchus contortus* adults in defined, semi-defined and complex cell-free media, *Proc. Helminthol. Soc. Wash.,* 53, 263, 1986.
62. **Leland, S. E.,** *In vitro* cultivation of *Cooperia punctata* from egg to egg, *J. Parasitol.,* 53, 1057, 1967.
63. **Zimmerman, G. L. and Leland, S. E.,** Completion of the life cycle of *Cooperia punctata in vitro, J. Parasitol.,* 57, 832, 1971.

64. **Stewart, T. B.,** The life history of *Cooperia punctata* a nematode parasite in cattle, *J. Parasitol.,* 40, 321, 1954.
65. **Marquardt, W. C.,** Separation of nematode eggs from fecal debris by gradient centrifugation, *J. Parasitol.,* 47, 248, 1961.
66. **Cauthen, G. E.,** A method of cultivating large numbers of *Haemonchus contortus* larvae from eggs in cattle feces, *Proc. Helminthol. Soc. Wash.,* 7, 82, 1940.
67. **Douvres, F. W.,** unpublished, 1987.
68. **Douvres, F. W.,** *In vitro* development of *Trichostrongylus colubriformis* from infective larvae to young adults, *J. Parasitol.,* 66, 466, 1980.
69. **Khan, Z. I. and Dorsman, W.,** *Trichostrongylus colubriformis*: cultivation of free-living stages, *J. Parasitol.,* 64, 1024, 1978.
70. **Dorsman, W. and Bijl, A. C.,** Cultivation of free-living stages of *Trichostrongylus colubriformis* in media without bacteria, animal tissue extract, or serum, *J. Parasitol.,* 71, 200, 1985.
71. **Glaser, R. W. and Stoll, N. R.,** Sterile culture of the free-living stages of the sheep stomach worm, *Haemonchus contortus, Parasitology,* 30, 324, 1938.
72. **Rapson, E. B., Jenkins, D. C., and Topley, P.,** *Trichostrongylus colubriformis*: *in vitro* culture of parasitic stages and their use for the evaluation of anthelmintics, *Res. Vet. Sci.,* 39, 90, 1985.
73. **Douvres, F. W.,** *In vitro* development of *Trichostrongylus axei* from infective larvae to young adults, *J. Parasitol.,* 65, 79, 1979.

Chapter 8

ACANTHOCEPHALA

J. D. Smyth

TABLE OF CONTENTS

I. BASIC PROBLEMS OF ACANTHOCEPHALA *IN VITRO* CULTURE

A. GENERAL COMMENTS

The Acanthocephala have proved to be exceptionally difficult to culture and no species has been successfully cultured through its life cycle *in vitro* to date. This is probably due to the fact that their *in vivo* development is relatively long — especially in species in poikilothermic hosts. Early studies on the group have been reviewed by Lackie[1] and more recently by Crompton and Nickol[2] and Crompton and Lassière.[3] The majority of culture experiments have been confined to *maintaining* the adult or larval stages for short or long periods, short-term maintenance (generally <12 h) being largely related to metabolic or other physiological studies and not related to achieving growth or differentiation.

Recent long-term studies on acanthellae[4] or cystacanths[5] have resulted in some growth being achieved, but in general, such culture attempts have also been relatively unsuccessful. In an early study, Jensen[6] grew the acanthella (from the crustacean *Hyalella aztica*) in a medium containing peptone broth, horse serum, yeast extract, and glucose and obtained some growth in size. Ovarian balls also developed within the female, but copulation was not observed and fertile eggs were not produced. The basic technique is summarized by Lackie.[1] In another early long-term experiment, Nicholas and Grigg[7] maintained *Moniliformis dubius* (= *moniliformis*) for 8 d with some viable eggs being produced during this period.

In this chapter, only recent techniques for the rodent acanthocephalean, *M. moniliformis*, which is widely used as an acanthocephalean model and readily maintained in the laboratory in rats and cockroaches (see Section II) are described. Other work has been succinctly reviewed by Crompton and Lassière.[3]

B. SHORT-TERM MAINTENANCE EXPERIMENTS

Some relevant short-term experiments, involving survival for 3 to 12 h, which will not be discussed further here, are those on *Moniliformis moniliformis*,[8,11] *Macracanthorhynchus hirudinaceus*,[12] *Polymorphus minutus*,[13] and *Paulisentis fractus*.[14]

II. *IN VITRO* CULTURE OF *MONILIFORMIS MONILIFORMIS* (= *DUBIUS*)

Cultured by: Lackie;[4] Tobias and Smith[5]

Definitive host: rat

Prepatent period: 5 to 6 weeks[15]

Common intermediate hosts: cockroaches (*Periplaneta americana, P. australasiae, Blatta orientalis, Blatella germanica*[16])

Experimental intermediate host: *P. americana* (In addition, three other species of Orthoptera and eight species of Coleoptera have been experimentally infected.[17])

A. GENERAL BIOLOGY AND LIFE CYCLE
1. Life Cycle

The general morphology of an adult is well-described in standard textbooks. The morphology of the male and female genitalia has been described in great detail by Asaolu,[18,19] as has the life cycle by Moore.[15] The adult worms live in the small intestine of rats and shelled acanthors, conveniently referred to as "eggs",[20] are passed in the feces, and are infective to cockroaches. Hatching occurs in the lower mid-gut of the insect with 24 to 48 h. The outer shell of the egg (Figure 1) is lost within 24 h, and after 48 h, the freed acanthors may be found in the gut lumen or actively penetrating the gut wall tissues. During migration

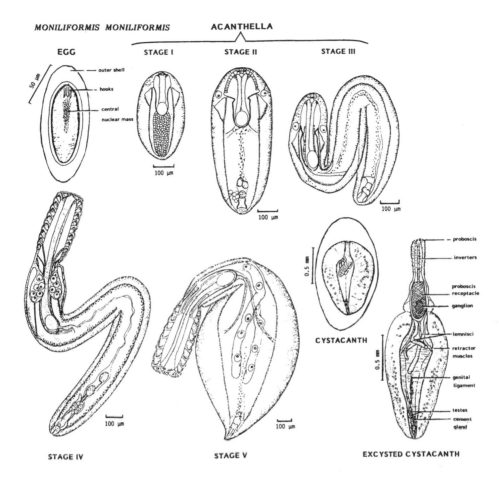

FIGURE 1. Stages in the development of *Moniliformis moniliformis* (= *dubius*) from egg to cystacanth. (Modified from Moore, D. V., *J. Parasitol.*, 32, 257, 1946 and King, D. and Robinson, E. S., *J. Parasitol.*, 53, 142, 1967. With permission.)

through the gut wall, the acanthors do considerable wandering and in about 10 to 12 d, the larvae appear as individual specks on the outer wall of the gut. Within a day or so, these larvae drop into the hemocoel where they lie free or become embedded in the adipose tissue. Here they go through a series of developmental stages (see below) eventually developing into an encysted acanthella or *cystacanth* (see Figure 1).

2. Stages in Larval Development

The following terminology[21] is used for the various stages:

- Stage I acanthor: a motile, hatched larva before penetration of the insect gut
- Stage II acanthor: stage covering the period of development in the gut wall and the short period in the hemocoel before structural differentiation begins
- Stage I acanthella (see Figure 1): stage marked by the initial differentiation of the central nuclear mass and the loss of body spines and the displacement of the rostellar hooks from their essential anterior position
- Stage II acanthella (see Figure 1): stage marked by the formation of the lemniscal nuclear ring
- Stage III acanthella (see Figure 1): stage characterized by rapid growth and bending of the body into a Z-shape

- Stage IV acanthella (see Figure 1): stage marked by the appearance of the lemnisci
- Stage V acanthella (see Figure 1): stage marked by rounding and flattening of the body
- Stage VI acanthella: stage marked by the invagination of the proboscis and neck (This stage is not yet infective.)
- *Cystacanth* (see Figure 1): final infective stage resulting from further morphological modifications of the previous stage (The cystacanth is surround by a "cyst" which is the product of the host's hemocytes and not of parasite origin.[21])

The time of development in the cockroach, at 27°C, is about 5 to 7 weeks.[21]

B. *IN VITRO* CULTURE TECHNIQUE

1. Cystacanths: Technique of Tobias and Schmidt[5]

The most successful of the early work with this species appears to be that of Tobias and Schmidt,[5] who in a brief research note reported cultivating the excysted cystacanths of *Moniliformis moniliformis* for 55 d, obtaining some limited growth, but little development of genitalia.

Source of material and sterilization — Intact cystacanths were dissected from infected *Periplaneta americana* and rinsed rapidly in 0.0625% Zephiran chloride and the cystacanth envelope (cyst) was teased away. The "uncovered" cystacanths were placed in sterile 0.85% NaCl and activated by the method of Graf and Kitzman[22] using bile salts.

Media and procedures — Activated juveniles were placed in pairs in 16 × 125 mm Kimax culture tubes in 12 ml of medium, gassed with 10% CO_2 in N_2 sealed with parafilm, and incubated at 37°C at a slight incline. The most successful medium was NCTC 199 with 10% inactivated calf serum and glucose alone or further supplemented with 20% hen egg yolk or 10% rat intestinal extract. The following antibiotics were added: 300 IU per milliliter penicillin; 50 mg/ml of streptomycin sulfate; and 50 mg/ml of tetracycline. The initial pH was 7.5 and the medium was renewed three times per week, 5 ml at a time.

Results — In all three media, worms survived for 55 d and somatic growth occurred, but no noticeable development of testes or ovarian balls resulted.

2. Acanthellae: Technique of Lackie[4]

The following technique of Lackie[4] has been quoted by Crompton and Lassière.[3]

Source of material and sterilization — Infected *Periplaneta americana* were swabbed with 70% alcohol and the crop was removed and discarded. The abdomen was opened in a 3-cm petri dish in sterile BSS and transferred to a 15-ml conical tube. Larvae were washed three times in Hills' BSS, with medium and 5% inactivated calf serum in the last wash (to prevent stickiness). Larvae were washed six times more, with one change of container tube.

Medium and procedure — The culture medium (Table 1) was based on one devised for the maintenance of cells from the cockroach, *Gromphadorhina laevigata*. Larvae were cultured in wells of a Nunc 24-well multidisk (Gibco Ltd.) with 750 µl of medium and 10% fetal calf serum. Stage I acanthellae (see Figure 1) were cultured at 50 per well; Stage III acanthellae or cystacanths were cultured at 20 to 30 per well; cultures were incubated at 25 to 28°C in the dark. Half of the medium was replaced every 2 to 3 d.

Results — Very little detailed information on the results of using this technique is available, but it is reported[3] that Stage II acanthellae developed to Stage V.

C. EVALUATION

The above techniques are in such a preliminary stage of development that it is impossible to give any kind of meaningful assessment of their value. Clearly, the whole field of acanthocephalean *in vitro* culture requires many more basic studies to be undertaken.

TABLE 1
Modified Cockroach Cell Medium D73 Used by Lackie[4] for Culturing the Acanthellae of *Moniliformis Moniliformis*

Amino acids (mg dl^{-1})		Salts (mg dl^{-1})	
L-Arginine-HCl	30	NaCl	600
L-Aspartic acid	25	KCl	80
L-Cysteine-HCl (BDH)	26	MgSO$_4$ · H$_2$O	18
L-Glutamic acid	95	Na$_2$HPO$_4$	5.0
Glycine	15	KH$_2$PO$_4$	6.0
L-Histidine-HCl	26	CaCl$_2$ · 2H$_2$O	14
L-Isoleucine	24		
L-Leucine	25	**Other components**	
L-Methionine	50	D-Glucose	100 mg dl^{-1}
L-Phenylalanine	20	HEPES	476 mg dl^{-1}
L-Proline	67	Streptomycin sulfate	10 mg dl^{-1}
L-Serine	8	Penicillin (sodium salt)	0.15 mg dl^{-1}
L-Threonine	20	Cyanocobalamine	0.5 μg
L-Tryptophan	20	Flow BME (basal medium Eagle) vitamins (× 100)	0.5 ml
L-Tyrosine	10	Gibco medium 199 (× 10) without bicarbonate and	4.0 ml
L-Valine	15	glutamine	
		Gentamicin sulfate	100 μg ml^{-1}

From Crompton, D. W. T. and Lassiere, O. L., in *In Vitro Methods for Parasite Cultivation*, Taylor, A. E. R. and Baker, J. R., Eds., Academic Press, London, 1987, 394. With permission.

REFERENCES

1. **Lackie, A. M.**, Acanthocephala, in *Methods of Cultivating Parasites In Vitro*, Taylor, A. E. R. and Baker, J. R., Eds., Academic Press, London, 1978, 279.
2. **Crompton, D. W. T. and Nickol, B. B.**, *Biology of the Acanthocephala*, Cambridge University Press, Cambridge, 1985.
3. **Crompton, D. W. T. and Lassière, O. L.**, Acanthocephala, in *In Vitro Methods for Parasite Cultivation*, Taylor, A. E. R. and Baker, J. R., Eds., Academic Press, London, 1987, 394.
4. **Lackie, A. M.**, unpublished work; as quoted in **Crompton, D. W. T. and Lassière, O. L.**, *In Vitro Methods for Parasite Cultivation*, Taylor, A. E. R. and Baker, J. R., Eds., Academic Press, London, 1987, 394.
5. **Tobias, R. C. and Schmidt, G. D.**, *In vitro* cultivation of *Moniliformis moniliformis*, *J. Parasitol.*, 63, 588, 1977.
6. **Jensen, T.**, The Life Cycle of the Fish Acanthocephalan, *Pomphorhynchus bulbocolli* (Linkins) Van Cleave, 1919, With Some Observations on Larval Development *In Vitro*, Ph.D. dissertion, University of Minnesota, Minneapolis, 1952.
7. **Nicholas, W. L. and Grigg, H.**, The *in vitro* culture of *Moniliformis dubius*, *Exp. Parasitol.*, 16, 332, 1965.
8. **Ward, P. V. F. and Crompton, D. W. T.**, The alcoholic fermentation of glucose by *Moniliformis dubius* (Acanthocephala), *in vitro*, *Proc. R. Soc. London Ser. B*, 172, 65, 1969.
9. **Körting, W. and Fairnbairn, D.**, Aerobic energy metabolism in *Moniliformis dubius* (Acanthocephala), *J. Parasitol.*, 58, 45, 1972.
10. **Crompton, D. W. and Ward, P. F. V.**, Selective metabolism of L-serine by *Moniliformis* (Acanthocephala) *in vitro*, *Parasitology*, 89, 133, 1984.
11. **Hovarth, K.**, Glycogen metabolism in larval *Moniliformis dubius*, *J. Parasitol.*, 57, 132, 1971.
12. **Wong, B. S., Miller, D. M., and Dunagan, T. T.**, Electrophysiology of acanthocephalan body wall muscles, *J. Exp. Biol.*, 82, 273, 1979.
13. **Crompton, D. W. T. and Ward, P. F. V.**, Lactic and succinic acids as excretory products of *Polymorphus minutus* (Acanthocephala) *in vitro*, *J. Exp. Biol.*, 46, 423, 1967.

14. **Hibbard, K. M. and Cable, R. M.,** The uptake and metabolism of tritiated glucose, tyrosine, and thymidine by adult *Paulisentis fractus* Van Cleave & Bangham, 1949 (Acanthocephala: Neoechinorhynchidae), *J. Parasitol.,* 54, 517, 1968.

15. **Moore, D. V.,** Studies on the life history and development of *Moniliformis dubius* Meyer, 1933, *J. Parasitol.,* 32, 257, 1946.

16. **Lackie, J. M.,** The host specificity of *Moniliformis dubius* (Acanthocephala) a parasite of cockroaches, *Int. J. Parasitol.,* 5, 301, 1975.

17. **Siddikov, B. Kh.,** The biology of *Moniliformis moniliformis* (Bremer, 1811), *Uzb. Biol. Zh.,* 5, 43, 1986, (in Russian).

18. **Asaolu, S. O.,** Morphology of the reproductive system of female *Moniliformis dubius* (Acanthocephala), *Parasitology,* 81, 433, 1980.

19. **Asaolu, S. O.,** Morphology of the reproductive system of the male *Moniliformis moniliformis* (Acanthocephala), *Parasitology,* 82, 297, 1981.

20. **Lackie, J. M.,** The course of infection and growth of *Moniliformis dubius* (Acanthocephala) in the intermediate host *Periplaneta americana, Parasitology,* 64, 95, 1972.

21. **King, D. and Robinson, E. S.,** Aspects of the development of *Moniliformis dubius, J. Parasitol.,* 53, 142, 1967.

22. **Graff, D. J. and Kitzman, W. B.,** Factors influencing the activation of acanthocephalan cystacanths, *J. Parasitol.,* 51, 424, 1965.

Chapter 9

CRYOPRESERVATION OF HELMINTH PARASITES

Eric R. James

TABLE OF CONTENTS

I. INTRODUCTION

Cryopreservation, as an integral component of the *in vitro* cultivation of parasitic helminths, has now reached the level of importance and usefulness it achieved some time ago with the protozoa. In the past few years, cryopreservation techniques have been described for a number of new (i.e., previously uncryopreserved) species and stages of helminths, and several reviews covering this subject have been published.[1-5]

The many different protocols for cryopreserving helminths can broadly be divided into those techniques which use slow cooling and those which use rapid cooling and those which incorporate cryoprotectants and those which do not. In all cases, however, the underlying cryobiological theory is the same. It is the interaction of the particular physical and chemical stresses, produced by freezing aqueous solutions, with the fundamental biological characteristics of a particular species or stage of helminth which determines the optimum cryopreservation technique for that helminth.

The goal in helminth cryopreservation is to achieve stable long-term maintenance with minimal loss of viability. The aim of this chapter is to act as a practical guide to helminth cryopreservation covering basic cryobiological theory and to provide an outline of specific techniques used for the cryopreservation of different helminths, culminating in an approach that one might take in designing a successful cryopreservation protocol.

II. CRYOPRESERVATION THEORY

Crystallization of water is the main threat to the survival of helminths during cryopreservation. Intracellular ice is generally lethal and the success of cryopreservation protocols is dependent on reducing, or in most cases entirely eliminating, its formation. Two basic strategies are available for reducing or preventing intracellular ice. The first is to reduce the water content of the organism. The second is to interfere with the ability of the intracellular water to crystallize. All cryopreservation protocols employ a combination of both strategies. Extracellular (or extraorganismal) ice is generally considered to be much less damaging and not infrequently can be beneficial.

A. CRYSTALLIZATION OF WATER

Ice nucleation within an aqueous solution is a random event occurring spontaneously once the temperature reaches or falls below the solution's freezing point. Once formed, the rate at which an ice crystal propagates through the solution depends on the viscosity of the solution and the degree of supercooling. Water which is still liquid below its freezing point is supercooled (or undercooled). The maximum limit of supercooling for pure water is $-38°C$.[6] At that temperature, individual water molecules can act as ice nucleation sites (homogenous nucleation). In practice, suspensions of cells or organisms do not supercool significantly since particles within the solution or the cell act as nucleation sites (heterologous nucleation).

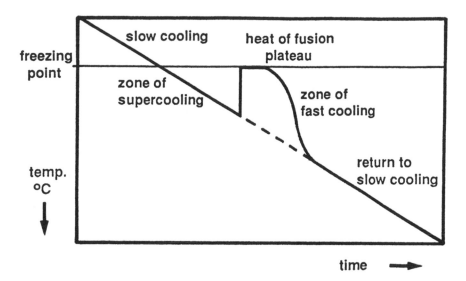

FIGURE 1. A sample cooled slowly often becomes supercooled below its freezing point. As soon as the sample freezes, however, the temperature rises as the latent heat of fusion is evolved. There will then be a marked temperature difference between the sample and the coolant in which the sample tube is sitting and the sample will cool fast to catch up. This zone of fast cooling is frequently damaging to cells.

Water liberates 79.7 cal/g heat of fusion as it crystallizes. If crystallization commences at or close to the freezing point of the solution, the latent heat of fusion will be evolved gradually. However, if the solution has become supercooled, then as soon as the first ice nucleation event occurs, crystallization will proceed rapidly. This causes a sudden release of the latent heat of fusion and a rapid warming of the solution up to, or almost to, its freezing point (Figure 1).

In most commercially available cooling machines, the sample being cryopreserved is in contact with a liquid or gas refrigerant which is itself being continuously cooled at a controlled rate. When all the latent heat of fusion has been evolved, the sample then has to cool faster to catch up with the liquid (or gas) in the cooling device. This sudden rise and subsequent fall in temperature is potentially very damaging to cells. To avoid this trauma, nucleation can be induced at the freezing point before supercooling occurs. There are many ways to achieve this. Tapping the sample container vigorously, touching the suspension momentarily with a much colder object, or seeding with an ice crystal are all methods to induce nucleation. Manufacturers of controlled cooling devices have incorporated various versions of these approaches for minimizing the trauma associated with crossing the freezing point. In those controlled rate freezers which use cold nitrogen gas as the refrigerant, a common method of overriding the latent heat of fusion is to generate a short pulse of cold gas as soon as nucleation is detected in a reference sample. In this way, the evolution of latent heat of fusion is effectively neutralized and a linear cooling rate can be maintained within the sample.

In an aqueous cell suspension, the intracellular solute concentration will generally be greater than the extracellular solute concentration, thus depressing the freezing point of the intracellular contents relative to the extracellular medium. Additionally, each cell forms a separate compartment while the extracellular medium is essentially a single large compartment. Thus, there is a much greater probability of ice formation occurring within the extracellular medium first rather than within any individual cell.

Ice crystals exclude practically all solutes (and dissolved gases) as they form. These solutes further depress the freezing point of the remaining unfrozen water. As cooling continues, so the solutes become concentrated into a progressively smaller volume of un-

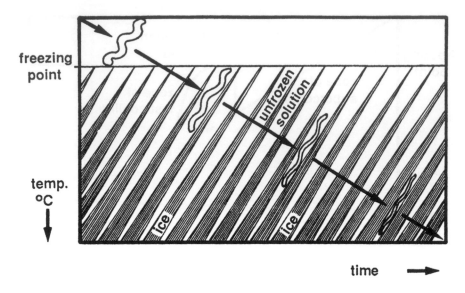

FIGURE 2.   When ice forms in an aqueous solution, all dissolved and suspended material is effectively excluded from the ice crystals. As cooling proceeds, so the volume of the unfrozen fraction containing the solutes and suspended material, including helminths, becomes progressively smaller.

frozen water. Helminths are also confined within this residual unfrozen fraction (Figure 2). The concentrated solutes may be chemically toxic and/or cause damage to the worms through excessive osmotic stress. During slow cooling in the presence of extracellular ice, helminths will lose water by osmosis, i.e., by the higher vapor pressure of liquid water inside the cell vs. the lower vapor pressure of ice external to the cell (Figure 3).

Classical cryobiological theory holds that for a given cell type, there is an optimal cooling rate for successful cryopreservation[7] (Figure 4). At cooling rates faster than optimal, cells are damaged by intracellular ice, while at suboptimal cooling rates, damage to cells occurs by extended exposure to toxic levels of solutes and/or by osmotic stress and excessive water loss — these latter stresses are termed solution effects.

B. CRYOPROTECTANTS

A wide variety of chemicals, including many alcohols, sugars, amides, and proteins, have cryoprotective properties. Several of these are synthesized naturally by animals and plants in response to low temperatures. Cryoprotectants are often arbitrarily divided into the relatively larger molecular weight compounds which act extracellularly and the smaller molecular weight compounds which permeate into cells. Glycerol and ethanediol can be either extracellular or intercellular cryoprotectants or both, since these compounds penetrate cells well at 37°C, but due to their high coefficients of viscosity, they penetrate poorly at 0°C. Their addition at 0°C produces partial dehydration of the organism.

Several successful cryopreservation techniques use sequential incubation steps (see Section III, Methods for Cryopreserving Helminths, below). The first incubation at 37°C in a low to moderate concentration induces penetration of the cryoprotectant. The second incubation step in a higher concentration at 0°C induces partial dehydration and consequent elevation of the intracellular concentration of cryoprotectant. Ethanediol has been used to best effect in this respect notably with *Schistosoma mansoni* schistosomula and with several species of microfilariae.[8-11] The combining of two or more cryoprotectants can have similar results. In this way, a single incubation in 10% Me_2SO together with 10% dextran as used by Nolan et al.[12] for *Strongyloides* would have simultaneously caused penetration of the

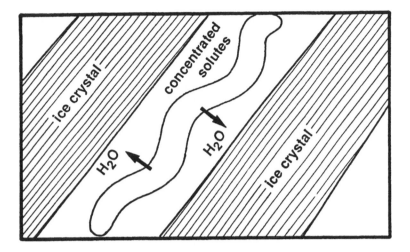

FIGURE 3.   Cells and organisms trapped within the unfrozen fraction between ice crystals are exposed to elevated concentrations of solutes. As the temperature decreases, the concentration of solutes in the unfrozen fraction increases. Water moves by osmosis out of the organism which becomes progressively dehydrated.

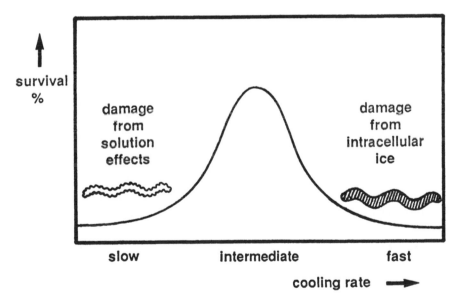

FIGURE 4.   At fast cooling rates, water cannot move out of the organism fast enough to keep pace with the reduction in temperature and it crystallizes inside the organism. Intracellular ice is almost always lethal. Slow cooling reduces the likelihood of intracellular ice forming. However, while the water content of the organism may be in equilibrium with the surrounding unfrozen fraction at slow cooling rates, the organism is damaged by spending an extended period of time in the concentrated solutes. Damage from solution effects is time dependent and thus these effects are lessened at faster cooling rates. These two factors interact and are both reduced at an intermediate cooling rate when optimal levels of survival are obtained.[7] The terms slow, intermediate, and fast are relative and are different for each cell type. For many cell types, 1°C min⁻¹ is an intermediate cooling rate; for erythrocytes, this rate is around 3000°C min⁻¹.

Me$_2$SO and partial dehydration resulting from the dextran. A similar mechanism occurs with the use of methanol and ethanediol together at 0°C for *Trichinella spiralis nativa*.[13]

Cryoprotectants operate by one or more of several different mechanisms most of which are not completely understood. The most frequently cited beneficial action of cryoprotectants is their colligative effect — when added to an aqueous solution, they depress the freezing point. This leads to a reduction in the concentration of other solutes, most importantly salt, and thus to a reduction in the cell shrinkage associated with cooling (Figure 5). The antifreeze proteins of antarctic fish appear to be protective by binding to the developing ice crystal lattice and preventing the addition of water molecules to the ice. Other common characteristics of many cryoprotectants include their ability to "stabilize" water through hydrogen bond formation (e.g., notably the alcohols), retarding ice crystal growth by increasing solution viscosity, (e.g., particularly PVP and glycerol), and scavenging of free radicals produced during cryopreservation (e.g., methanol, ethanediol, Me$_2$SO). In the presence of high concentrations of certain cryoprotectants, or when slow cooling proceeds until the cryoprotectants in the residual unfrozen fraction (also containing the helminth) become sufficiently concentrated, then crystallization is inhibited and the sample vitrifies (see below).

Generally, each cryoprotectant combines several of these protective mechanisms. The cumulative effect of cryoprotectant addition is to enhance survival at the optimum cooling rate, shift this optimum to a slower rate and also to reduce the rate at which vitrification occurs enhancing survival at rapid cooling rates (Figure 6).

C. VITRIFICATION

Water vitrifies when it fails to organize into crystals and retains its amorphous molecular arrangement as it solidifies. Viscosity and cooling rate interact in the process of vitrification. Pure water can only be vitrified by cooling extremely rapidly — approximately 10^9°C sec^{-1}. Many cryoprotectants significantly enhance the viscosity of water, e.g., an aqueous solution of 40% v/v polyvinylpyrrolidone will not crystallize at any cooling rate; 40% glycerol will vitrify at cooling rates in excess of around 100°C min^{-1}; and 40% ethanediol will vitrify at cooling rates in excess of around 1000°C min^{-1}.

During slow cooling to -60°C, the removal of water from solution by crystallization will effectively concentrate solutions of glycerol or ethanediol to 58 or 52% v/v, respectively, in the unfrozen fractions (Figure 7). Thus, if the cells are in approximate osmotic equilibrium with this residual unfrozen fraction, their intracellular contents will eventually become vitrified even at cooling rates as low as 1°C min^{-1} (Figure 8).

Many cells and organisms, including certain of the parasitic helminths, do not tolerate slow cooling well, but can be cryopreserved effectively if high concentrations of cryoprotective additives are used in combination with rapid cooling. The appropriate combination of cryoprotectant concentration and cooling rate varies for each helminth. Technically, cooling rates in excess of around 10,000°C min^{-1} are impractical and thus the lowest useful concentration of cryoprotectant to induce vitrification by rapid cooling is about 35% v/v. At slower cooling rates, a concomitant increase in the cryoprotectant concentration is required for the solution to vitrify (see Figure 8).

D. COOLING RATES

The optimum cooling rate varies for different cell types basically in relation to the cell's size, the permeability of its limiting membrane, and its tolerance to cryoprotectants, salts, and dehydration. Samples of around 1ml when plunged into liquid nitrogen will cool at 100 to 500°C min^{-1}, depending on the wall thickness of the tube containing the sample and the material from which it is constructed. Slower cooling rates can be achieved by insulating the tube or cooling it in vapor phase nitrogen — such as in the Union Carbide BF-6 apparatus. Various techniques have been described for achieving rates of around 1°C min^{-1}, which

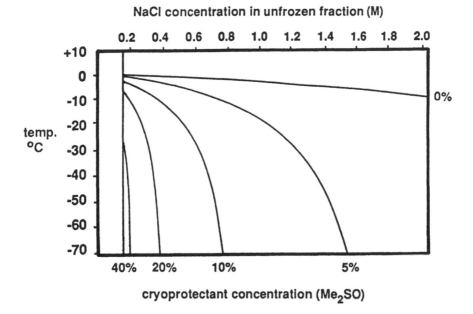

FIGURE 5. The addition of a cryoprotectant to an aqueous solution lowers its freezing point. Therefore at any given temperature, the volume of the unfrozen fraction will be larger (and the concentration of the dissolved solutes will be lower) than in the absence of the cryoprotectant. (Based on data of Farrant.[14])

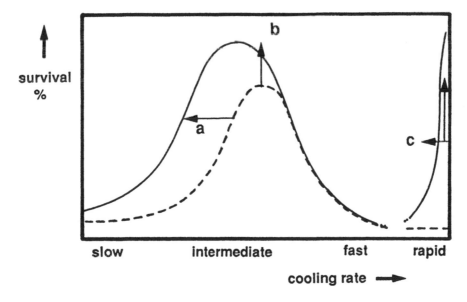

FIGURE 6. Cryoprotectants increase survival levels at intermediate cooling rates and generally shift the optimum cooling rate to a lower rate. Additionally, by increasing the viscosity of the solution they reduce the cooling rate at which vitrification occurs. Many cell types can survive rapid cooling if they vitrify internally (see Figure 19).

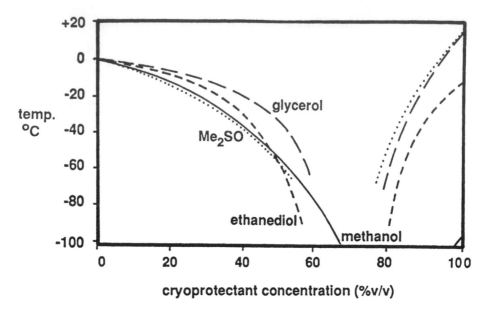

FIGURE 7.   Freezing point determinations for saline (lactalbumin hydrolysate with Earle's salts) containing different concentrations (percent volume/volume) of the cryoprotectants methanol, Me₂SO, ethanediol and glycerol.

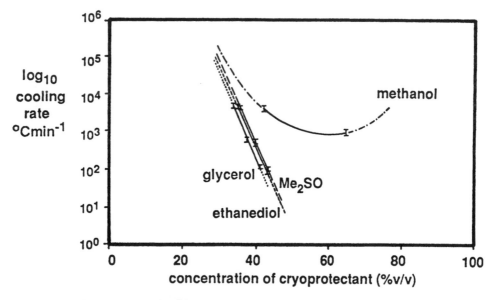

FIGURE 8.   Relationship between the critical cooling rate required to induce vitrification and the concentration of cryoprotectant (percent volume/volume in 0.85% saline). (Data of James, E. R. and Renfro, J.)

seems to have become a favorite rate for many workers wishing to cryopreserve a range of different types of cells. In some cases, 1°C min⁻¹ (or so) has been shown by careful experimentation to be optimal, but in many cases this rate has been picked arbitrarily without any evaluation of other cooling rates.

There are several controlled rate cooling apparatuses available commercially capable of generating cooling rates of less than 0.1°C min⁻¹ up to around 100°C min⁻¹. Some programmable cooling devices can produce an almost infinite variety or combination of cooling rates and these are excellent for helping to define an optimum cryopreservation protocol during the experimental phase. For routine cryopreservation, they are expensive to purchase and operate, but do give near perfect reproducibility.

At the other end of the scale, rapid cooling is simple and does not require expensive equipment. To achieve rapid cooling rates, the sample volumes and/or the sample container have to be reduced in size. The contents ($\sim$20 μl) of a sealed capillary tube plunged into liquid nitrogen cool at around 3000°C min^{-1}. A similar-sized sample spread out to 4 $\times$ 10 mm on a No. 2 glass cover slip plunged into liquid nitrogen cools at approximately 5000°C min^{-1}. A 20 μl sample, not supported on a glass sliver, prefrozen at -20°C or so, may cool as fast as 10,000°C min^{-1} when plunged into liquid nitrogen. These cooling rates can be almost doubled by using freezing point nitrogen instead of liquid nitrogen which is at its boiling point of -196°C. Freezing point nitrogen at -210°C can be produced by allowing some of the nitrogen to boil off under a partial vacuum. This can be achieved in certain freeze-drying apparatuses, but extreme care is required for this as it is a potentially dangerous procedure. Frozen nitrogen is a soft, white solid, but above the solid phase, the liquid is also at, or close to, -210°C. Small samples plunged into this liquid will not cause boiling around the sample as occurs with nitrogen at -196°C. In the absence of this insulating gas envelope, cooling rates of 10,000 to 20,000°C min^{-1} are attainable — which is close to the resolution of most electromechanical temperature sensing equipment.

To achieve even faster cooling rates, samples have to be reduced to small droplets or spray mists and shot into liquid nitrogen or onto the surface of a metal target precooled to -196°C or so. Obviously these methods are impractical for cryopreserving helminths (Figure 9).

E. STEPPED COOLING

Hybrid cooling techniques which involve slow cooling down to a set intermediate temperature, followed by rapid cooling, usually into liquid nitrogen, can be extremely useful (Figure 10). A holding period at the intermediate temperature may also be included. Monitoring survival from these samples in conjunction with the survival of control samples warmed from various points during the slow cooling step and/or the holding period at the intermediate temperature enables considerable information to be gathered: the tolerance of the helminth to cryoprotectants and dehydration, its preference for slow or rapid cooling, and its potential for vitrification.

Stepped cooling experiments gave the first indication that *Schistosoma mansoni* schistosomula could be cryopreserved[15,16] and has been very useful in defining cryopreservation protocols for microfilariae (mf) and third stage larvae (L3s) of the filarial worms,[17] the L3s of *Dictyocaulus*[18] and infective larvae of *Trichinella*,[19] the first stage juveniles (J1s) and infective juveniles (IJs) of *Steinernema* (= *Neoaplectana*),[20] and the cysticerci of *Taenia*.[21]

F. STORAGE TEMPERATURE

Four media are commonly used for cryogenic storage of living organisms — dry ice at -69°C, freezers at temperatures ranging from -70°C to about -140°C, vapor phase nitrogen at around -140°C, and liquid nitrogen at -196°C. A variety of liquid, and vapor phase, nitrogen storage vessels are available from a number of manufacturers. These can be purchased with inventory systems for the glass capillaries preferred by many protozoa cryobanks, plastic straws used predominantly for sperm and embryo storage, or for the screw-top plastic cryotubes which are tending to become the industry standard for most cryobanking operations.

Samples cryopreserved by slow cooling and stored in dry ice (-69°C) or in the higher temperature freezers (above -90°C) are relatively stable, but will deteriorate with time as these temperatures are above the glass point and the recrystallization point (Figure 11).

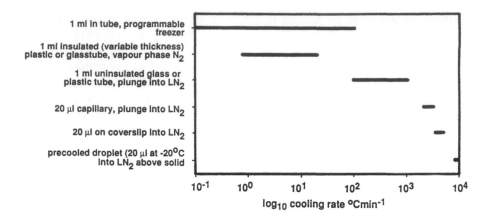

FIGURE 9.   Cooling rates produced by various methods.

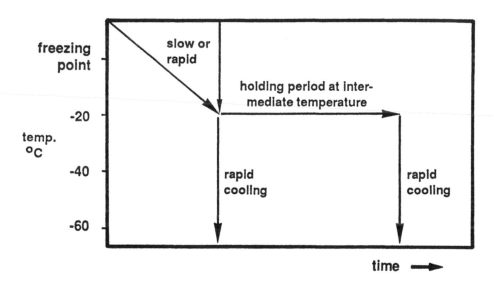

FIGURE 10.   There are two basic methods of producing stepped cooling: (1) slow cooling to an intermediate temperature followed by rapid cooling and (2) rapid cooling to an intermediate temperature
followed by a holding period at that temperature and then a second rapid cooling step. Partial cellular
dehydration occurs during the slow cooling step in method 1 or during the holding period in method 2.
The intracellular contents are vitrified during the subsequent rapid cooling step.

Vitrified samples produced by techniques which use high cryoprotectant concentrations and
rapid cooling are particularly unstable at the higher temperatures and should only be stored
in liquid or vapor phase nitrogen. It is also extremely important to limit the time a cryopreserved sample spends outside vapor phase or liquid nitrogen when it is being transported
or when other samples are being retrieved from a common storage vessel. So long as these
precautions are followed, a cryopreserved sample stored in liquid nitrogen will be equally
viable when thawed after 32 s or after 32,000 years.[23]

G. WARMING AND DILUTION

The method used to recover an organism from low temperature storage can be as
important as the protocol used for cooling, but rarely has this aspect of cryopreservation
been given much consideration. During warming, samples pass through the glass point at
around -135 to $-100°C$ (depending on the cryoprotectant concentration), the devitrification

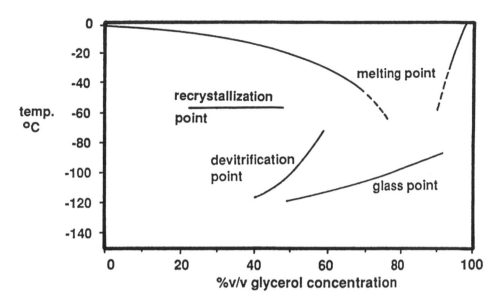

FIGURE 11. Phase diagram for glycerol indicating temperatures of melting, recrystallization, devitrification, and glass point transition for different concentrations (%w/w). (Redrawn from Luyet, B. J. and Rasmussen, D., *Biodynamica*, 10, 167, 1968. With permission.)

point at about −120 to −70°C and the recrystallization point at about −60°C (see Figure 11). Ice crystals which form above the devitrification point during warming are generally considered to be minute and innocuous,[24] but these will aggregate into large damaging crystals above the recrystallization point.

Samples which have vitrified during either slow or rapid cooling must be warmed rapidly to avoid damage due to devitrification and recrystallization. However, some cells or organisms which have been cooled slowly may only survive warming at slow rates. Continuous slow cooling to very low temperatures can produce considerable cellular dehydration. Some cells have to be rehydrated gradually to avoid dilution stresses — which is only accomplished by slow warming. However, there is no evidence so far that any helminth prefers to be warmed slowly.

Following thawing, care may also have to be taken in diluting the sample to remove the cryoprotectant. If dilution proceeds too rapidly, the cell/organism may swell and burst as water enters the cell faster than the cryoprotectant can exit. For some cell types, e.g., erythrocytes, this is probably the single most critical step in cryopreservation. However, except for one study with *Onchocerca* mf,[25] there does not appear to be much evidence that helminths are prone to dilution stress. Dilution stress can be reduced by incorporating an osmotic buffering agent such as glucose or serum in the diluting medium. This will reduce the amount of water entering the cell/organism while the intracellular cryoprotectant is leaving the cell.

H. VIABILITY ASSAYS

The quickest method for assaying helminth viability following cryopreservation is usually by motility. However, while this is convenient for determining trends or comparing different treatments, it is not definitive. Metabolic assays such as vital stains or uptake of radiolabels have not been used to assay helminth survival following cryopreservation, although several published methods are available. Obviously infectivity and/or subsequent development *in vivo* or *in vitro* are the only true measures of survival, and to be useful, these assays should be quantitative and should always incorporate the relevant normal untreated and cryoprotectant-treated noncryopreserved controls.

III. METHODS FOR CRYOPRESERVING HELMINTHS

Cryopreservation of multicellular organisms is intrinsically difficult since a cryopreservation protocol which is optimal for one cell type, e.g., muscle cells, may be damaging to other cell types, e.g., nerve cells or cells of the reproductive organs. However, some helminths can be cryopreserved using remarkably simple techniques.

Perhaps it is not surprising that the infective larvae of the gastrointestinal (GI) nematodes of ruminants were among the first to be successfully cryopreserved.[26-34] The L3s are free-living and may be exposed to extremes of low temperature and desiccation during the course of their dispersal and search for new hosts. They have been found able to tolerate temperatures as low as $-31°C$ in the field.[35] These nematodes retain their shed L2 cuticle as a sheath protecting them from desiccation. This sheath, however, not only interferes with water movement across the larval surface, but also with the uptake of cryoprotectants. In order to be cryopreserved, the larvae must first be exsheathed. This phenomenon is not limited to L3s or GI nematodes. Schistosome schistosomula newly transformed from cercariae gradually lose the water-impervious glycocalyx over 90 min of *in vitro* culture and only then can be cryopreserved successfully.[8] Also, the infective larvae of *Trichinella spiralis* have two additional inert surface tegumental layers which must first be removed before they can be cryopreserved.[19,36]

All cryopreservation protocols, whether they involve slow or rapid cooling or internal or external cryoprotectants, induce some degree of dehydration of the organism. This partial dehydration is beneficial and is often crucial for cryopreservation success. The bulk of the dehydration which occurs with slow cooling is induced as a result of extracellular ice crystal formation (see above). In techniques which employ rapid cooling in combination with high cryoprotectant concentrations, the shrinkage is induced entirely by the cryoprotectant in the absence of intracellular ice. For some organisms, e.g., *Steinernema* ($=$ *Neoaplectana*)[37] and *Trichostrongylus*,[38] optimum survival from cryopreservation only follows prior dehydration by exposure to lowered relative humidity in a desiccation chamber. For ease of reference, the cryopreservation methods which are outlined below have been divided into the GI/lung nematodes, tissue nematodes, entomogenous nematodes, trematodes, and cestodes.

A. NEMATODES OF THE GI TRACT AND LUNGS

As regards cryopreservation, interest in this group of nematodes has been mainly with the L3 stage. Many of these helminths are inherently resistant to low temperatures and can be successfully cryopreserved without using any cryoprotective additive (which might suggest that they contain their own natural cryoprotectants). For a few species, the addition of cryoprotectants enhances levels of survival, and for *Strongyloides*, the use of cryoprotectants appears to be mandatory. Coles et al.[32] compared fast cooling (rate not given, but probably $\sim70°C$ min^{-1}) with slow cooling ($1°C$ min^{-1}) for eight different species of GI nematodes and obtained consistently better results with the slower rate. Successful techniques report the use of slow cooling rates or of stepped cooling protocols, which suggests that some desiccation is necessary for cryopreservation to be successful.

1. Single Cryoprotectant, Slow Cooling: Eggs/L1s

Angiostrongylus cantonensis **eggs,** *Strongyloides stercoralis* **L1s**

Pretreatment	None
No./volume	35/! ml to 50,000/1 ml
Cryoprotectant	1 M Me$_2$SO at 37°C, then 1.5 M Me$_2$SO at 0°C for 15 min;[39] 10% Me$_2$SO + 10% dextran, 60 min at room temperature[12]

Cooling rate	0.3°C min^{-1};[39] 0.8°C min^{-1} to −40°C, then 10°C min^{-1} to −70°C[12]
Warming rate	5°C min^{-1};[39] ~70°C min^{-1} (see Reference 12)
Dilution	1:10 in RPMI
Viability	95.4% grew *in vitro* to L3 (These were infective to puppies.)
References	Nolan et al.,[12] Uga et al.[39]

Note: During the first Me$_2$SO incubation in the protocol described by Uga et al.,[39] the sample was cooled at 1°C min^{-1} to 0°C. The *Angiostrongylus* eggs were cooled only to −50°C. Extended storage at −50°C was not reported and neither was storage at −196°C; however, survival from temperatures lower than −50°C is unlikely with the cooling protocol as reported.[39]

2. No Cryoprotectant, Slow Cooling: L3s

Cooperia, Haemonchus, Marshallagia, Nematodirus, Nematospiroides, Oesophagostomum, Ostertagia, Trichostrongylus

Pretreatment	Exsheath in 0.72% NaOCl, 5 min, 20°C; or CO$_2$ for 30 min
No./volume	~50,000 to 150,000/1 ml in glass or plastic tube
Cryoprotectant	None
Cooling rate	1°C min^{-1};[27,32] 10°C min^{-1},[34] in vapor phase nitrogen
Warming rate	~100°C min^{-1} in warm water at 20°C; also ~1,000°C min^{-1} (see Reference 27)
Dilution	n.a.
Viability	Motility: 15%[33] (*Marshallagia*); 25%[28] (*Cooperia*); 38—61%[28] (*Ostertagia*); 86% (*Haemonchus*); 6,[27] −87,[28] −100%[33] (*Trichostrongylus*); 19—91%[28] (*Oesophagostomum*); 94%[33] (*Nematodirus*)
	Infectivity: 11%[32] (*O. circumcincta*); 12%[32] (*Nematodirus*); 25%[32] (*Haemonchus*); 7 vs. 27% for unfrozen controls[26] (*N. filicollis*), 25 vs. 10%[32] (*Ostertagia* — unfrozen controls were lower); 30 vs. 36%[26] *N. battus*); 52 vs. 43%[29] (*Haemonchus*)
References	Parfitt;[26] Isenstein and Herlich;[27] Campbell and Thompson;[28] Campbell et al.;[29] Kelly and Campbell;[30] Van Wyk and Gerber;[31] Coles et al.;[32] Van Wyk et al.;[33] Ramp et al.[34]

Note: For this group of nematodes, cryoprotectant additives have been found to confer no benefit or to be detrimental to survival.[30,32,40] The cryopreservation of a mixed population of horse strongylids (70% *Trichonema* spp.) in desiccated feces has also been reported by Bemrick.[42] For the use of desiccation in cryopreservation protocols, see the section on entomogenous nematodes.

3. No Cryoprotectant, Stepped Cooling: L3s

Dictyocaulus viviparus

Pretreatment	Exsheath in 0.1% NaOCl for 10 min at 20°C
No./volume	~1,000 in 20-µl droplet
Cryoprotectant	None
Cooling rate	1°C min^{-1} to −10°C on an aluminium support, followed by plunge into liquid nitrogen ~10,000°C min^{-1}
Warming rate	6,800°C min^{-1} by agitating droplet in 2 ml of 37°C saline
Dilution	n.a.
Viability	Motility: up to 73%
	Infectivity: 0.8 to 1.1% for cattle
References	James and Peacock[18]

Note: The use of cryoprotectants with this parasite is also detrimental.

4. Single Cryoprotectant, Slow Cooling: L3s

Ancylostoma caninum, Nippostrongylus brasiliensis, Strongyloides stercoralis

Pretreatment	Exsheath in 0.72% NaOCl (*Nippostrongylus, Ancylostoma*) or CO$_2$ for 60 s[30,43] (*Ancylostoma*) or none (*Strongyloides*)
No./volume	20,000—200,000/1 ml

Cryoprotectant	10% glycerol 20 h at 26°C;[44] 10% Me$_2$SO, incubation temp and time not given[45] (*Nippostrongylus, Ancylostoma*); 10% Me$_2$SO + 10% dextran, 90 min at room temperature[12] (*Strongyloides*)
Cooling rate	Slow — rate not given (*Nippostrongylus, Ancylostoma*); ~10°C min^{-1} using vapor phase nitrogen[12] (*Strongyloides*)
Warming rate	Fast — rate not given (*Nippostrongylus, Ancylostoma*), but probably ~70°C min^{-1} by immersion in water at 40°C until thawed[12]
Dilution	Not given (*Nippostrongylus, Ancylostoma*); 1:10 in RPMI[12] (*Strongyloides*)
Viability	Motility: 47—54%,[45] 48%,[44] 58%[43] (*Ancylostoma*); 10—64%[30] (*Nippostrongylus*), 73.5%[12] (*Strongyloides*)
	Infectivity: 25—31%[45] compared to unfrozen controls (*Ancylostoma*) and 28—32%[30] (*Nippostrongylus*); confirmed, but not quantified[12] (*Strongyloides*)
References	Nolan et al.;[12] Kelly and Campbell;[30] Vetter and Klaver-Wesseling;[43] Miller and Cunningham;[44] Kelly et al.[45]

Note: Although successful cryopreservation of *Nippostrongylus* has been reported in the absence of a cryoprotectant,[30] cryoprotectant use does enhance survival, and it appears that *Strongyloides* cannot be cryopreserved in the absence of cryoprotectant.[42]

B. TISSUE NEMATODES
1. No Cryoprotectant, Slow Cooling: L1s (Embryos)

Dracunculus medinensis

Pretreatment	None
No./volume	2,000/1 ml
Cryoprotectant	None — Me$_2$SO and glycerol are deleterious
Cooling rate	5°C min^{-1} to −78°C
Warming rate	Rate not given — in water bath at 37°C — probably ~100°C min^{-1}
Dilution	n.a.
Viability	Motility: 60—100%
	Infectivity: 1—5 vs. 80—90% for unfrozen controls develop to L3 in *Cyclops*
References	Bandyopadhyay and Chowdhury;[46] Muller;[47] Muller[48]

Note: Survival of this parasite using the method outlined here is only barely adequate for strain maintenance since there is a considerable difference between motility following thawing and development to L3 in the intermediate host. The authors did not state what cryoprotectant concentrations were evaluated nor the incubation temperature(s) and time(s).

2. Single Cryoprotectant, Slow Cooling: L3s

Brugia malayi, Acanthocheilonema (= *Dipetalonema*) *viteae, Dirofilaria immitis*)

Pretreatment	Free from vector tissues
No./volume	Number of L3s not given, 1ml/2-ml tube[49]
Cryoprotectant	5% Me$_2$SO + 20% newborn calf serum in Ham's F12(K) medium[49] or 6% Me$_2$SO + 15% newborn calf serum in RPMI 1640 or 9% Me$_2$SO + 4 mM PVP
Cooling rate	0.5 to 1.0°C min^{-1} in vapor phase LN$_2$ to −70 or −80°C, then plunge into LN$_2$
Warming rate	Fast in a water bath at 37 or 50°C
Dilution	1:5.5 to 1:10 in saline or Ham's F12(K) medium
Viability	Motility: up to 97%; infectivity: 0.9 to 7.2%[49,50]
References	Lok et al.;[49] Lowrie;[50] Wang and Huijun[51]

Note: *A. viteae* L3s have also been successfully cryopreserved within the tick vector by McCall et al.[52] using a protocol similar to that outlined above for free L3s — up to 76% of cryopreserved L3s subsequently recovered from tick tissue and injected into jirds were recovered as adult worms 81 days later.

3. Single Cryoprotectant and Slow Cooling: L1s (NBL)/L2s

Trichinella spp. NBL, *Toxocara canis* L2s

Pretreatment	None
No./volume	1,000/1.8 ml;[54,55] 1 ml[53]
Cryoprotectant	5% Me₂SO at 22°C to −2°C over 30 min;[53] 10% Me₂SO at 37°C for 15 min[54,55]
Cooling rate	0.6°C min⁻¹ to −70°C, then plunge into liquid nitrogen at 140 to 170°C min⁻¹ (see References 54 and 55); 1.1°C min⁻¹ to −28°C, then 1.7°C to −80°C then plunge[53]
Warming rate	90 to 100°C min⁻¹ in 37°C water bath
Dilution	1:10 with Basal Medium Eagles with Earle's[54,55] or Hanks' salts[53]
Viability	Motility: 48—58% initially — *Toxocara*;[53] 19—39% after 35 weeks of *in vitro* culture;[53] 80% *Trichinella*[54,55]
	Infectivity: 33% (*T. s. spiralis*), 21% (*T. s. nativa*), 2% (*T. pseudospiralis*), 0% (*T. s. nelsoni*) relative to unfrozen controls;[54,55] infectivity of *Toxocara* L2s not quantified[53]
References	Ramp et al.;[53] Rossi and Pozio;[54] Pozio et al.[55]

Note: *Trichinella* newborn larvae (NBL) are harvested from 24-h *in vitro* culture of adult female worms. The L2 of *Toxocara* are isolated from eggs.

4. Single Cryoprotectant, Stepped Cooling: L3s

Brugia pahangi

Pretreatment	Free from vector tissue
No./volume	Number of worms not given; 40-µl aliquots
Cryoprotectant	20% methanol + 20% newborn calf serum at 0°C for 10 min
Cooling rate	5°C min⁻¹ to −21°C, then plunge (10,000°C min⁻¹) into LN₂
Warming rate	~10,000°C min⁻¹ by dropping pellet into 2 ml of 37°C M199
Dilution	1:50 in M199
Viability	~42% normally motile; 6.4% infective compared to unfrozen controls
References	Ham and James[56]

5. Double Cryoprotectant Incubation, Rapid Cooling: L1s (MSL)

Trichinella spp.

Pretreatment	Agitation in 10% dog bile at 37°C for 60 min
No./volume	~1,000/20 µl on a glass cover slip (5 × 40 mm)
Cryoprotectant	20% ethanediol at 37°C for 10 min, then 33% ethanediol + 33% methanol at 0°C for 15 min (*T. s. spiralis*, *T. s. nelsoni*, *T. pseudospiralis*) or single incubation at 0°C in 33% ethanediol + 33% methanol for 30 min (*T. s. nativa*)
Cooling rate	5,100°C min⁻¹ by plunging into liquid nitrogen
Warming rate	~10,000°C min⁻¹ by dropping into 2 ml of saline at room temperature
Dilution	1:100 with 0.85% saline
Viability	Infectivity: recovery of adult worms following per os infection as percent of controls −7.8% (*T. s. spiralis*); 73.2% (*T. s. nativa*); 2.3% (*T. s. nelsoni*); 1.2% (*T. pseudospiralis*)
References	James et al.;[13] Jackson-Gegan and James[19]

Note: The MSL are obtained by standard pepsin/HCl digestion of host muscle.

6. Single Cryoprotectant, Slow (or Stepped) Cooling: mf

Brugia malayi, B. pahangi, Chandlerella quiscali, Dirofilaria corynoides, D. immitis, D. repens, Loa loa, Mansonella ozzardi, Onchocerca gutturosa, Wuchereria bancrofti

Pretreatment	None
No./volume	Number of mf not given per 1 ml in 1.2-ml tube[50,58] or sealed ampule;[59] 3,000/1 ml;[59] 100,000/1 ml[60]

Cryoprotectant	5% Me_2SO + 4 mM PVP[60] (*Loa loa, Mansonella perstans*); 6% Me_2SO + 4 mM PVP[50] (*D. corynoides*); 5%[59] or 10%[61] Me_2SO + 5% glucose and 0.9% lactose[59,61] for 30 min at 0°C;[61] 12.5% Me_2SO + 37.5% fetal calf serum[58] (*M. ozzardi*); 5% Me_2SO;[62] 9% Me_2SO[50] (*B. malayi*); 16% HES 15—30 min, 0°C[57] (Incubation temperatures and times not given by many authors.)
Cooling rate	0.5—0.8°C min^{-1} (*B. malayi, W. bancrofti*);[50] 1°C min^{-1} to −30°C then 2°C min^{-1} to −50°C then into LN_2[61] (*B. pahangi*); 1.6°C min^{-1} to −40°C then 0.6°C min^{-1} to −60°C[59] (*C. quiscali*); 1.8°C min^{-1};[60] 1.2—2.2°C min^{-1};[57] 2—5°C min^{-1} (*D. corynoides*)[50]
Warming rate	~100—200°C min^{-1} in a 37°C water bath;[50,57,59,60] slow in air to −50°C then fast[58]
Dilution	1:12 in Hanks' Balanced Salt Solution;[50] 1:9 in RPMI 1640 + 10% FCS;[58] 1:50 in RPMI 1640;[60] 1:9 in Hayne's saline + 5% blood[61]
Viability	Motility: 99% after 1 h, 92% after 18 h[60]
	Infectivity: 22—32 vs. 39% for unfrozen controls[50] (*B. malayi, D. corynoides*), 25—47% (*W. bancrofti*),[50] 78% of control mf 7 d postinjection in chickens
References	Lowrie;[50] Minjas and Townson;[57] Lawrence;[58] Granath and Huizinga;[59] Cesbron et al.;[60] Obiamiwe and Macdonald;[61] Ogunba[62]

The cryoprotectant of choice with this protocol is Me_2SO. It appears that mf benefit from the presence of an extracellular nonpermeating cryoprotective agent in the slow cooling and also the rapid cooling protocols (see next method below). Polyvinylpyrrolidone (PVP), serum albumin, glucose, lactose, and hydroxy ethyl starch (HES) can all act as external cryoprotectants. Their mechanisms of action are undefined, but probably include scavenging of free radicals, partial desiccation, increasing extracellular solution viscosity, and colligative effects. In one study,[57] HES was used alone, though it is difficult to evaluate this technique as unfrozen controls were not incorporated into the infectivity study. In a related technique, mf of *Onchocerca volvulus* have been successfully cryopreserved within nodule tissue.[63]

In most versions of this technique, samples containing mf are cooled slowly down to −60 or −70°C before transferring into liquid nitrogen for storage. However, transfer can be carried out at higher temperatures when the sample volume is small (~20 μl). Ham et al.[17] successfully cryopreserved the mf of *O. volvulus, O. gutturosa*, and *O. cervicalis* using 5% Me_2SO or 5% methanol in combination with 20% serum, a cooling rate of 1°C min^{-1} to between −14 and −20°C followed by plunging into liquid nitrogen at a rate of around 10,000°C min^{-1}. Ham and James[25] also determined that higher levels of survival could be attained if serum at a concentration of 60% was incorporated into the thawing/diluting medium. Survival using the stepped cooling method was similar to that of the slow cooling method with 85% of mf showing normal motility following cryopreservation, 30% migrating to the ears in mice proxy hosts, and 29 to 46% developing further in the insect vector.

7. Double Cryoprotectant Incubation, Rapid Cooling: mf/Adults

Mf of *Brugia pahangi*, *Onchocerca* spp., *Wucheria bancrofti*, Adults of *Onchocerca*

Pretreatment	None
No./volume	1,000—250,000/20 μl on glass coverslip (5 × 40 mm)
Cryoprotectant	10% ethanediol at 37°C for 15 min, then 40% ethanediol at 0°C 12—15 s (*Onchocerca* spp. mf); for *B. pahangi* mf, the second incubation step is extended to 45 s; for *W. bancrofti*, 60 s; and for *Onchocerca* spp. adults, 10 min
Cooling rate	~5,100°C min^{-1}
Warming rate	~10,000°C min^{-1}
Dilution	1:50 or 1:100 in M199, Hanks', or RPMI 1640 supplemented with 20% serum
Viability	Motility: *Onchocerca* mf, 77—95%;[9,64] *Brugia* mf, 91%; *Onchocerca* adults, 100% up to 7 d
	Infectivity: *Onchocerca* mf in proxy hosts, 72—79%; *Onchocerca* mf, 78%, and *Brugia* mf, 81% relative to unfrozen controls developed in respective vectors; *Onchocerca* adults, 58% survived 11 d i.p. in CBA mice
References	Ham et al.;[9] Owen and Anantaraman;[11] El Sheikh and Ham;[64] Townson;[65] Townson and Ham[66]

This technique is derived from one originally described for *Schistosoma mansoni* schistosomula[8] (see below). It is a very simple technique to perform, requires very little equipment, and has been used effectively under field conditions to collect material. Another major advantage of this method is mf can be stored more compactly — up to around 250,000 mf per sliver with 10 slivers per standard 3.5-ml cryotube. Mf can be stored at lower density also so that it is possible to thaw up small numbers as required for use. The first incubation step in ethanediol at 37°C causes cryoprotectant to penetrate into the worms. During the second incubation step at 0°C very little extra ethanediol will penetrate, but the increase in concentration to 40% causes partial dehydration and also therefore elevates the internal ethanediol concentration to 40%. The rapid cooling step is fast enough to prevent crystallization of either the extracellular solution or the intracellular contents. However, because the worms are vitrified they must be stored below the devitrification point (approximately −130°C) and therefore cannot be transported in dry ice. Levels of survival with this method are high and reproducible.

C. ENTOMOGENOUS NEMATODES
1. Single Cryoprotectant, Rapid Cooling: J1s

Steinernema feltiae (= *Neoapleactana carpocapsae*)

Pretreatment	None
No./volume	~1,000/20 μl
Cryoprotectant	60% methanol, 45 s, 0°C
Cooling rate	5,100°C min^{-1}
Warming rate	~10,000°C min^{-1}
Dilution	1:100 in water at 20°C
Viability	Motility: 12.3%
References	Smith et al.[20]

Note: The J1s are harvested from eggs hatched in axenic culture.

2. Single Cryoprotectant, Rapid Cooling: IJs

Steinernema feltiae

Pretreatment	Surface sterilized in 0.5% hyamine for 30 min, pelleted on Whatman #1 filter paper, and desiccated to 97% RH at 25°C over 24 h
No./volume	100,000 IJs/0.1 ml
Cryoprotectant	70% methanol, 10 min 0°C
Cooling rate	330—4,800°C min^{-1}
Warming rate	~600°C min^{-1}
Dilution	1:20 saline at 20°C
Viability	Motility: 84%; infectivity: growth, development, and reproduction *in vitro* demonstrated, but not quantified
References	James and Popiel[37]

Note: IJs are harvested from *in vitro* culture.

D. TREMATODES
1. Single Cryoprotectant, Slow Cooling: Sporocysts

Schistosoma mansoni

Pretreatment	Fragments of parasitized hepatopancreas and ovotestis excised from *S. mansoni*-infected *Biomphalaria glabrata* suspended in Chernin's Balanced Salt Solution (BSS)
No./volume	"Small quantities" of tissue in 1.2-ml plastic tube
Cryoprotectant	15—20% Me$_2$SO, 22—24°C for 30 min, then 4°C for 30 min

Cooling rate	Not given; tubes placed at −20°C for 30 min, then at −70°C
Warming rate	Not given — fast by pipetting 37°C BSS into tube
Dilution	Fragments washed twice in BSS
Viability	"Slightly lower than for unfrozen sporocysts"
References	Cohen and Eveland[67]

2. Double Cryoprotectant Incubation, Rapid Cooling: Schistosomula

Schistosoma spp. schistosomula are derived from cercariae by *in vitro* transformation. This technique evolved from a stepped cooling method with 17.5% methanol as the cryoprotectant, through a method using a 40% methanol incubation for 10 s and rapid cooling. As with the double incubation/rapid cooling method described above for mf, the first incubation step allows cryoprotectant to penetrate the worms, while the second incubation step induces partial desiccation and simultaneous elevation of the internal cryoprotectant concentration to 35%. Glycerol and ethanediol followed by glycerol also work, but not as effectively as ethanediol alone, resulting in lower infectivity of the cryopreserved schistosomula. The technique was devised for *S. mansoni*, but has also been used to cryopreserve *S. haematobium*, *S. mattheei*, *S. bovis*, and *S. japonicum* schistosomula, although the viability levels of these other species have not been quantitated.

Schistosoma mansoni

Pretreatment	Culture for 90 min at 37°C to stimulate loss of surface glycocalyx
No./volume	~5,000 per 20 μl on glass sliver (5 × 40 mm)
Cryoprotectant	10% ethanediol 10 min at 37°C, then 35% ethanediol for 10 min at 0°C
Cooling rate	5,000°C min^{-1}
Warming rate	10,000°C min^{-1} by dropping glass sliver into 2 ml of Earle's lactalbumin medium (ELAC) prewarmed to 42°C
Dilution	1:100 in ELAC
Viability	Motility: 44—66% normally motile
	Infectivity: 47.4% of unfrozen controls
References	James[8]

E. CESTODES
1. Single Cryoprotectant, Slow Cooling: Metacestodes

Echinococcus multilocularis

Pretreatment	Separated from host tissue and cut into 0.2 g blocks suspended in Tyrode's Saline
No./volume	0.2 g parasite block in 1 ml in plastic tubes
Cryoprotectant	10% glycerol in Eagles Minimum Essential Medium with Earle's salts (EMEM)
Cooling rate	0.7°C min^{-1} from +20 to 0°C, then 1°C min^{-1} to −20°C, then 1.7°C min^{-1} to −80°C, then to −196°C at 65°C min^{-1}
Warming rate	105°C min^{-1} in 37°C water bath
Dilution	1:10 in EMEM, then washed 2× in EMEM
Viability	Infectivity: 100% of *Meriones* injected i.p. contained viable metacestodes at 8 weeks
References	Eckert and Ramp[68]

2. Single Cryoprotectant, Stepped Cooling: Cysticerci

Taenia crassiceps

Pretreatment	None
No./volume	Number of cysticerci per 50-μl droplet not given; on aluminium support
Cryoprotectant	20% methanol in ELAC for 2 min at 0°C
Cooling rate	5°C min^{-1} to −35°C then plunge to −196°C (~5,000°C min^{-1})
Warming rate	~10,000°C min^{-1} by agitating droplet in 2 ml of ELAC at 37°C
Dilution	1:40 in ELAC
Viability	Infectivity: 2 of 5 mice following infection with 50 cysticerci positive after 3 months
References	Ham[21]

IV. STRATEGY FOR DESIGNING A CRYOPRESERVATION TECHNIQUE

No two helminths are alike in their tolerance of the various physical and chemical stresses induced during cryopreservation. However, when beginning to develop a cryopreservation technique for a new, previously uncryopreserved helminth, it is always worth trying a technique that has been shown to work for a related species. This approach is very unlikely to give an optimum result first time, but it is obviously easier to refine an existing method than to have to develop an entirely new approach.

Assuming that some degree of survival is obtained using a technique already described, parameters which should be evaluated further are

1. Varying the concentration of the cryoprotectant used
2. Varying the time of incubation in the cryoprotectant
3. The temperature of incubation

Other parameters that may be considered for investigation are

4. A range of cooling rates
5. Speeding up the rate of warming
6. The method of diluting out of the cryoprotectant

Dilution may be preceded by warming or may occur simultaneously with warming. All cryopreserved samples should be stored below the glass point, i.e., below about $-130°C$, which generally requires access to liquid nitrogen. If the simple approach of modifying an existing technique fails to yield viable organisms, then the experimenter should follow a series of steps designed to investigate the parasite's tolerance to cryoprotectants, to dehydration and to low temperatures:

A. STEP ONE: SURFACE CHANGES

In all cases, whether cryoprotectants are present or absent in the cryopreservation protocol, helminths which retain an impervious surface appear to be impossible to cryopreserve. The first step is therefore to determine the nature of the tegument and whether removal of a sheath or outer layer is possible and whether this increases transmembrane solute movement.

B. STEP TWO: CRYOPROTECTANTS

A range of cryoprotectants should be evaluated for toxicity and dehydration potential. The four most commonly used compounds are methanol, Me_2SO, ethanediol, and glycerol. The list of compounds which may protect parasites during cryopreservation is extensive,[69] but other compounds which may be considered are propan-1,2-diol, trigol, dimethylformamide, methylformamide, PVP, and HES. The additives should be tested at varying concentration for different periods of time at several temperatures, most usually 0, 20, and 37°C (e.g., Figure 12). If the organism being evaluated can only tolerate low concentrations of cryoprotectant (<10%) for short periods (<5 min) at low temperatures (0°C), it is most likely that slow cooling will be the most productive route to follow. Tolerance of higher concentrations or longer time periods or a higher incubation temperature should suggest that slow cooling might be successful, but that rapid cooling in combination with higher cryoprotectant concentrations will eventually prove to be more effective. A parasite may appear to tolerate cryoprotectants and dehydration stress well yet be incapable of being cryopreserved. This probably means that its surface layers are impermeable to the cryoprotectants used and step one should be reinvestigated.

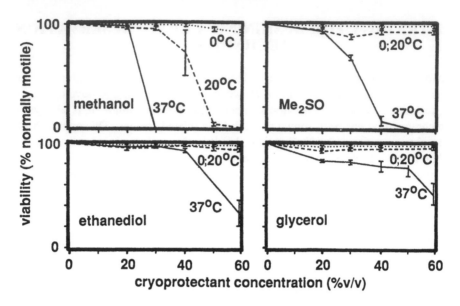

FIGURE 12.   Tolerance of *Trichinella spiralis spiralis* muscle stage larvae (MSL) to the cryoprotectants methanol, Me₂SO, ethanediol, and glycerol at 0, 20, and 37°C. Methanol and ethanediol (1:1:1 with saline) was found to be optimal for cryopreservation of *T. spiralis*. (Redrawn from Jackson-Gegan, M. and James, E. R., *Am. J. Trop. Med. Hyg.*, 38, 558, 1988. With permission.)

C. STEP THREE: STEPPED COOLING

Pick a low concentration of one of the best tolerated cryoprotective additives (5 to 20% v/v initially) with a moderate incubation period (10 to 30 min) — results from step two will suggest what these parameters should be. Cool small samples (20 to 50 μl or so) at ~1 to −70°C min⁻¹, plunging one sample or several replicates into liquid nitrogen every 2 to 5°C. Control samples should be thawed directly from the same intermediate temperatures (Figure 13).

Occasionally, survival may be high in samples plunged into liquid nitrogen from all intermediate temperatures. If this is the case, then slow cooling is the way to go. Rarely, as with *Dictyocaulus*,[18] this approach may determine that stepped cooling itself leads to optimal survival (Figure 14).

The results from stepped cooling should indicate at the very least the lowest temperature down to which the helminth can be cooled before viability is lost. Repeat using several different cooling rates and different starting concentrations of cryoprotectant. With one of the permutations of cryoprotectant and cooling rate, it should be possible to identify a window of survival from plunge (Figure 15). This will immediately indicate what is likely to be the correct final cryoprotectant concentration to use in a rapid cooling protocol. If, for example, survival peaked with plunge from −30°C and the cryoprotectant used was methanol, then the concentration required for rapid cooling will be 40% v/v (see Figure 7). Commercial programmable freezing machines can help considerably with this phase of the study,[70] but it is possible to construct cheap apparatuses that will give fairly reproducible nonlinear cooling rates of small samples.[71] If there was no survival window and viability declined to zero at a relatively high sub-zero temperature with all cryoprotectants and cooling rates tested, then rapid cooling is most likely the only option.

D. STEP FOUR: RAPID COOLING

Select a standard rapid cooling method. Several have been described. One of the simplest is to score 20 × 40 mm No. 2 glass coverslips with a diamond pen and break into 5 × 40

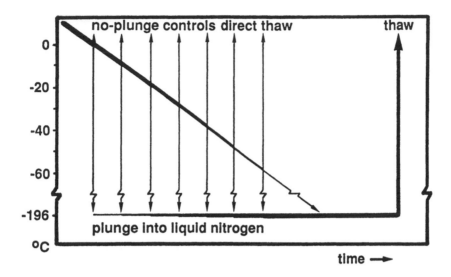

FIGURE 13. Schematic for a stepped cooling experiment. Samples are cooled slowly to different intermediate temperatures and then either thawed directly (controls) or plunged into liquid nitrogen (plunge) before thawing.

mm slivers which are supported on an aluminum block resting on crushed ice. A 20-μl sample placed on this sliver spread out to approximately 4 × 10 mm will cool at around 5100°C min^{-1} when plunged into liquid nitrogen. Obviously for large sample volumes and more discrete sample drops, the cooling rate will be slower. Faster cooling rates allow more flexibility in the concentrations of cryoprotectants which can be evaluated.

The aim with rapid cooling is to vitrify the intracellular contents of the helminth. The constituents of the cytosol contribute to the viscosity of the intracellular contents and hence the critical cryoprotectant concentration for achieving intracellular vitrification will be slower than indicated in Figure 8. To achieve vitrification, the right combination of cryoprotectant and level of dehydration which the parasite can tolerate has to be determined. Different concentrations of each cryoprotectant, incubated for increasing time periods, should be evaluated for their effect on survival both with and without plunge into liquid nitrogen (Figure 16).

It may be necessary to divide the incubation in cryoprotectant into two or more steps in order to separate and better control the processes of cryoprotectant penetration and de-hydration. For example, the addition of 35% ethanediol in a single step proved too damaging for *S. mansoni* schistosomula.[8] By incubating first in a low concentration of cryoprotectant at a high temperature — 10% at 37°C for 10 min — which induced the cryoprotectant to penetrate, the damaging effects of the final higher concentration of cryoprotectant were considerably ameliorated (Figure 17). This sequence may achieve partial dehydration during the second incubation step if the cryoprotectant is one of the higher molecular weight and/or more viscous compounds.

Occasionally, similar results can be achieved by adding a mixture of a low molecular weight penetrating cryoprotectant together with a higher molecular weight nonpenetrating additive, e.g., with *T. s. nativa*.[13] In this case, methanol and ethanediol with saline in a 1:1:1 proportion was optimal with a 30 min incubation at 0°C (Figure 18).

The sequence of dehydration and penetration can also be reversed. Very recently, with *Steinernema*, it has been found that optimal survival of IJs is obtained if the addition of 70% methanol as the cryoprotectant is preceded by desiccation for 24 h to a RH of 97%.

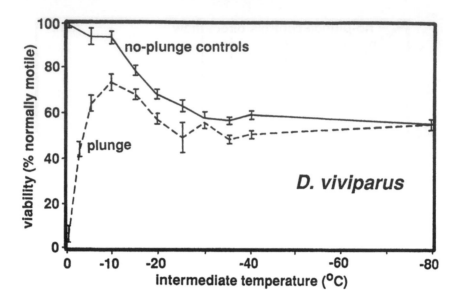

FIGURE 14.   Stepped cooling of *Dictyocaulus viviparus* infective (L3) larvae. Samples of larvae were cooled to various intermediate temperatures before plunge into liquid nitrogen. Control samples were thawed directly from the intermediate temperatures. (Redrawn from James and Peacock.[18])

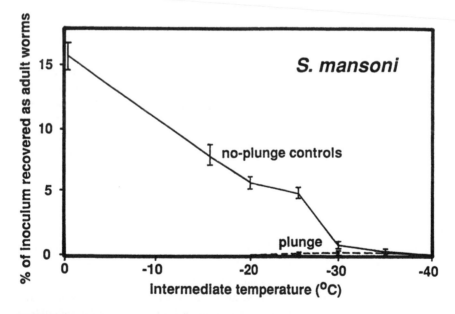

FIGURE 15.   Stepped cooling of *Schistosoma mansoni* schistosomula. Samples were cooled to different intermediate temperatures followed by plunge into liquid nitrogen. Control samples were thawed directly from the intermediate temperatures. A small ''window of survival'' occurred between −25 and −35°C. (Redrawn from James and Farrant.[15])

E. STEP FIVE: WARMING AND DILUTING

The method used to recover the parasite from liquid nitrogen should not be overlooked as an area where it is possible to achieve an increase in viability. Almost without exception, fast warming is more beneficial than slow warming. Protocols which induce vitrification by rapid cooling must be warmed rapidly to avoid or minimize devitrification and recrystallization (see Section II.C above and Figure 19). Various warming rates can be achieved by

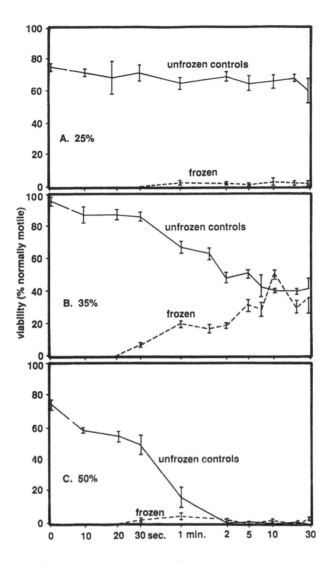

FIGURE 16. Cryopreservation of *Schistosoma mansoni* schistosomula by double incubation in ethanediol. All samples were incubated in 10% v/v ethanediol at 37°C initially for 10 min. The second
incubation was carried out at 0°C in 25% (A), 35% (B), or 50% (C)
ethanediol for increasing lengths of time. There is little decline in
viability of the schistosomula incubated in 25% ethanediol over 60
min, but this concentration is not sufficient to protect against plunge
into liquid nitrogen (A). Viability declined to <50% over 60-min
incubation in 35%, but this concentration protected against subsequent plunge into liquid nitrogen with a peak of survival at 10 min.
(B) Incubation in 50% ethanediol caused significant loss of viability
of control unplunged schistosomula by 1 min (C). These data suggest
that the optimum protocol (B) induces both partial penetration of the
ethanediol and partial dehydration. The cooling rate of the plunge
step was 5100°C min⁻¹ which would be sufficient to induce vitrification (see Figure 8). (Data from James.[8])

cryopreserving different sized samples and by insulating the sample container for the slower
rates.

Rapid warming may be simple enough to achieve but may pose rehydration problems
for the parasite particularly if a high concentration of cryoprotectant is used. Dilution stress

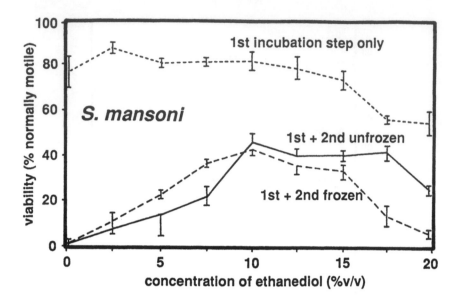

FIGURE 17. Cryopreservation of *Schistosoma mansoni* schistosomula using the ethanediol double incubation method. Here the second incubation step was constant for all samples — 35% v/v for 10 min at 0°C — while the concentration of ethanediol in the first incubation step at 37°C for 10 min was varied. Controls consisted of unfrozen samples exposed to the first incubation step only or to both incubation steps. Viability of the first incubation controls declined as the concentration of ethanediol increased, while optimum protection against the second incubation in 35% ethanediol and against plunge into liquid nitrogen was obtained with 10% ethanediol in the first incubation. (Data from James.[8])

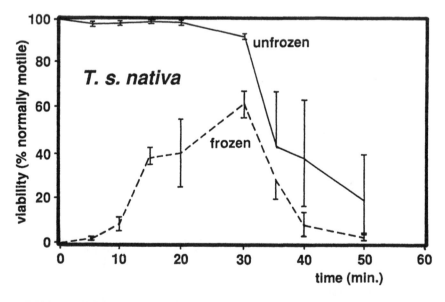

FIGURE 18. Cryopreservation of *Trichinella spiralis nativa* MSL using a mixture of 1:1:1 methanol to ethanediol to saline. Protection is dependent on the time of incubation in the cryoprotectant cocktail with peak survival occurring after 30 min of incubation at 0°C. (Data from James et al.[13])

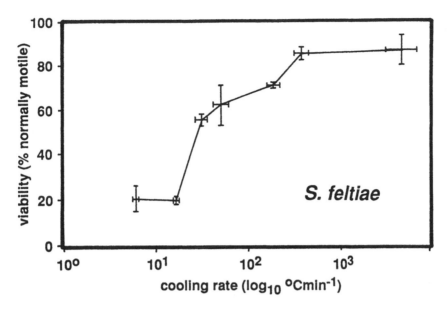

FIGURE 19.   Cryopreservation of desiccated *Steinernema feltiae* IJs using 70% v/v methanol. Optimal survival occurred at relatively rapid cooling rates of 330 to 4800°C min⁻¹, when the samples would have vitrified, and declined precipitously at slower cooling rates. (Data from James and Popiel.[37])

occurs particularly when a small volume sample is transferred into warm medium and immediately agitated. This can be alleviated by incorporating either a low concentration of the cryoprotectant used, or a nonpenetrating compound such as glucose or HES, into the recovery medium. The parasites can then be transferred to unsupplemented medium after a suitable equilibration period is spent in this lower concentration of cryoprotectant. Stepwise dilution or osmotic buffering is very necessary for certain of the protozoa but for the helminths has only been reported for the mf of *Onchocerca*[25] so far.

This stepwise approach to trouble shooting and designing a *de novo* cryopreservation technique is not comprehensive. It has, however, been based on experience with cryopreserving a number of helminth parasites, particularly schistosomula, mf, L3s of *Dictyocaulus*, L3s of *Steinernema*, and MSL of *Trichinella*, all of which, with the exception of mf, resisted cryopreservation by more traditional methods. Any experimenter who encounters problems while following the above approach will, however, gain a feel for the science of cryobiology and potentially successful new avenues of investigation are more than likely to suggest themselves.

REFERENCES

1. **Van Wyk, J. A., Gerber, H. M., and van Aardt, W. P.,** Cryopreservation of the infective larvae of the common nematodes of ruminants, *Onderstepoort J. Vet. Res.,* 44, 173, 1977.
2. **James, E. R.,** Protozoa and helminth parasites of man and animals, in *Low Temperature Preservation in Medicine and Biology,* Ashwood-Smith, M. F. and Farrant, J., Eds., Pitmans, Tunbridge Wells, U.K., 1980, 155.
3. **James, E. R.,** Cryopreservation of helminths, *Parasitol. Today,* 1, 134, 1985.
4. **Eckert, J.,** Cryopreservation of parasites, *Experientia,* 44, 873, 1988.
5. **James, E. R.,** Cryopreservation of parasites, in *Parasitology in Focus, Facts and Trends,* Mehlhorn, H., Ed., Springer Verlag, Heidelberg, 1988, 684.

6. **Angell, C. A., Shuppert, J., and Tucker, J. C.,** Cooperative behaviour in supercooled water: heat capacity, expansivity and PMR chemical shift anomalies from 0 to −38°C, *Phys. Chem.,* 77, 3092, 1973.

7. **Mazur, P., Leibo, S. P., and Chu, E. H. Y.,** A two-factor hypothesis of freezing injury, *Exp. Cell Res.,* 71, 345, 1972.

8. **James, E. R.,** *Schistosoma mansoni:* cryopreservation of schistosomula by two-step addition of ethanediol and rapid cooling, *Exp. Parasitol.,* 52, 105, 1981.

9. **Ham, P. J., Townson, S., James, E. R., and Bianco, A. E.,** An improved technique for the cryopreservation of *Onchocerca* microfilariae, *Parasitology,* 83, 139, 1981.

10. **Ham, P. J. and Townson, S.,** Improved development of *Brugia* microfilariae following cryopreservation in liquid nitrogen using a technique suitable for field conditions, *Trans. Roy. Soc. Trop. Med. Hyg.,* 80, 150, 1986.

11. **Owen, D. G. and Anantaraman, M.,** Successful cryopreservation of *Wuchereria bancrofti* microfilariae, *Trans. Roy. Soc. Trop. Med. Hyg.,* 76, 232, 1982.

12. **Nolan, T. J., Aikens, L., and Schad, G. A.,** Cryopreservation of first-stage and infective third-stage larvae of *Strongyloides stercoralis, J. Parasitol.,* 74, 387, 1988.

13. **James, E. R., Jackson-Gegan, M., Rawls, J. T., Smith, B., Hodgson-Smith, A., Callahan, H. L., and Murrell, K. D.,** Maintenance of *Trichinella* spp. isolates by cryopreservation, in *ICT 7, Proc. 7th Int. Congr. on Trichinellosis,* Kim, C. W., Ed., 1990, 24.

14. **Farrant, J.,** Mechanisms of cell damage during freezing and thawing and its prevention, *Nature,* 205, 1284, 1965.

15. **James, E. R. and Farrant, J.,** Studies on preservation of schistosomula of *Schistosoma mansoni* and *S. mattheei, Cryobiology,* 13, 625, 1976.

16. **James, E. R. and Farrant, J.,** Recovery of infective *Schistosoma mansoni* schistosomula from liquid nitrogen: a step towards a live schistosomiasis vaccine, *Trans. R. Soc. Trop. Med. Hyg.,* 71, 498, 1977.

17. **Ham, P. J., James, E. R., and Bianco, A. E.,** The successful cryopreservation of *Onchocerca* spp. microfilariae at −196°C and their subsequent development in the insect host, *Parasitology,* 47, 384, 1979.

18. **James, E. R. and Peacock, R.,** Studies on the cryopreservation of *Dictyocaulus viviparus* (Nematoda) third-stage larvae, *J. Helminthol.,* 60, 65, 1986.

19. **Jackson-Gegan, M. and James, E. R.,** Cryopreservation of *Trichinella spiralis* muscle stage larvae, *Am. J. Trop. Med. Hyg.,* 38, 558, 1988.

20. **Smith, B. S., Hodgson-Smith, A., Popiel, I., Minter, D. M., and James, E. R.,** Cryopreservation of the entomogenous nematode parasite *Steinernema feltiae* (= *Neoaplectana carpocapsae*), *Cryobiology,* 27, 319, 1990.

21. **Ham, P. J.,** Recovery of *Taenia crassiceps* (ERS Tol derived strain) from sub-zero temperatures, *J. Helminthol.,* 56, 131, 1982.

22. **Luyet, B. J. and Rasmussen, D.,** Study by differential thermal analysis of the temperatures of instability of rapidly cooled solutions of glycerol, ethylene, glycol, sucrose and glucose, *Biodynamica,* 10, 167, 1968.

23. **Ashwood-Smith, M. J.,** Low temperature preservation of cells, tissues and organs, in *Low Temperature Preservation in Medicine and Biology,* Ashwood-Smith, M. J. and Farrant, J., Eds., Pitmans, Tunbridge Wells, U.K., 1980, 19.

24. **Rall, W. F., Reid, D. S., and Farrant, J.,** Innocuous biological freezing during warming, *Nature,* 286, 511, 1980.

25. **Ham, P. J. and James, E. R.,** Protection of cryopreserved *Onchocerca* microfilariae (Nematoda) from dilution shock by the use of serum, *Cryobiology,* 19, 448, 1982.

26. **Parfitt, J. W.,** Deep freeze preservation of nematode larvae, *Res. Vet. Sci.,* 12, 488, 1971.

27. **Isenstein, R. S. and Herlich, H.,** Cryopreservation of infective third-stage larvae of *Trichostrongylus axei* and *T. colubriformis, Proc. Helminthol. Soc. Wash.,* 39, 140, 1972.

28. **Campbell, W. C. and Thompson, B.,** Survival of nematode larvae after freezing over liquid nitrogen, *Aust. Vet. J.,* 49, 110, 1973.

29. **Campbell, W. C., Blair, L. S., and Egerton, J. R.,** Unimpaired infectivity of the nematode *Haemonchus contortus* after freezing for 44 weeks in the presence of liquid nitrogen, *J. Parasitol.,* 59, 425, 1973.

30. **Kelly, K. D. and Campbell, W. C.,** Survival of *Nippostrongylus brasiliensis* larvae after freezing over liquid nitrogen, *Int. J. Parasitol.,* 4, 173, 1974.

31. **Van Wyk, J. A. and Gerber, H. M.,** Survival and development of larvae of the common nematodes of ruminants after long term cryopreservation and investigation of different routes of infestation, *Onderstepoort J. Vet. Res.,* 47, 129, 1980.

32. **Coles, G. C., Simpkin, K. G., and Brisco, M. G.,** Routine cryopreservation of ruminant nematode larvae, *Res. Vet. Sci.,* 28, 391, 1980.

33. **Van Wyk, J. A., Gerber, H. M., and Alves, R. M. R.,** Methods of infesting sheep with gastro-intestinal nematodes after cryopreservation: dosing of larvae in gelatin capsules compared to dosing of larvae in water suspension, *Onderstepoort J. Vet. Res.,* 51, 217, 1984.

34. **Ramp, Th., Eckert, J., and Christen, C.**, Erfahrungen mit der Gefrierkonservierung dritter Larvenstadien von Trichostrongyliden der Wiederkauer, *Schweiz. Arch. Tierheilkd.*, 128, 79, 1986.
35. **Marquardt, W. C., Fritts, D. H., Senger, C. M., and Seghetti, L.**, The effect of weather on the development and survival of the free-living stages of *Nematodirus spathiger* (Nematoda: Trichostrongylidae), *J. Parasitol.*, 55, 431, 1969.
36. **Stewart, G. L., Despommier, D. D., Burnham, J., and Raines, K. M.**, *Trichinella spiralis*: behavior, structure and biochemistry of larvae following exposure to components of the host enteric environment, *Exp. Parasitol.*, 63, 195, 1987.
37. **James, E. R. and Popiel, I.**, Cryopreservation of *Steinernema feltiae* (= *Neoaplectana carpocapsae*) third-stage larvae using desiccation, 18M methanol and rapid cooling, *Exp. Parasitol.*, in press.
38. **Andersen, F. L. and Levine, N. D.**, Effect of desiccation on survival of the free-living stages of Trichostrongylus colubriformis, *J. Parasitol.*, 54, 117, 1968.
39. **Uga, S., Araki, K., Matsumura, T., and Iwamura, N.**, Studies of the cryopreservation of eggs of *Angiostrongylus cantonensis*, *J. Helminthol.*, 57, 297, 1983.
40. **Campbell, W. C., Blair, L. S., and Egerton, J. R.**, Motility and infectivity of *Haemonchus contortus* larvae after freezing, *Vet. Rec.*, 91, 13, 1972.
41. **Bemrick, W. J.**, Tolerance of equine strongylid larvae to desiccation and freezing, *Cryobiology*, 15, 214, 1978.
42. **Weinman, D. and McAllister, J.**, Prolonged storage of human pathogenic protozoa with conservation of virulence: observations on the storage of helminths and leptospiras, *Am. J. Hyg.*, 45, 102, 1947.
43. **Vetter, J. C. M. and Klaver-Wesseling, J. C. M.**, Unimpaired infectivity of *Ancylostoma ceylanicum* after storage in liquid nitrogen for one year, *J. Parasitol.*, 63, 700, 1977.
44. **Miller, T. A. and Cunningham, M. P.**, Freezing of infective larvae of the dog hookworm, in *Annu. Rep. East African Trypanosomiasis Research Organization*, 1965, 19.
45. **Kelly, J. D., Campbell, W. C., and Whitlock, H. V.**, Infectivity of *Ancylostoma caninum* larvae after freezing over liquid nitrogen, *Aust. Vet. J.*, 52, 141, 1976.
46. **Bandyopadhayay, A. K. and Chowdhury, A. B.**, Preliminary observations on the effect of prolonged hypothermia on *Dracunculus medinensis*, *Bull. Calcutta Soc. Trop. Med.*, 13, 49, 1965.
47. **Muller, R.**, The development of *Dracunculus medinensis* in the intermediate host after deep freezing, *Trans. R. Soc. Trop. Med. Hyg.*, 61, 451, 1967.
48. **Muller, R.**, Development of *Dracunculus medinensis* after freezing, *Nature*, 226, 662, 1970.
49. **Lok, J. B., Mika-Grieve, M., and Grieve, R. B.**, Cryopreservation of *Dirofilaria immitis* microfilariae and third-stage larvae, *J. Helminthol.*, 57, 319, 1983.
50. **Lowrie, R. C., Jr.**, Cryopreservation of third-stage larvae of *Brugia malayi* and *Dipetalonema viteae*, *Am. J. Trop. Med. Hyg.*, 32, 767, 1983.
51. **Wang, S.-H. and Huijun, Z.**, Cryopreservation of infective-stage larvae of *Brugia malayi*, Southeast Asian *J. Trop. Med. Public Health*, 18, 488, 1987.
52. **McCall, J. W., Jun, J., and Thompson, P. E.**, Cryopreservation of infective larvae of *Dipetalonema viteae*, *J. Parasitol.*, 61, 340, 1975.
53. **Ramp, T., Eckert, J., and Gottstein, B.**, Cryopreservation and long-term *in vitro* maintenance of second-stage larvae of *Toxocara canis*, *Parasitol. Res.*, 73, 165, 1987.
54. **Rossi, P. and Pozio, E.**, Cryopreservation of *Trichinella* newborn larvae, *J. Parasitol.*, 74, 510, 1988.
55. **Pozio, E., Rossi, P., and Scrimitore, E.**, Studies on the cryopreservation of *Trichinella* species, *Exp. Parasitol.*, 67, 182, 1988.
56. **Ham, P. J. and James, E. R.**, Successful cryopreservation of *Brugia pahangi* third-stage larvae in liquid nitrogen, *Trans. R. Soc. Trop. Med. Hyg.*, 77, 815, 1983.
57. **Minjas, J. N. and Townson, H.**, The successful cryopreservation of microfilariae with hydroxyethyl starch as cryoprotectant, *Ann. Trop. Med. Parasitol.*, 74, 571, 1980.
58. **Lawrence, D. N.**, Extended cryopreservation of *Mansonella ozzardi* microfilaria concentrated from human peripheral blood, *Am. J. Trop. Med. Hyg.*, 29, 313, 1980.
59. **Granath, W. O. and Huizinga, H. W.**, *Chandlerella quiscali* (Onchocercidae: Filarioidea): cryopreservation and post freezing viability of microfilariae in chickens, *Exp. Parasitol.*, 46, 239, 1978.
60. **Cesbron, J.-Y., Taelman, H., Henry, D., Myelle, L., and Capron, A.**, Extended cryopreservation of *Loa loa* and *Mansonella perstans* microfilariae from human peripheral blood, *Trans. R. Soc. Trop. Med. Hyg.*, 80, 563, 1986.
61. **Obiamiwe, B. A. and Macdonald, W. W.**, The preservation of *Brugia pahangi* microfilariae at sub-zero temperatures and their subsequent development to the adult stage, *Ann. Trop. Med. Parasitol.*, 65, 547, 1971.
62. **Ogunba, E. O.**, Preservation of frozen *Brugia pahangi* using dimethylsulphoxide, *J. Parasitol.*, 55, 1101, 1969.
63. **Schiller, E. L., Turner, V. M., Marroquin, H. F., and D'Antonio, R.**, The cryopreservation and *in vitro* cultivation of larval *Onchocerca volvulus*, *Am. J. Trop. Med. Hyg.*, 28, 997, 1979.

64. **El Sheikh, H. and Ham, P. J.,** Human onchocerciasis: cryopreservation of isolated microfilariae, *Lancet,* 1, 450, 1982.

65. **Townson, S.,** The maintenance of *Onchocerca gutturosa* adult worms at low temperatures, *Tropenmed. Parasitol.,* 36, 28, 1985.

66. **Townson, S. and Ham, P. J.,** Successful cryopreservation of adult male *Onchocerca gutturosa* in liquid nitrogen, *Tropenmed. Parasitol.,* 37, 117, 1986.

67. **Cohen, L. M. and Eveland, L. K.,** Cryopreservation of *Schistosoma mansoni* sporocysts, *J. Parasitol.,* 70, 592, 1984.

68. **Eckert, J. and Ramp, Th.,** Cryopreservation of *Echinococcus multilocularis* metacestodes and subsequent proliferation in rodents (Meriones), *Z. Parasitenkd.,* 71, 777, 1985.

69. **O'Connell, K. M., Hunter, S. H., Fromentin, M., Frank, O., and Baker, M.,** Cryoprotectants for *Crithidia fasciculata* stored at − 20°C with notes on *Trypanosoma gambiense* and *T. conorhini, J. Protozool.,* 15, 719, 1968.

70. **Fuller, B. J. and James, E. R.,** Cryopreservation of isolated rat hepatocytes in pellet form, *Cryo-Letters,* 6, 49, 1985.

71. **James, E. R.,** A portable apparatus for controlled slow, or 2-step cooling of small volumes, *Cryo-Letters,* 1, 17, 1979.

Chapter 10

IN VITRO CULTURE AS A BIOLOGICAL TOOL

J. D. Smyth

TABLE OF CONTENTS

I. APPLICATIONS: GENERAL SURVEY

A. GENERAL COMMENTS

It is self-evident that once a larval or adult helminth can be cultured *in vitro* — even for short periods — the horizons for research on the basic biology of the organism are greatly extended. However, it is only within recent years, in parallel with the commercial availability of standard media, sterile plastic ware, and wide-spectrum antibiotics, that these techniques for helminths are coming into more general use. This is particularly true in the fields of immunobiology and chemotherapy where either application in the search for effective vaccines and/or drugs against helminths tends to dominate the *in vitro* culture field.

Yet it must be emphasized that *in vitro* techniques have a much wider application than the production of excretory/secretory (E/S) antigens or the evaluation of anthelminthics, in that they are also invaluable for studies on fundamental biological phenomenon many of which are common to most animal groups and thus have a much wider significance. The areas in which *in vitro* culture can make a meaningful contribution can be summarized as follows:

- Replacement of definitive or intermediate hosts
- Identification of unknown larval stages
- Immunobiology
- Biochemistry
- Physiology
- Chemotherapy
- Cell or tissue differentiation

It is not intended here to attempt to review all the applications of *in vitro* culture in the above areas, but only to present a few examples to illustrate the principles involved.

B. REPLACEMENT OF LABORATORY HOSTS

Although some helminths, e.g., *Hymenolepsis diminuta* (Chapter 5, Section V) can now be cultured through their entire life cycle *in vitro*, the techniques involved are still too elaborate to be used routinely to replace the intermediate and/or definitive hosts. On the other hand, several cestodes with progenetic plerocercoids and digenetic trematodes with progenetic metacercariae can be matured by relatively simple *in vitro* culture systems, thus entirely replacing the definitive host (usually a bird) in laboratory investigations. Typical examples of these are *Microphallus similis* (Chapter 3, Section V) or *Microphalloides japonicus* (Chapter 3, Section VI) among the trematodes and *Schistocephalus solidus* (Chapter 5, Section II) and *Ligula intestinalis* (Chapter 5, Section III) among the cestodes. This situation really represents a spectacular advance in research potential by providing simple experimental models, without the use of laboratory animals, on which even relatively junior students can carry out significant biochemical and physiological experiments. Although fertilized eggs, capable of hatching and continuing the life cycle are only produced in a few instances, e.g., *S. solidus*, development is often sufficiently normal for much valuable data to be obtained.

C. IDENTIFICATION OF UNKNOWN LARVAL STAGES

Although *in vitro* techniques have been little used to identify unknown species of larval helminths, there are several interesting examples of this in the case of fish parasites. Thus, the metacercaria of the strigeid *Cotylurus erraticus* was first found in Ireland in trout, but its species unidentified as the adult had not been found. This metacercaria proved to be unusually easy to culture to maturity *in vitro* (in spite of being nonprogenetic) and the adult which developed from the larva was readily identified as *C. erraticus* (Chapter 4, Section I). A comparable result has been obtained with unknown nematode larvae in fish. Thus, larvae found in the Chilean hake, *Merluccius gayi*, were tentatively identified as unknown species of *Anisakis* and *Phocanema*.[1] The adult worms were obtained by culturing (Chapter 7, Section III.B) and the species identified as *Anisakis simplex* and *Phocanema decipiens*. This is likely to prove to be a valuable taxonomic technique which could be more widely exploited once its value is appreciated and the appropriate techniques are developed.

D. IMMUNOBIOLOGY
1. Vaccine Development

Although *in vitro* techniques have not been used directly for immunodiagnosis of parasitic helminths, they have been extensively used indirectly as a method of producing antigens from a wide range of species of trematodes (especially *Fasciola*), cestodes (especially *Echinococcus* spp. and *Taenia* spp.), and nematodes (especially *Ascaris lumbricoides* and *Trichinella spiralis*). Although it was hoped that the development of *in vitro* techniques — particularly the production of E/S antigens — would lead to the development of reliable vaccines against parasitic helminths, to date this has proved not to be the case, although some level of limited protection is usually produced by the use of such antigens.

A vast literature exists on the production of antigens by helminth groups and this has been comprehensively reviewed by Rickard and Howell.[2] Many of the culture techniques described in this text are suitable for the production of E/S antigens or can be adapted for this purpose. Unfortunately, a major difficulty is that most of the culture media used contain natural products such as serum which must be omitted if antigens, uncontaminated by other natural products, are to be obtained. This factor greatly limits the application of many of the culture techniques, in that organisms survive for much shorter periods in serum-free media and growth and differentiation may be severely limited.

2. Neurobiology

A recent application of *in vitro* techniques has been the identification of neurosecretory cells in the nervous system of cestodes. Although presumptive neurosecretory cells have been described in trematodes[3] and cestodes,[4] they have only been "identified" by the presence of granules which stain positively with paraldehyde fuchsin (PAF). It is now widely recognized that this strain is nonspecific so that results obtained by its use must be regarded as being equivocal. If, however, it could be shown that the activity within the cells was related to an induced pattern of physiological activity, the case for identifying such cells as neurosecretory cells would be greatly strengthened. This has been unequivocally demonstrated in the nerve cells of *Diphyllobothrium dendriticum*[4-6] by *in vitro* methods. It was shown that peptidergic neurons existed in the nervous system, being tentatively identified by using (in addition to PAF) chromalum-hematoxylin/phloxine, paraldehyde-thionin, rescorcin/fuchsin and Alcian blue/Alcian yellow. When the plerocercoids obtained from a fish were cultured *in vitro* at 38°C (i.e., the temperature of the adult bird host) increased cellular activity began within 5 min and was marked after 1 h. This result suggested that the rise in temperature induced a physiological "switch" in what appeared to be true neurosecretory cells. The later application of immunocytological methods has since unequivocally supported this view[5,6] and at least nine vertebrate neuropeptides have been unequivocally identified. This field has been further reviewed by Smyth and McManus.[4]

E. BIOCHEMISTRY AND PHYSIOLOGY

The maintenance of parasitic helminths *in vitro*, for even short periods, means that they can be readily exposed to most routine biochemical techniques. As extensive literature based on both long-term and short-term methods of cultivation has been built up and this has been extensively reviewed[3,4,7] and will not be further discussed here. Many of these *in vitro* techniques are also suitable for evaluation of anthelminthics and as such have been widely used.[8,9] However, the effect of drugs should be interpreted with caution and adequate controls should be utilized, as *in vitro* systems are notoriously affected by the physicochemical conditions of the medium and small differences in factors, such as pCO_2 or Eh, may easily affect survival *in vitro* and give misleading results of apparent drug action.

In addition to obtaining routine data on the basic metabolism of helminths, additional information can often be obtained by examining the physiological or biochemical "switches" which may occur when a larva/larva or larva/adult transformation is taking place. One useful model for examining the biochemical changes occurring during such switches has been the plerocercoid larva of *Schistocephalus solidus* which is readily transformed into an adult by a simple rise in temperature (Chapter 5, Section II). The biochemistry of this transformation has been examined in some detail by Korting and Barrett[10] and Beis and Barrett.[11]

A larval/larval transformation which has received much attention is the cercaria/schistosomulum transformation which can readily be carried out *in vitro* by a number of mechanical means (Chapter 4, Section VI.B). Although the biochemistry of this transformation has not been much studied, the physiological changes, which particularly lend themselves to *in vitro* study, have been thoroughly investigated.

F. CELL AND TISSUE DIFFERENTIATION
1. Larval/Adult Transformation

Since many parasitic helminths, especially trematodes and cestodes, undergo dramatic morphological changes during their life cycles, it is self-evident that the successful achievement *in vitro* of larva/larva or larva/adult transformations opens up fascinating fields for fundamental studies of cell and tissue differentiation, as well as of the physiological aspects mentioned above. Thus, in the case of the cestodes, *Schistocephalus solidus* and *Ligula intestinalis*, the ease with which the plerocercoid/adult transformation can be carried out *in vitro* has enabled the processes of spermatogenesis and eggshell formation to be extensively studied,[12-14] and in the latter instance, these studies have thrown much light on the process of quinone tanning by which the eggshell in pseudophyllidean cestodes and trematodes is formed and stabilized.[3,4]

2. Asexual/Sexual Differentiation

Perhaps the most interesting aspect of differentiation which has emerged from recent *in vitro* studies is that of the factors controlling asexual/sexual differentiation during cestode life cycles. *In vitro* systems are now available for *Echinococcus granulosus* and *E. multilocularis* and for *Mesocestoides corti* and *M. lineatus* which enable the larval stage — protoscolex or tetrathyridium — to enter either the asexual phase or the sexual phase of development. One of the surprising results from such studies is that all these species appear to utilize different "triggers" and/or physiological requirements to initiate sexual differentiation. Thus, *E. granulosus* (Chapter 5, Section XI.E.) requires the presence of a suitable protein substrate in a diphasic culture system to stimulate sexual differentiation, whereas *E. multilocularis* (Chapter 5, Section XII.C.) will differentiate sexually in a liquid monophasic medium. The stimulus to trigger sexual differentiation in *M. corti* (Chapter 5, Section IX.E.) appears to be an anaerobic environment, whereas *M. lineatus* (Chapter 5, Section X.B.) will only differentiate sexually in the presence of trypsin. These *in vitro* systems provide powerful tools for the study of this aspect of differentiation which is a fundamental phenomenon throughout the animal kingdom and one which is little understood.

II. FUTURE HORIZONS

As mentioned earlier, the commercial availability of ready-made media, sterile plastic ware, and an increasing range of wide-spectrum antibiotics has revolutionized the field of *in vitro* culture and has greatly stimulated progress in this area of biological research. Nevertheless, the rationale of many of the successful culture systems remains unknown and their development still contains a high element of empiricism.

In spite of the standardization of the known media components, variability in the natural constituents of media (i.e., those from animal sources), such as serum or yeast extract, still presents a problem well known to research workers. It is widely recognized that the growth properties of serum, in particular, may vary greatly from batch to batch — organisms growing well in one batch and failing to grow in another. Although some suppliers test the growth properties of their sera against standard cell lines, this may not screen for the properties required for a particular species of helminth. This uncertainty undoubtedly discourages workers to enter this field and this is a problem which needs continuous attention and further research. Although some serum "substitutes" are available, they appear to be generally unsatisfactory for this kind of work.

The long-term aim of workers in this field must be to be able to maintain the life cycles of many species of helminths with relatively simple, standardized systems. Much progress has been made within the last decade, particularly with trematodes, cestodes, and intestinal nematodes, but systems for filaria, acanthocephala, monogenetic trematodes, and the intramolluscan stages of digenetic trematodes still remain somewhat poorly developed.

This text has confined itself to the *in vitro* cultivation of whole organisms, but a future area which could prove to be promising is that of cell culture. The area of cell culture remains relatively unexplored, and although some interesting experiments have been carried out with *Fasciola hepatica*,[15] *Echinococcus granulosus*,[16] and *E. multilocularis*, the establishment of a cell line which is unequivocally of helminth origin (i.e., and not derived inadvertently from host tissue) remains to be demonstrated.[7] This is clearly a most challenging field for further work.

REFERENCES

1. **Carvajal, J., Barros, C., Santander, G., and Alacalde, C.,** *In vitro* culture of larval anisakid parasites of the Chilean hake *Merluccius gayi, J. Parasitol.,* 67, 958, 1981.
2. **Rickard, M. D. and Howell, M. J.,** Helminth vaccines, in *In Vitro Methods for Parasite Cultivation,* Taylor, A. E. R. and Baker, J. R., Eds., Academic Press, London, 1987, 407.
3. **Smyth, J. D. and Halton, D. W.,** *The Physiology of Trematodes,* Cambridge University Press, Cambridge, 1983.
4. **Smyth, J. D. and McManus, D. P.,** *The Physiology and Biochemistry of Cestodes,* Cambridge University Press, Cambridge, 1989.
5. **Gustafsson, M. K. S. and Wikgren, M. C.,** Peptidergic and aminergic neurons in adult *Diphyllobothrium dendriticum* Nitzsch, 1824 (Cestoda, Pseudophyllidea), *Z. Parasitenkd.,* 64, 121, 1981.
6. **Gustafsson, M. K. S.,** Activation of the peptidergic neurosecretory system in *Diphyllobothrium dendriticum* (Cestoda), *Cell Tissue Res.,* 220, 473, 1981.
7. **Barrett, J.,** *Biochemistry of Parasitic Helminths,* Macmillan, London, 1981.
8. **Rapson, E. B., Jenkins, D. C., and Topley, P.,** *Trichostrongylus colubriformis:* in vitro culture of parasitic stages and their use for the evaluation of anthelmintics, *Res. Vet. Sci.,* 39, 90, 1985.
9. **Jenkins, D. C., Armitage, R., and Carrington, T. S.,** A new primary screening test for anthelminthics utilizing the parasitic stages of *Nippostrongylus brasiliensis, in vitro, Z. Parasitenkd.,* 63, 261, 1980.
10. **Korting, W. and Barrett, J.,** Carbohydrate catabolism in the plerocercoids of *Schistocephalus solidus* (Cestoda: Pseudophyllidea), *Int. J. Parasitol.,* 7, 411, 1977.

11. **Beis, I. and Barrett, J.**, The contents of adenine nucleotides and glycolytic and tricarboxylic acid cycle intermediates in activated and non-activated plerocercoids of *Schistocephalus solidus* (Cestoda: Pseudophyllidea), *Int. J. Parasitol.*, 9, 465, 1979.
12. **Smyth, J. D.**, Studies on tapeworm physiology. VI. Effect of temperature on the maturation *in vitro* of *Schistocephalus solidus*, *J. Exp. Biol.*, 29, 304, 1952.
13. **Smyth, J. D.**, Studies on tapeworm physiology. IX. Histochemical study of egg-shell formation in *Schistocephalus solidus (Pseudophyllidea)*, *Exp. Parasitol.*, 5, 519, 1956.
14. **Smyth, J. D. and Clegg, J. A.**, Egg-shell formation in trematodes and cestodes, *Exp. Parasitol.*, 8, 286, 1959.
15. **Howell, M. J.**, An approach to the production of helminth antigens *in vitro*: the formation of hybrid cells between *Fasciola hepatica* and a rat fibroblast cell line, *Int. J. Parasitol.*, 11, 235, 1981.
16. **Fiori, P. L., Monaco, G., Scappaticci, S., Pugliese, A., Canu, N., and Cappuccinelli, P.**, Establishment of cell cultures from hydatid cysts of *Echinococcus granulosus*, *Int. J. Parasitol.*, 18, 297, 1988.
17. **Howell, M. J. and Matthaei, K.**, Points in question: *in vitro* culture of host or parasite cells?, *Int. J. Parasitol.*, 18, 883, 1988.

INDEX

V

W